Viral Genome Methods

Viral Genome Methods

Edited by

Kenneth W. Adolph, Ph.D.
Department of Biochemistry
University of Minnesota Medical School
Minneapolis, Minnesota

CRC Press
Boca Raton New York London Tokyo

Library of Congress Cataloging-in-Publication Data

Viral genome methods / edited by Kenneth W. Adolph.
p. cm.
Includes bibliographical references and index.
ISBN 0-8493-4412-3 (alk. paper)
1. Viral genetics--Laboratory manuals. 2. Molecular virology--Laboratory manuals. I. Adolph, Kenneth W., 1944-.
QH434.V56 1996
576′.6484--dc20 96-11626
CIP

International Standard Book Number 0-8493-4412-3
Library of Congress Card Number 96-11626
Printed in the United States of America 1 2 3 4 5 6 7 8 9 0
Printed on acid-free paper

Preface

Viral Genome Methods is a practical guide to the application of important molecular biology and genetics techniques to research on viruses. The chapters are written by experts who are recognized authorities in their research areas and often originated the new techniques that are described. The central parts of the chapters are the experimental protocols which are presented so as to be readily used at the laboratory bench. These step-by-step protocols are intended to be concise and easy to follow. The procedures should be reproducible in the hands of researchers with different levels of expertise. Suggestions to successfully apply the procedures are included, along with recommended materials and suppliers. A special feature of the chapters is that, in addition to the protocols, important background information and representative results of applying the methods are given. Viruses with roles in human disease are mainly discussed, but it should be emphasized that the experimental procedures should be easy to adapt to different viruses and research questions.

Molecular biology and genetics approaches now pervade viral research. A working knowledge of these gene-centered approaches is therefore essential. DNA sequencing, cloning, and related techniques are basic to investigations of viral protein structure and function, since these investigations are founded upon, or are incomplete without, knowledge of gene structure and regulation. Similarly, cell biology studies concerned with the process of viral infection exploit the power of molecular biology and genetics approaches. *Viral Genome Methods* therefore aims to be a useful and convenient source of some of the newest molecular techniques of viral research.

The volume consists of two sections, which deal with "Retroviruses/HIV/RNA Viruses" and "DNA Viruses". The emphasis is upon viruses important in human disease. In the first section, the human immunodeficiency virus (HIV) receives the greatest coverage. Chapters present methods to study HIV genetic variation, gene structure, reverse transcriptase function, and Tat and Nef viral protein effects. Procedures are also given for retrovirus-mediated gene transfer, and for quantitating human retroviruses and simian immunodeficiency virus (SIV). Among other significant RNA viruses, an example discussed is vesicular stomatitis virus. In the second section, methods are outlined relating to DNA viruses including human cytomegalovirus,

Epstein-Barr virus, herpes simplex virus, adeno-associated virus, and adenovirus. The protocols are concerned with viral gene expression and regulation, DNA replication, and the molecular biology of infection.

The volume is intended for researchers in virology and related biomedical fields. Thus, microbiologists, molecular biologists, biochemists, pathologists, geneticists, and cell biologists will find the work a valuable aid to their research. Established investigators, postdoctoral researchers, graduate students, and technicians should benefit from the volume.

Kenneth W. Adolph

The Editor

Kenneth W. Adolph, Ph.D., is currently Associate Professor in the Department of Biochemistry of the University of Minnesota Medical School in Minneapolis. He received his Ph.D. from the Department of Biophysics of the University of Chicago. His B.S. and M.S. degrees were from the University of Wisconsin–Milwaukee. After receiving the Ph.D., he was a postdoctoral fellow at the Medical Research Council Laboratory of Molecular Biology in Cambridge, England, and at the Rosenstiel Basic Medical Sciences Research Center, Brandeis University. He was then a post-doctoral fellow in the Department of Biochemical Sciences, Princeton University. More recently, he was a visiting scientist in the Department of Biochemistry of the University of Washington in Seattle. He is a member of the American Society for Biochemistry and Molecular Biology.

Kenneth W. Adolph has research interests in gene structure and regulation, chromosome organization, nonhistone chromosomal proteins, and virus assembly. Research projects have been concerned with the structure and regulation of the human thrombospondin genes, electron microscopy studies of chromosome organization through the cell division cycle, the synthesis and post-translational modifications of nonhistone chromosomal proteins, and the assembly and structure of plant viruses.

Contributors

Akio Adachi, Ph.D.
Laboratory of Gene Analysis
Department of Viral Oncology
Institute for Virus Research
Kyoto University
Kyoto, Japan

Estuardo Aguilar-Cordova, Ph.D.
Department of Pediatrics
Baylor College of Medicine
Houston, Texas

David G. Anders, Ph.D.
The David Axelrod Institute
Wadsworth Center for Laboratories and Research
and Department of Biomedical Sciences
University at Albany School of Public Health
Albany, New York

Shirley M. Bartido, Ph.D.
Department of Biology
New York University
New York, New York

William Bebrin, D.M.D., D.M.Sc.
Department of Biological Chemistry and Molecular Pharmacology
and Committee on Virology
Harvard Medical School
Boston, Massachusetts

John W. Belmont, M.D., Ph.D.
Department of Molecular and Human Genetics
Baylor College of Medicine
Houston, Texas

Bruce A. Bunnell, Ph.D.
Clinical Gene Therapy Branch
National Center for Human Genome Research
National Institutes of Health
Bethesda, Maryland

Barrie J. Carter, Ph.D.
Targeted Genetics Corporation
Seattle, Washington

Javier Chinen, M.D.
Department of Microbiology and Immunology
Baylor College of Medicine
Houston, Texas

Connie Chow, B.Sc.
Committee on Virology
Harvard Medical School
Boston, Massachusetts

Donald M. Coen, Ph.D.
Department of Biological Chemistry and Molecular Pharmacology
Harvard Medical School
Boston, Massachusetts

Frank E. J. Coenjaerts, Ph.D.
Laboratory for Physiological Chemistry
Utrecht University
Utrecht, The Netherlands

Brian C. Corliss, B.S.
Gladstone Institute of Virology and Immunology
University of California, San Francisco
San Francisco, California

Swapan K. De, Ph.D.
Laboratory of Molecular Biology
National Institute of Mental Health
National Institutes of Health
Bethesda, Maryland

Job Dekker, M.Sc.
Laboratory for Physiological Chemistry
Utrecht University
Utrecht, The Netherlands

Paul Digard, Ph.D.
Division of Virology
Department of Pathology
University of Cambridge
Cambridge, United Kingdom

William C. Drosopoulos, Ph.D.
Department of Microbiology and Immunology
Albert Einstein College of Medicine
Bronx, New York

Mark B. Feinberg, M.D., Ph.D.
Gladstone Institute of Virology and Immunology
University of California, San Francisco
San Francisco, California

Jessica L. Guthrie, B.S.
Gladstone Institute of Virology and Immunology
University of California, San Francisco
San Francisco, California

Andrea C. Iskenderian, M.S.
The David Axelrod Institute
Wadsworth Center for Laboratories and Research
and Department of Biomedical Sciences
University at Albany School of Public Health
Albany, New York

Marcia L. Kalish, Ph.D.
HIV Laboratory Investigations Branch
Division of AIDS, STD and TB Laboratory Research
National Center for Infectious Diseases
Centers for Disease Control and Prevention
Atlanta, Georgia

Meiko Kawamura, Ph.D.
Laboratory of Gene Analysis
Department of Viral Oncology
Institute for Virus Research
Kyoto University
Kyoto, Japan

Yvonne Kew, M.S.
Department of Microbiology and Immunology
Albert Einstein College of Medicine
Bronx, New York

Jon W. Marsh, Ph.D.
Laboratory of Molecular Biology
National Institute of Mental Health
National Institutes of Health
Bethesda, Maryland

Richard A. Morgan, Ph.D.
Clinical Gene Therapy Branch
National Center for Human Genome Research
National Institutes of Health
Bethesda, Maryland

Shelli A. Oien
Department of Biology
New York University
New York, New York

Joseph S. Pagano, M.D.
UNC-Lineberger Comprehensive Cancer Center
and Departments of Medicine, and Microbiology and Immunology
University of North Carolina at Chapel Hill
Chapel Hill, North Carolina

Gregory S. Pari, Ph.D.
Hybridon, Inc.
Worcester, Massachusetts

Vinayaka R. Prasad, Ph.D.
Department of Microbiology and Immunology
Albert Einstein College of Medicine
Bronx, New York

Suraiya Rasheed, Ph.D., FRCPath.
Laboratory of Viral Oncology and AIDS Research
University of Southern California
School of Medicine
Los Angeles, California

Carol S. Reiss, Ph.D.
Department of Biology
and Center for Neural Science
and Kaplan Comprehensive Cancer Center
New York University
New York, New York

Hiroyuki Sakai, Ph.D.
Laboratory of Gene Analysis
Department of Viral Oncology
Institute for Virus Research
Kyoto University
Kyoto, Japan

Gerald Schochetman, Ph.D.
HIV Laboratory Investigations Branch
Division of AIDS, STD and TB Laboratory Research
National Center for Infectious Diseases
Centers for Disease Control and Prevention
Atlanta, Georgia

Nirupama Deshmane Sista, Ph.D.
UNC-Lineberger Comprehensive Cancer Center
University of North Carolina at Chapel Hill
Chapel Hill, North Carolina

Silvija I. Staprans, Ph.D.
Gladstone Institute of Virology and Immunology
University of California, San Francisco
San Francisco, California

Shambavi Subbarao, Ph.D.
HIV Laboratory Investigations Branch
Division of AIDS, STD and TB Laboratory Research
National Center for Infectious Diseases
Centers for Disease Control and Prevention
Atlanta, Georgia

Kenzo Tokunaga, Ph.D.
Laboratory of Gene Analysis
Department of Viral Oncology
Institute for Virus Research
Kyoto University
Kyoto, Japan

Peter C. van der Vliet, Ph.D.
Laboratory for Physiological Chemistry
Utrecht University
Utrecht, The Netherlands

Contents

Section I

Retroviruses/HIV/ RNA Viruses

Chapter 1

Retrovirus-Mediated Gene Transfer

*Bruce A. Bunnell and Richard A. Morgan**

Contents

I. Introduction

Retroviral-based gene transfer vectors are currently one of the most widely used and highly effective gene transfer systems available. Retroviruses are a family of RNA viruses that possess the ability to efficiently and precisely integrate a DNA replication intermediate into dividing cells. Sophisticated molecular biology techniques have led to the development of a family of vectors that use the effective replication and integration mechanisms of retroviruses to stably transfer genes into a wide variety of cell types, such as hepatocytes and peripheral blood lymphocytes.[1,2] Retroviral vector technology has extensive applications, ranging from the potential clinical treatment of genetic disorders and inhibition of infectious agents such as human immunodeficiency virus type 1 (HIV-1), to basic applications such as mapping mammalian genomes and examining mammalian development.[3-5] The purpose of this chapter is to describe the technology and protocols used in retroviral-mediated gene transfer. The topics discussed in this chapter include (1) retroviral vectors, (2) retroviral packaging cell lines, and (3) generation and characterization of high-titer retrovirus supernatants.

II. Life Cycle of the Retrovirus

Retroviruses are found in several invertebrate and vertebrate species, including most mammals.[6] They are divided into three subfamilies based primarily on their association with disease: oncovirinae, lentivirinae, and spumavirinae. The most commonly studied retroviruses are the simple oncoviruses, such as Rous sarcoma virus (RSV) and the Moloney murine leukemia virus (Mo-MLV). Each retrovirus particle (a virion) contains two copies of single-stranded (positive sense) RNA within a protein core surrounded by a lipid envelope. The most interesting property of the retrovirus is the existence of an RNA-dependent DNA polymerase termed reverse transcriptase.[7,8]

The retrovirus life cycle begins by the binding of a retrovirus envelope protein to a specific receptor on the cell surface. Depending on the virus, the virion then gains access to the cell through endocytosis or by direct fusion between the virus and cellular membrane. Following entry, the viral core complex passes through the cytoplasm and eventually enters the nucleus. During this migration, the reverse transcriptase reaction progresses through a complex intermolecular reaction to create a double-stranded DNA copy of the input viral RNA. Within the nucleus, the DNA copy is inserted into a host cell chromosome by the action of the integrase protein (the completion of the reverse transcription and integration process requires cell division). The integrated DNA form of the reverse transcribed RNA is called a provirus.

As described in subsequent sections, the retrovirus used most widely in gene transfer systems is the Moloney murine leukemia virus (Mo-MLV). A diagram of the provirus form of Mo-MLV is shown in Figure 1.1. The Mo-MLV genome is 8264 bp in length and is organized with its protein-coding domains flanked by cis-acting nucleic acid sequences. Duplicated at either end of the provirus are long terminal repeats (LTR). Within the LTR are the sequences necessary for the integration and the regulation of transcription. The LTR is approximately 600 bp in length and contains a duplicated transcription enhancer element followed by a promoter region, and a polyadenylation (poly A) signal. (Following the 5′ LTR is a region (–P) of RNA where the tRNA primer binds and reverse transcription is initiated. Next is the splice donor (SD) sequence used in the production of the subgenomic envelope messenger RNA. The region between the splice donor and the start of the first protein coding domain (termed Ψ) is deemed necessary for encapsulation of the genomic RNA.

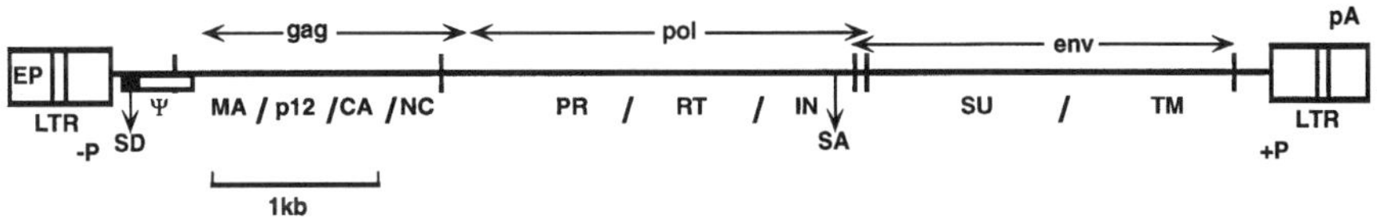

FIGURE 1.1
Genomic organization of the Moloney murine leukemia virus. The virus structural genes are labeled *gag, pol,* and *env.* The viral proteins derived from the structural genes are depicted below the genome (see the text for details). The regulatory elements are denoted as follows: E, enhancer; P, promoter; –P, tRNA primer binding site; SD, splice donor; Ψ, packaging signal; SA, splice acceptor; +P, positive strand initiation site; pA, polyadenylation signal. The E, P, and pA sites are found in both LTRs but are not duplicated in the figure, for clarity. The size in kilobase pairs (kb) is indicated.

The majority of the structural gene products are encoded by the first large open reading frame, called the *gag* (group antigen) gene. An initial gag polyprotein is processed to yield four proteins: p15 (matrix, MA), p12, p30 (capsid, CA), and p10 (nucleocapsid, NC). The next region, the *pol* (polymerase) gene, is organized into protease (PR), reverse transcriptase/RNase H (RT), and the integrase (IN) coding domains. These proteins are all derived from a gag/pol fusion protein that results from read-through translation of the gag stop codon. Before the end of the pol coding region is found the splice acceptor (SA) site that is used in the production of the subgenomic envelope (env) mRNA (the pol and env coding regions overlap and are out of frame relative to one another). The *env* gene is synthesized as a polyprotein that is then cleaved by a cellular protease to yield the gp70 (surface protein, SU) and the p15E (transmembrane protein, TM). Between the end of the *env* gene and the start site of the 3′ LTR is a purine-rich region that serves as the site for the initiation of plus strand cDNA synthesis (+P). As mentioned, the 3′ LTR duplicates all the information of the 5′ LTR and contains the sequence for mRNA poly A addition.

III. Retroviral Vector Designs

A. Early Vector Designs

Retroviral-mediated gene transfer takes advantage of the inherent separation of the protein-coding domains from the cis-acting regulatory elements to create an efficient gene transfer system. The first retroviral vectors were produced by inserting intact genes into naturally occurring, deleted retrovirus genomes. These defective viruses were then rescued with replication competent helper virus. As examples of this type of approach, several groups reported that the herpes simplex virus thymidine kinase (HSV-tk) gene could be inserted into both defective avian and murine retroviruses.[9-12] A nondefective helper virus was then used to demonstrate transfer of the tk gene-containing retrovirus to naive cells. Because these vectors required helper virus for infection, their use was limited. The discovery that retroviruses have a defined packaging element within the 5′ region of their RNAs prompted a greatly expanded effort by many laboratories to develop designs for retroviral vectors.[13-15]

In the design of the next generation of retroviral vectors, selectable marker genes were inserted immediately after the packaging element by deleting most of the remaining viral structural genes. Two examples of these early vectors were the construct pMSVgpt reported by Mann and co-workers[14] and the pLPL vector described by Miller et al.[16] The vector pMSVgpt contained the bacterial *gpt* gene (permitting growth in the presence of mycophenolic acid, xanthine, and aminopterin), while the pLPL vector contained the *hprt* gene (this gene can be used to select $hprt^+$ cells by growth in hypoxanthine, aminopterin, and thymidine). Analysis of these vectors revealed that the titers (the number of drug-resistant colony-forming units per milliliter) produced were approximately tenfold less than wild-type helper virus. During the analysis of deletion mutants of Mo-MLV-based vectors, Armentano et al.[17] found one retroviral vector construct (called N2) that was capable of generating titers near wild-type levels. In the N2 vector, the bacterial neomycin phosphotransferase II gene (Neo®) was inserted into a deleted version of the Mo-MLV that, in addition to the Ψ site, also contained sequences from the *gag* gene. The ability of gag sequences to enhance viral titer was verified in subsequent studies, and this larger packaging element is termed Ψ^+.[18-20] Vectors containing the Ψ^+ element can routinely generate titers between 10^6 and 10^7 cfu/ml.

B. Single Gene Vectors

Refinements in the current design of retroviral vectors that express single genes are aimed at both increasing gene expression and decreasing the potential for the generation of replication competent virus (Figure 1.2). The LNL6 vector improved on the design of the N2 vector by modifying the 5′ and Ψ^+ regions.[18] First, the *gag* gene sequence was altered by mutating the start codon to a stop codon. This prevents initiation of translation of any potential replication competent recombinants (this also has the potential for reducing the number of premature translation starts that may lessen the potential for proper translation of the inserted gene). In addition, the LTR and 5′ untranslated region from Mo-MLV were replaced with sequences from Moloney murine sarcoma virus (Mo-MSV). These substitutions both negate the potential for the production of a Mo-MLV-specific precursor gag protein as well as lessen the homology between the vector and the packaging genome. The LNL6 vector was further modified by the Miller group (and termed LN) by deletion of the Neo® gene 3′ untranslated region and all sequences derived from the Mo-MLV *env*

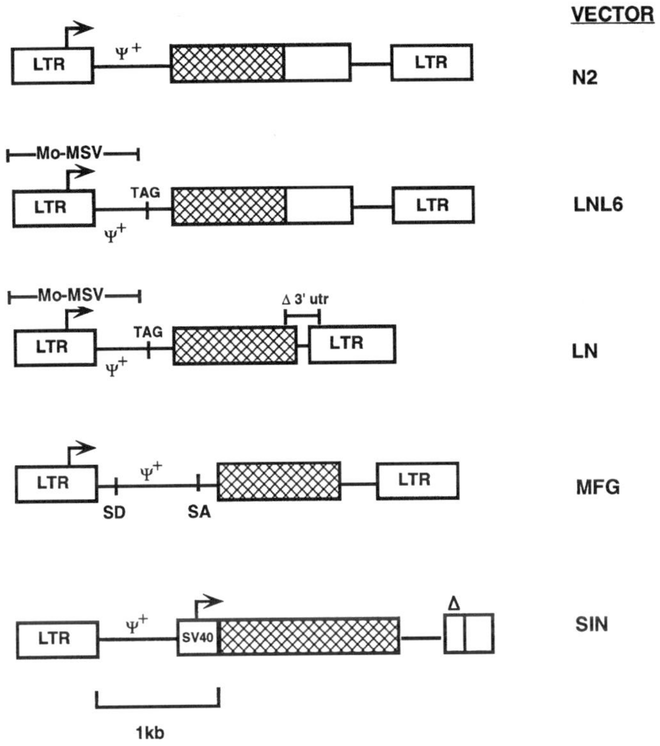

FIGURE 1.2
Single gene retroviral vectors. Five types of retroviral vectors are depicted and are described from top to bottom. All of the vectors were designed to express single genes. The N2 vector is an example of a Ψ^+ vector which uses the retroviral LTR to drive expression of a gene inserted 3′ to the extended packaging signal. The LNL6 vector is an example of a Ψ^+ modification 1 vector which has been modified to prevent translation initiation at the *gag* gene start codon, and has a substitution of Mo-MSV sequences for Mo-MLV sequences as indicated. The Ψ^+ modification 2 vector family, which includes the LN vector, contains further deletions in the 3′ end sequences derived from the Mo-MLV *env* gene. The MFG vector is representative of the splicing class of vectors. The splicing vectors use native splice donor and acceptor regions from the Mo-MLV to express an inserted gene in a similar manner to the natural RNA coding for the *env* gene. The self-inactivating vectors, SIN, contain deletions in the 3′ LTR that are translated to the 5′ LTR during reverse transcription. The gene expression in SIN vectors is mediated by an internal promoter, such as the SV-40 promoter. See the text for further details and references.

gene.[21] Recently vectors similar to the LN design but containing multiple cloning sites for gene insertion have been described (Figure 1.3).[22,23]

In both the N2 vector and the LNL6 vector, the retroviral splice donor sequence is retained (for the N2 virus the presence of this element may be required for gene expression). The importance of retaining the splice donor (SD) sequence in the LNL6-like vectors has been investigated by two groups.[20,21] In these reports, it was demonstrated that, while some SD mutations do not affect viral titer or gene expression, some other SD alterations can adversely influence titer. These results suggest that the packaging element may extend into the SD region. Additional vectors

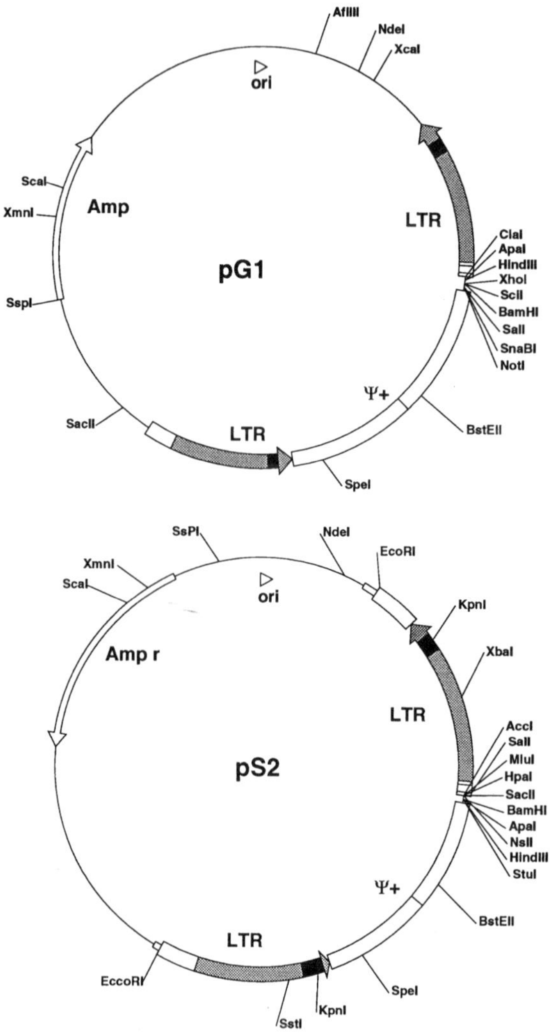

FIGURE 1.3

Single gene vectors containing multiple cloning sites. Depicted are two retroviral vectors containing multiple cloning sites which are similar to the LN vectors. The pG1 vector was constructed by the insertion of a linker containing multiple unique cloning sites between the 5′ and 3′ LTRs. The pS2 vector was generated by replacement of the neomycin gene in the N2 vector with a synthetic multiple cloning site. The purpose of generating such vectors is to simplify the cloning of the gene to be expressed in the context of the retroviral vector.

designed to express individual genes by a spliced mRNA have been developed. A vector termed MFG, developed by Mulligan and colleagues, mimics the normal pattern of mRNA splicing in the wild-type Mo-MLV.[24,25] In this vector, the gene of interest is inserted such that translation begins at the position of the native viral envelope gene, mediated by insertion of sequences from the Mo-MLV pol/env gene boundary (this region contains the natural splice acceptor and envelope protein translation start signal).

An alternative to expression of inserted genes from the retroviral LTR is to drive the expression of an internal promoter. The potential advantage of this design is that nonviral promoter/enhancer elements may permit tissue-specific gene expression. One theoretical reason to avoid using the retroviral LTR as a promoter is that, in certain cell types, the viral LTR can either have limited activity or is turned off by cellular functions.[26,27] An alternative to the use of internal promoters is to use a modified LTR. Several groups have reported that transcriptional silencing may be overcome by either using natural variants of the retroviral LTR, or by making chimeric enhancer elements within the context of the normal LTR enhancer.[28] Examples of hybrid LTRs include vectors containing substitutions of the Mo-MLV enhancer with enhancer elements from polyoma virus or the immunoglobin heavy chain gene.[29,30] The ability to make these biologically active chimeric viruses can be attributed to the plasticity of the retroviral genome.

In some circumstances, it may be advantageous to have a vector containing an internal promoter in the context of an inactivated LTR. Such an arrangement may increase the likelihood that the internal promoter will work independently and in a tissue-specific manner. The potential negative influence of the LTR promoter on the expression of an internal promoter has been observed in some systems and is termed promoter suppression. Vectors that inactivate the retroviral LTR have been termed suicide vectors or self-inactivating (SIN) vectors, and are constructed by making deletions in the U3 region of the 3′ LTR, usually in the enhancer.[23,31-33] As a consequence of the process of reverse transcription, any changes in the 3′ U3 region are copied to the 5′ end in the generation of the proviral LTR. This results in an inactive 5′ LTR and no, or very little, LTR-mediated gene expression.

C. Multi-Gene Vectors

In the majority of applications it is often advantageous to have more than one gene expressed from a retroviral vector. Of obvious utility is to have a selectable marker gene co-expressed in the same cell as the investigational gene; this permits enrichment for gene-containing cells. Several types of retroviral vectors have been constructed that use different mechanisms for achieving the expression of multiple genes (Figure 1.4). These include alternative splicing vectors, vectors containing internal promoters, vectors containing internal ribosome entry sites, and vectors that express a double copy of a transcription unit within the retroviral LTR. The first vector to drive the expression of two independent genes was the pZIP-NEOXV(X) construct.[34] This vector is an example of an alternative splicing vector, where the retroviral LTR is used to directly promote the expression of a gene of interest while a selectable marker gene (in this case the Neo® gene) is produced via splicing.

The most common design for the construction of multi-gene retroviral vectors is to use an internal promoter element. Variations on this theme are extensive and include the use of internal promoters derived from strong viral promoters, the use of cellular promoters, and the use of intact (intron containing) cellular genes. As

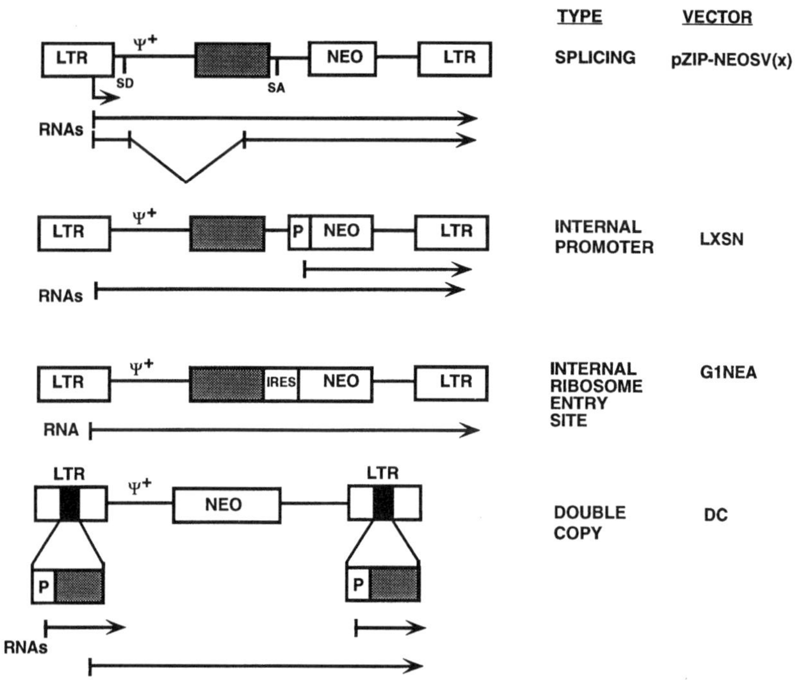

FIGURE 1.4
Multi-gene expressing retroviral vectors. Depicted are the general designs of retroviral vectors used to express multiple genes (described from top to bottom). The splicing vectors use the LTR to express both genes, one directly and one via RNA splicing (pZIP-NEOXV(X) is an example of this type of vector). Internal promoters use two (or more) promoters to express independent genes (LXSN is an example of this type of vector). The G1NEA vector is a vector containing an internal ribosome entry site (IRES). These vectors use the LTR to drive the expression of a single gene, the second cistron of which is translated via internal ribosome binding. The double copy vectors use a second promoter to express an additional gene, but this promoter/gene combination is inserted into the 3′ LTR, which is then copied to the 5′ LTR during reverse transcription. See the text for more details and references.

with splicing vectors, most internal promoter vectors express a selectable marker as one of the two genes. Three commonly used examples of this arrangement are the pDO-L,[35] the LXSN,[21] and the pBabe vectors.[20] All three vectors use the LTR to drive the expression of a gene of interest and an internal SV40 virus early region promoter to express a selectable marker. The pDO-L vectors use the minimal packaging element Ψ while the LXSN and pBabe use the extended packaging element Ψ^+. The differences between the LXSN and pBabe vectors include an inactivated splice donor signal and additional selectable marker genes in the pBabe vectors.

A new class of retroviral vectors has been reported recently that can be used to express multiple genes without the use of internal promoters or alternative splicing.[36,37] These vectors use internal ribosome entry sites (IRES) to facilitate the internal translation initiation of multiple protein coding domains. The IRES elements used in these vectors are naturally occurring sequences found in the 5′ untranslated regions of picornaviruses (e.g., the human poliovirus or the mouse encephalomyocarditis virus). Insertion of an IRES between protein coding sequences (between the stop codon of the first gene and the start codon of the second gene) produces a bicistronic mRNA. The types of genes expressed in these bicistronic vectors include both reporter genes (e.g., chloramphenicol acetyl transferase, CAT) and clinically relevant genes (e.g., adenosine deaminase, ADA). In our experience with these vectors, we have found them to be useful for expression of multiple genes; furthermore, constructs containing up to three genes can easily be assembled.[37]

A final category of vectors that can be used to express multiple genes are the double copy vectors.[38-40] The concept for these vectors relies on experience with SIN vectors. It was demonstrated that manipulation in the U3 region of the 3′ LTR is transferred to both LTRs following transduction; in addition to deletions, insertions into the U3 region are also propagated to both LTRs. The site of these insertions is in the extreme 5′ part of the U3 region, upstream of the enhancer. This insertion site places the double copy promoter outside of the retroviral transcription unit, leaving the normal LTR promoter intact and functional. Examples of vectors that have been constructed based on this design include a double copy vector that produces ADA and a double copy vector producing the HIV TAR region using the tRNA Met pol III promoter. Both of these vectors drive the *neo* selectable marker gene from the standard LTR promoter.

IV. Retroviral Packaging Cell Lines

Retrovirus packaging cell lines provide all the proteins required to assemble a functional retroviral vector (Figure 1.5). A variety of packaging cell lines based on mouse, primate, and avian retroviruses are available (Table 1.1). One of the essential principles behind the development of packaging cell lines is the capability to pseudotype vector-derived virus with envelope components of retroviruses that have a specific host range. This property is due to the ability of the murine retrovirus gag and pol proteins to interact with different envelope proteins (including non-murine envelopes). Most retroviral packaging cell lines contain the Mo-MLV *gag* and *pol* genes and a variable *env* gene. Both the ecotropic (infecting rat- and mouse-derived cells) and amphotropic (infecting a broad range of cell types, including human) viral envelope genes are used extensively in the production of retroviral vector packaging cells.

The development of packaging cell lines has progressed through three generations, with each generation increasing in complexity. The first generation packaging cell lines (e.g., Ψ-2) were constructed by deletion of the retroviral packaging elements (Ψ elements) from the replication competent retroviral genome.[14] The principle behind the cell line was that the virus genome synthesized all of the native proteins, but the genomic RNA was inefficiently packaged into virions; however, the recombinant retroviral vector genome (containing the packaging element) would efficiently be packaged into virions. While packaging cell lines based on this design effectively generate high-titer retrovirus preparations, they can easily generate replication competent retrovirus, as a single recombination event can transfer the Ψ element to the packaging genome.

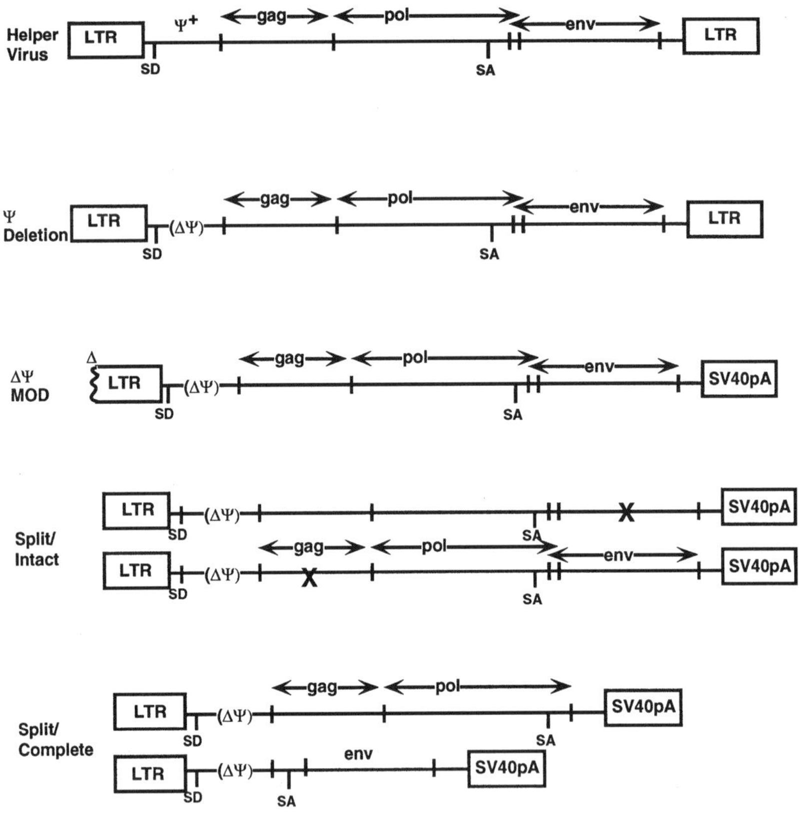

FIGURE 1.5
Retroviral packaging cell lines. Depicted are examples of four of the most common types of genomes used to construct retroviral packaging cell lines (they are described from top to bottom). The Ψ deletion cell lines were the first generation of packaging cell lines created and were produced by simple deletion of the packaging element. The second-generation packaging cell lines (ΔΨ mod) were created by removal of the 3′ LTR and deletion of the 5′ LTR sufficient to inhibit integration. The split gene/intact packaging cell lines are one of the new classes of packaging cell lines that use two modified genomes in combination. In this case, the genes are inactivated by insertions or deletions of *gag* and *env* genes, respectively. The split gene/complete cell lines also use two genomes to provide packaging cell function, but in this case most of the *env* gene has been deleted from the *gag/pol* expression plasmid and the *env* expression plasmid only contains *env* sequences. See the text for greater details and references.

Further retroviral genome modifications led to the development of second-generation packaging cell lines that contained multiple deletions in the 5′ and 3′ regulatory elements in addition to the deletion of the Ψ elements. One of the most widely used of the second-generation packaging cell lines is PA317.[41] In addition to the deletion of the Ψ site, the native viral genes beyond the 3′ end of the envelope gene are deleted and replaced with SV40 polyadenylation signals and, in 5′ LTR, elements involved in integration are removed. The resulting packaging cell line requires two recombination events to occur to generate replication competent retrovirus, and this greatly reduces the potential for helper virus production.

TABLE 1.1
Retrovirus Packaging Cell Lines

Envelope	Name of cell line	Type	Ref.
Ecotropic	Ψ2	Ψ Deletion	14
	Clone 32	Split/partial	52
	Ψ-CRE	Split/inactivation	42
	GP+E 86	Split/complete	43
	PE501	ΔΨ/Modifications	21
	WE	Split/complete	20
	ampli-GPE	Split/complete	53
	SV-Ψ-MLV	COS-7/Transient	46
Amphotropic	Ψ-AM	Ψ Deletion	2
	T19-14X	ΔΨ/Modification	54
	VT19-17-HZ	ΔΨ/Modification	54
	PA12	Ψ Deletion	55
	PA317	ΔΨ/Modification	41
	ΨCRIP	Split/inactivation	42
	GP-envAM12	Split/complete	44
	SV-Ψ-A-MLV	COS-7/Transient	46
	BOSC 23	293/Transient	47
	p*kat* ampac	293/Transient	46
Avian	C3A2	Split/partial	51
	Q4dh	Ψ Deletion	56
	pHF-g	ΔΨ/Modifications	57
	DSN	Split/complete	58
	Isolde	Split/complete	59
	SV-RSV-env	COS7/Transient	46
Primate	PG13	Split/complete	45
	PG53	Split/complete	45

The third and latest generation of stable packaging cell lines utilizes two modified genomes that combine to provide the necessary viral proteins. ΨCRIP and ΨCRE are examples of third-generation cell lines in which the gag/pol and env protein-coding regions are expressed independently.[42] In addition, the 3′ LTR is deleted from the two modified genomes. The expression vectors are sequentially introduced into cell lines and both plasmids selected. Further refinement of this design involves the physical separation of gag/pol and env coding regions on two independent expression vectors prior to transfection into cell lines.[43-45] This latest generation of packaging cell lines requires three recombination events to produce wild-type virus.

All of the aforementioned cell lines are utilized for the generation of stable retrovirus-producing cell clones, a process that can be protracted. Recently, several vector systems that allow for the production of high-titer retroviral supernatants without the generation of stable producer cell lines have been described. One of the first of these systems utilizes packaging plasmids that contain SV40 origins of replication that are cotransfected with the retroviral vector into COS-7 cells.[46] The gag/pol and env coding regions are physically separated on two packaging plasmids. The packaging plasmids and the recombinant retroviral vector are cotransfected into COS-7 cells. The COS-7 cells express the SV40 large T antigen, which permits replication of the packaging plasmids to a high copy number. The system produces retroviral supernatants free of helper virus, with titers ranging from 1×10^5 to 1×10^6 infectious units per milliliter.

Another high-titer transient retrovirus production system uses the highly transfectable human cell line, 293.[47] The retroviral packaging cell line BOSC23 was derived from the cell line 293tsA1609neo, which is a human embryonic kidney cell line that was transformed with adenovirus type 5. The BOSC23 packaging cell line was generated by the sequential transfection of two expression vectors containing physically separated gag/pol and env coding regions. High-titer retroviral supernatant that is free of helper virus can be generated by the calcium phosphate-mediated transfection of standard retroviral vectors into the BOSC23 cells. Under optimal conditions this system has generated retrovirus supernatants ranging from 1×10^6 to 1×10^7 infectious units per milliliter.

A final example of the transient transfection systems, the *kat* transduction system, incorporates the strong human cytomegalovirus (hCMV) immediate early promoter/enhancer regions into vectors to produce high steady-state levels of transcripts upon transfection into 293 cells.[48] In this system the vectors encoding the packaging functions contain a hybrid 5′ LTR that consists of the enhancer/promoter elements from the immediate early region of the hCMV fused directly to the U3 region of Mo-MSV. Both intact and split genome packaging vectors can be generated in this manner. The *kat* retroviral vectors are modified to contain a direct fusion of the hCMV promoter/enhancer to the R region of a retroviral LTR. When the *kat* retroviral and packaging vectors are cotransfected into 293 cells, the hCMV promoter elements are trans-activated by the E1A and E1B gene products localized within the 293 cells to produce high levels of transcript from the retroviral and packaging vectors. Viral titers have been observed as much as 10- to 5-fold higher than transient transfection into 3T3-based retroviral packaging cell lines. Although the utility of these novel transient high-efficiency systems has not been sufficiently tested, they may prove to be an excellent tool for screening multiple vectors and possibly allow for the transient expression proteins that are difficult to express in a stable manner.

V. Production of Retrovirus Supernatant

The first step in retroviral-mediated gene transfer is the transfection of retroviral vector plasmid DNA into a packaging cell line. The transfected vector DNA is then transcribed into an RNA that is packaged and excreted from the cell. The cell culture medium containing retroviral vector particles is generally referred to as retroviral vector supernatant. Retrovirus supernatant production can be either transient or stable in nature. The transient production of retrovirus supernatant can be very beneficial for rapidly screening multiple retroviral vectors or as an intermediate step in generating stable retrovirus-producing cell lines. However, stable retrovirus-producing cell lines are the preferred method of retrovirus vector production for most applications. The DNA transfection technique used in this laboratory for the generation of retrovirus-producing cell lines is calcium phosphate-mediated DNA coprecipitation.

Materials for Transfection

2.0 *M* $CaCl_2$	
29.4 g/100 ml	
2X HBS solution	
6 m*M* dextrose	1.19 g
137 m*M* NaCl	8.01 g
5 m*M* KCl	0.37 g
0.7 m*M* Na_2HPO_4	0.10 g
21 m*M* Hepes (free acid)	5.47
	1000 ml
pH solution to 7.05 with 1 *N* NaOH	
Store in aliquots at –20°C	

The packaging cell lines commonly used in our laboratory are PA317 for amphotropic vector production, GP+E 86 for ecotropic vector production, and PG13 for gibbon ape leukemia virus vector production. The PA317 and PG13 cell lines are maintained in Dulbecco's modified Eagle medium (DMEM) with 4.5 g/l glucose supplemented with 10% (v/v) fetal bovine serum (termed D-10). The GP+E 86 packaging cell line is maintained in the same medium supplemented with 10% newborn calf serum. All cell lines are cultured at 37°C in a humidified incubator in an atmosphere of 5% CO_2/air.

A. Transient Virus Production

Day 1: Plate the log-phase retrovirus packaging cell line (PA317, GP+E 86, or PG13) at 1×10^6 cells/10 cm dish, and incubate overnight.

Day 2: Replace medium with fresh D-10 4 to 6 h prior to transfection. Prepare the calcium phosphate DNA precipitate with the plasmid DNA as follows: the DNA should be suspended in 10 m*M* Tris-HCl, pH 7.4. In a sterile 12 × 75 mm polypropylene tube, mix 40 µg of plasmid DNA with 5 µl of 2 *M* $CaCl_2$ and sterile water to a final volume of 500 µl. Add the DNA-$CaCl_2$ solution in a dropwise manner to 500 µl of 2X HBSS while gently vortexing the tube. Incubate the solution at room temperature for 45 min, at which point a fine precipitate should be present. Add this precipitate directly to the plates and gently swirl to distribute the precipitate. Incubate the plates at 37°C for a minimum of 6 h to a maximum of 16 h.

Day 3: Aspirate the medium from the plate, wash the cells four times with PBS^- and replace with 10 ml of D-10.

Day 4: Collect the virus-containing medium and filter through a 0.45-μm cellulose acetate filter to remove cells and cellular debris. The retroviral supernatant can either be used to transduce cells immediately or frozen long-term at –70°C. Supernatants frozen at –70°C are stable indefinitely.

B. Stable Retroviral Vector-Producing Cell Lines

The generation of stable retroviral vector-producer cell lines is the preferred method for the production of retroviral vectors. The distinct advantage of stable transfection of packaging cell lines is the generation of an unlimited supply of a retroviral vector from a well-defined source. There are two different methods for the production of stable vector-producing cell lines; trans-infection and micro-ping-pong. While both techniques can yield high-titer producer cell clones, micro-ping-pong generally yields higher-titer cell populations than trans-infection.

1. Trans-Infection

Day 1: Plate 1×10^6 log-phase GP+E 86 cells into a 10 cm dish, and incubate overnight.

Day 2: Transfect the cells with the calcium phosphate DNA precipitate as previously described.

Day 3: Aspirate the medium from the plates of transfected GP+E 86, wash four times with PBS^-, and replace with 10 ml of D-10. Seed log-phase PA317 cells at 2×10^5 cells/10 cm dish. Make five PA317 plates per plate of transfected GP+E 86 cells.

Day 4: Collect the medium from the GP+E 86 plates and filter through a 0.45-μm filter to remove cells and debris. Split the GP+E 86 cells 1:5 into D-10 plus the drug of choice for selection. For example: 0.6 mg/ml G418 (active concentration), 4 m*M* histidinol, or 0.4 mg/ml hygromycin B. These cells can be utilized as a new source of ecotropic virus, if no PA317 cells grow out of the transinfection/selection process. Use 5.0 ml, 0.5 ml (2 times), and 0.1 ml (2 times) of GP+E 86 cell supernatant to infect the PA317 plates. Adjust the indicated amounts of supernatant to a final volume of 7 ml D-10 containing 8 μg/ml polybrene. Aspirate the D-10 medium from the PA317 cells and add the virus dilutions, and incubate overnight.

Day 5: Aspirate medium and add 10 ml of D-10 containing one of the selection agents previously mentioned, plus HAT (100 μ*M* hypoxanthine, 400 nm aminopterin, 16 μ*M* thymidine) to select for the packaging genome. After a period of 10 to 14 d, drug-resistant colonies should appear in the PA317 plates. Using cloning rings, pick a minimum of 12 colonies from each transduction in which the colonies are well separated. Cloning rings are prepared by coating one end of the ring with vacuum grease, placing the rings greased side down into a glass Petri dish, and sterilizing by autoclaving. To pick clones, aspirate the medium from the tissue culture dish and, with the plate inverted in the tissue culture hood, circle individual colonies with a marking pen on the bottom of the dish. Return the tissue culture dish to its upright position and place cloning rings greased side down over the previously marked positions. To each ring, add 50 μl of trypsin/ethylene diamine tetraacetic acid (EDTA) and incubate for 5 min at room temperature.

Add 50 μl of medium to a cloning ring, and vigorously pipette the medium to dislodge the cells from the dish. Place the cells into a single well of a 24-well tissue culture plate containing 0.5 ml of D-10 with selective agent plus HT (hypoxanthine, thymidine). As the individual clones reach confluence, expand the cells to 6-well (using D-10 plus selective agent plus HT) plates and finally into 10 cm dishes (with D-10 plus selective agent). Each clone should then be analyzed for viral titer, presence of replication competent helper virus and structure of the vector genome by Southern blot. All of these procedures will be discussed below.

2. Micro-Ping-Pong

The micro-ping-pong technique involves co-cultivation of amphotropic and ecotropic producer cell lines, followed by transfection with the retroviral vector DNA. The virus produced from the transfection of an ecotropic packaging cell can infect an amphotropic packaging cell, and vice versa. This ultimately leads to superinfection of both cell types. The result of the superinfection is that the transduced cells contain multiple copies of integrated provirus, thus generating a higher titer. The potential problems with this system are twofold: first, if the retroviral vector is unstable and prone to rearrangements, micro-ping-pong will select for the rearranged vectors. Second, this procedure can rapidly amplify replication competent retrovirus that may be at undetectable levels in the original cell lines.

Day 1: Plate log-phase cells into 10 cm dishes at the following densities: 3×10^5 GP+E 86 cells plus 2×10^5 PA317 per plate, and incubate overnight.

Day 2: Transfect the cells with the retroviral vector, as previously described.

Day 3: Aspirate the medium from the plates, wash four times with PBS^- and replace with 10 ml of D-10.

Days 4–9: Culture cells and split every 2 to 3 d out to 1:5.

Day 10: After a minimum period of 7 d post-transfection, split the cells 1:10 into D-10 containing HAT (100 μ*M* hypoxanthine, 400 nm aminopterin, and 16 μ*M* thymidine) and the selection agent found in the retroviral vector. The HAT selects for the PA317 that contains the herpes simplex virus thymidine kinase (tk) gene, while the GP+E 86 ecotropic producer cells lack this resistance gene. After clones form, pick individual colonies as previously discussed. The clones are placed in 6-well plates and selected using D-10 containing HT (100 μ*M* hypoxanthine and 16 μ*M* thymidine) and the selection agent such as G418. As the cells are split into 10 cm dishes, they are selected in D-10 containing the selection drug and an aliquot of the cells should be frozen away until the individual clones are analyzed for titer, helper virus, and retroviral genome organization by Southern blotting.

VI. Collection and Titration of Retrovirus Supernatant

Infectious retrovirus particles can be harvested from the producer cells simply by collecting medium off the tissue culture dishes. As the producer cells reach confluence, the medium is replaced with fresh D-10. Recent evidence suggests that the titer of retroviral producers can be increased by harvesting supernatant after incubation at 32°C, as opposed to 37°C, but supernatant can be harvested at either temperature.[49] The medium may be changed as frequently as every 8 h or the cells can be incubated overnight prior to collection. Supernatant can be harvested from confluent producer cells four to six times over a 24-h period from the same dish. Upon collection, centrifuge the medium at 3000 × *g* for 5 min at 4°C to remove cells and cellular debris, and store at 4°C (virus is stable at this temperature for several days). At the end of the collection period, aliquot the pooled supernatant into 50-ml tubes and store at –80°C. A small aliquot (<1 ml) of each virus is stored for titer determination.

A vital part of characterization of a producer cell line is the determination of viral titer, or the number of infectious virions per milliliter of supernatant. The protocol involves the transduction of target cells with limiting dilutions of the retroviral supernatant and the selection of clones of productively transduced cells. The recipient cells commonly used are 3T3 cells for amphotropic and ecotropic viruses or HeLa cells for gibbon ape leukemia viruses. The protocol is as follows:

Day 1: Split the recipient cells into 6-well plates at 1×10^5 cells per well, and incubate overnight.

Day 2: Generate enough D-10 plus 8 mg/ml polybrene to make all of the appropriate dilutions of retroviral supernatant. The polybrene stock (hexadimethrine bromide) is made up as a 1000X stock (8 mg/ml in PBS^-) and filtered through a 0.45-nm filter, and stored at 4°C. Thaw supernatants at room temperature or quickly at 37°C and immediately place on ice. Using the D-10/polybrene, prepare log-fold dilutions of viral supernatant from 1×10^{-1} to 1×10^{-5}. Aliquot enough of the virus (need 2 ml virus/well) such that samples can be analyzed in duplicate or triplicate. Aspirate the medium from the wells containing recipient cells, replace with the diluted virus, and incubate overnight at 32°C.

Day 3: Aspirate the inoculum and replace with 3 ml of D-10 containing the selective agent (0.6 mg/ml G418, 4 m*M* histidinol, or 0.4 mg/ml hygromycin B). Incubate plates at 37°C for 8 to 10 d, such that well-formed colonies can be seen on the plate.

Days 11–13: Stain colonies by removing the selective media, washing in PBS^-, and adding methylene blue stain. The methylene blue stain is prepared by adding 0.6 g methylene blue powder to 100 ml of methanol. Cells are stained at room temperature for 15 min, then wash plates with water and allow to air dry. The viral titer, expressed as colony-forming units per milliliter of supernatant, is determined by counting the number of colonies and multiplying by 1/dilution factor (e.g., 50 colonies $\times 1/10^{-5} = 5 \times 10^6$ cfu/ml).

VI. Assays for Replication Competent Helper Virus

The most important assay for the characterization of retroviral producer cell lines is the assay for replication competent retrovirus (helper virus). The presence of helper virus in a producer cell population renders that producer cell unusable for gene transfer purposes. The marker rescue assay is currently the most sensitive technique available for the detection of replication competent retrovirus.[42] This assay detects the mobilization of a retroviral vector from cells by a viral supernatant. Mobilization of the integrated vector would be the direct result of a transfer of the packaging function from the helper virus, such that the cells would begin to produce virions containing the genome produced from the integrated vector. The integrated vector contains a selectable marker, such as the *neo* gene, that acts as the indicator for the assay. An essential part of the assay is to be certain that the target cells chosen are capable of being infected by the helper virus. For example, if the helper virus being tested is produced by the GP+E 86 cell line, then human cells cannot be used for the assay, because the ecotropic virus will not infect human cells.

To initiate the assay, the target cell line must first be generated. This can be accomplished by the calcium phosphate-mediated transfection of 3T3 or HeLa cells with a retroviral vector containing the *neo* gene, such as LXSN. The cells are selected in G418 and resistant clones are pooled and collected. Another protocol for the generation of the target cell line involves transducting 3T3 or HeLa cells with a helper-free vector supernatant, selecting in G418, and generating a population of cells. The retrovirus transduced cells are passaged for 14 d to allow ample time for production of helper virus, which should not be present. The cells are assayed for the production of any helper virus by infecting 3T3 cells with supernatant collected from the cells, and selecting for *neo*. Ideally, no G418-resistant colonies should be detected. Cells containing the integrated retroviral vector are utilized for the marker rescue assay as follows:

Day 1: Plate the vector-containing target cells at 1×10^6 cells per 10 cm dish, and incubate overnight.

Day 2: Infect the target cells with 2 ml of test virus (filtered through a 0.45-μm filter), and add 7 ml of D-10 containing 8 μg/ml polybrene. Positive control plates can be set up by infecting the target cells with a small amount of retrovirus known to be replication competent.

Days 3–16: Passage cells for 14 d to allow for the replication and spread of the helper virus. Split cells to 1:10 dilutions with trypsin/EDTA as needed, using care not to cross-contaminate any cultures, as some may begin to produce significant amounts of helper virus.

Day 17: Change medium on target cells to fresh D-10. Plate naive 3T3 cells into 6-well plates at 1×10^5 cells per well.

Day 18: Collect the medium from target cells and add polybrene to 8 μg/ml and filter medium through a 0.45-μm filter. Use 1- to 10-ml fractions to infect the 3T3 cells at 37°C overnight.

Day 19: Begin selection by adding D-10 containing 0.6 mg/ml G418. Select in G418 for 8 to 10 d.

Days 28–30: Stain colonies as described previously with methylene blue and count. Detection of any colonies indicates that the *neo* vector was rescued by the replication competent retrovirus.

VIII. Analysis for Rearrangement of the Provirus

The successful generation of a retroviral producer cell or the productive transduction of target cells requires the integration of an unrearranged provirus. Therefore, analysis of the integrity of the retroviral vector is an important step in the analysis of producer cell production, as well as transduction of target cells. The structure of the provirus is analyzed by Southern blot analysis of genomic DNA isolated from either producer or target cells. The analysis is performed by standard Southern blot protocol,[50] using an enzyme that cuts the vector once within each LTR, such that the structure of the integrated viral vector is retained. For most Mo-MLV-based vectors, enzymes such as Kpn I or Sst I are ideal for this purpose. The probe for the blot can be either the investigational gene inserted into the vector or the *neo* gene. Rearranged provirus can be detected by the presence of multiple, smaller or larger fragments on the blot than the size of the unrearranged vector. The size of the unrearranged provirus can be determined by adding a control lane to the blot consisting of the retroviral vector DNA plasmid (30 pg) used to generate the producer cell.

IX. General Transduction Protocol

The exposure of cells directly to the retroviral supernatant is the method of choice for the transduction of many cell types. This can be accomplished as follows:

- Plate the target cells at a low density such that they will divide during the transduction period. Cell division is a requirement for successful retroviral transduction.
- Thaw the retroviral supernatant quickly at 37°C and then place immediately on ice (virus has a short half-life at 37°C).
- Supplement the supernatant with 8 μg/ml polybrene or 5 μg/ml protamine sulfate. Protamine sulfate is less toxic to some cells than polybrene.
- Sterile-filter the supernatant through a 0.45-μm filter. *Never* use a 0.22-μm filter, as the retrovirus does not pass efficiently through the filter.
- Use a multiplicity of infection ≥1 for the transduction of target cells. For suspension cells, pellet them by centrifugation and resuspend directly in the vector supernatant. Suspension cells can also be transduced by direct co-cultivation with producer cells. The cells can either be directly co-cultivated or separated by a porous membrane that allows for the diffusion of the retrovirus from the producer cells to the target cells. For adherent cells, aspirate the medium and replace with the retroviral supernatant. Incubate for 6 to 16 h at 37°C. There is some recent evidence to indicate that retroviral transductions performed at 32°C provide an increase in gene transfer, probably due to increased half-life of the virus.[49] For some cell types, it may be advantageous to repeat the transduction several times on consecutive days to achieve optimal gene transfer.
- Add fresh medium and culture cells for 24 to 48 h before applying selection.

- Selection protocols should be optimized prior to transduction to determine the amount of the selective agent necessary to kill all of the target cells within 10 to 14 d. For G418 this can vary from 0.2 mg/ml up to 3 mg/ml. Apply selection to transduced cells.
- If cells are to be cultured for extended periods of time, selection should be reapplied for 1-week intervals every 4 weeks.
- Analyze cells for gene transfer.

Acknowledgments

The authors acknowledge Theresa Lumsden for her assistance in the preparation of this manuscript. We would also like to thank Drs. Paula Gregory, David Bodine, and Linda Muul for their helpful discussions and critical review of this manuscript.

References

1. Wilson, J. M., Danos, O., Grossman, M., Raulet, D. H., and Mulligan, R. C., *Proc. Natl. Acad. Sci. U.S.A.*, 87, 439 (1990).
2. Cone, R. D. and Mulligan, R. C., *Proc. Natl. Acad. Sci. U.S.A.*, 81, 6349 (1984).
3. Blaese, R. M. and Anderson, W. F., *Hum. Gene Ther.*, 1, 327 (1990).
4. VandenDriessche, T., Chuah, M. K. L., and Morgan, R. A., *AIDS Update*, 7, 1 (1994).
5. Morgan, R. A. and Anderson, W. F., *Annu. Rev. Biochem.*, 62, 191 (1993).
6. Weiss, R. A., Teich, N., Varmus, H. E., and Coffin, J. M., Eds., *Molecular Biology of Tumor Viruses: RNA Tumor Viruses*, 2nd ed, Cold Spring Harbor Laboratory Press, Cold Spring Harbor, New York, 1985.
7. Temin, H. M. and Mizutani, S., *Nature*, 226, 1211 (1970).
8. Baltimore, D., *Nature*, 226, 1209 (1970).
9. Shimotohno, K. and Temin, H. M., *Cell*, 26, 67 (1981).
10. Wei, C. M., Gibson, M., Spear, P. G., and Skolnick, E. M., *J. Virol.*, 39, 935 (1981).
11. Tabin, D. J., Hoffman, J. W., Goff, S. P., and Weinberg, R. A., *Mol. Cell. Biol.*, 2, 426 (1982).
12. Joyner, A. L. and Bernstein, A., *Mol. Cell. Biol.*, 3, 2191 (1983).
13. Shank, P. R. and Linial, M., *J. Virol.*, 63, 513 (1989).
14. Mann, R., Mulligan, R. C., and Baltimore, D., *Cell*, 33, 153 (1983).
15. Watanabe, S. and Temin, H. M., *Proc. Natl. Acad. Sci. U.S.A.*, 79, 5986 (1982).
16. Miller, A. D., Jolly, D. J., Friedmann, T., and Verma, I. M., *Proc. Natl. Acad. Sci. U.S.A.*, 80, 4709 (1983).
17. Armentano, D., Yu, S. F., Kantoff, P. W., von Ruden, T., Anderson, W. F., and Gilboa, E., *J. Virol.*, 61, 1647 (1987).
18. Bender, M. A., Palmer, T. D., Gelinas, R. E., and Miller, A. D., *J. Virol.*, 61, 1639 (1987).
19. Adam, M. A. and Miller, A. D., *J. Virol.*, 62, 3802 (1988).
20. Morganstern, J. P. and Land, H., *Nucleic Acids Res.*, 18, 3587 (1990).

21. Miller, A. D. and Roseman, G. J., *BioTechniques,* 7, 980 (1989).
22. McLachlin, J. R., Mittereder, N., Daucher, M. B., Kadan, M., and Eglitis, M. A., *Virology,* 195, 1 (1993).
23. Faustinella, F., Kwon, H., Serrano, F., Belmont, J. W., Caskey, C. T., and Aguilar-Cordova, E., *Hum. Gene Ther.,* 5, 307 (1994).
24. Ohashi, T., Boggs, S., Robbins, P., Bahnson, A., Patrene, K., Wei, F.-S., Wei, J. F., Li, J., Lucht, L., Fei, Y., Clark, S., Kimak, M., He, H., Mowery-Rushton, P., and Barranger, J. A., *Proc. Natl. Acad. Sci. U.S.A.,* 89, 11332 (1992).
25. Jaffee, E. M., Dranoff, G., Cohen, L. K., Hauda, K. M., Mulligan, R. C., and Pardol, D., *Cancer Res.,* 53, 2221 (1993).
26. Hoeben, R. C., Migchielsen, A. A. J., vander Jagt, R. C. M., van Ormondt, H., and Vander Eb, A. J., *J. Virol.,* 65, 904 (1991).
27. Palmer, T. D., Rosman, G. J., Osborne, W. R. A., and Miller, A. D., *Proc. Natl. Acad. Sci. U.S.A.,* 88, 1330 (1991).
28. Couture, L. A., Mullen, C. A., and Morgan, R. A., *Hum. Gene Ther.,* 5 667 (1994).
29. Moore, K. A., Scarpa, M., Kooyer, S., Utter, A., Caskey, C. T., and Belmont, J. W., *Hum. Gene. Ther.,* 2, 307 (1991).
30. Valerio, D., Einerhard, M. P. W., Wamsley, P. M., Bakx, T. A., Li, C. L., and Verma, I. M., *Gene,* 84, 419 (1989).
31. Yu, S.-F., von Ruder, T., Kantoff, P. W., Garber, C., Seiberg, M., Ruthers, U., Anderson, W. F., Wagner, E. F., and Gilboa, E., *Proc. Natl. Acad. Sci. U.S.A.,* 83, 3194 (1986).
32. Cone, R. D., Reilly, E. B., Gisen, H. N., and Mulligan, R. C., *Science,* 236, 954 (1987).
33. Guild, B. C., Finer, M. H., Housman, D. E., and Mulligan, R. C., *J. Virol.,* 62, 3795 (1988).
34. Cepko, C. L., Roberts, B. E., and Mulligan, R. C., *Cell,* 37, 1053 (1984).
35. Korman, A. J., Frantz, J. D., Strominger, J. L., and Mulligan, R. C., *Proc. Natl. Acad. Sci. U.S.A.,* 84, 2150 (1987).
36. Adams, M. A., Ramesh, N., Miller, A. D., and Osborne, W. R. A., *J. Virol.,* 65, 4985 (1991).
37. Morgan, R. A., Couture, L., Elroy-Stein, O., Ragheb, J., Moss, B., and Anderson, W. F., *Nucleic Acids Res.,* 20, 1293 (1992).
38. Hantzopolous, P. A., Sullenger, B. A., Ungers, G., and Gilboa, E., *Proc. Natl. Acad. Sci. U.S.A.,* 86, 3519 (1989).
39. Sullenger, B. A., Gallardo, H. F., Ungers, G. E., and Gilboa, E., *Cell,* 63, 601 (1990).
40. Sullenger, B. A., Lee, T. C., Smith, C. A., Ungers, G. E., and Gilboa, E., *Mol. Cell. Biol.,* 10, 6512 (1990).
41. Miller, A. D. and Buttimore, C., *Mol. Cell. Biol.,* 6, 2895 (1986).
42. Danos, O. and Mulligan, R. C., *Proc. Natl. Acad. Sci. U.S.A.,* 85, 6460 (1988).
43. Markowitz, D., Goff, S., and Bank, A., *J. Virol.,* 62, 1120 (1988).
44. Markowitz, D., Goff, S., and Bank, A., *Virology,* 167, 400 (1988).
45. Miller, A. D., Garcia, J. V., von Suhr, N., Lynch, M., Wilson, C., and Eden, M. V., *J. Virol.,* 65, 2220 (1991).
46. Landau, N. R. and Littman, D. R., *J. Virol.,* 66, 5110 (1992).
47. Pear, W. S., Nolan, G. P., Scott, M. L., and Baltimore, D., *Proc. Natl. Acad. Sci. U.S.A.,* 90, 8392 (1993).
48. Finer, M. H., Dull, T. J., Qin, L., Farson, D., and Roberts, M., *Blood,* 83, 43 (1994).

49. Kotani, H., Newton, P. B. III, Zhang, S., Chiang, Y. L., Otto, E., Weaver, L., Blaese, R. M., Anderson, W. F., and McGarrity, G. J., *Hum. Gene Ther.,* 5, 19 (1994).
50. Southern, E. M., *J. Mol. Biol.,* 98, 503 (1975).
51. Watanabe, S. and Temin, H. M., *Mol. Cell. Biol.,* 3, 2241 (1983).
52. Bosselman, R. A., Hsu, R.-Y., Bruszewski, J., Hu, F., Martin, F., and Nicholson, M., *Mol. Cell. Biol.,* 7, 1797 (1987).
53. Takahra, Y., Hamada, K., and Housman, D. E., *J. Virol.,* 66, 3725 (1992).
54. Sorge, J., Wright, D., Erdman, V. D., and Cutting, A. E., *Mol. Cell. Biol.,* 4, 1730 (1984).
55. Miller, A. D., Law, M.-F., and Verma, I. M., *Mol. Cell. Biol.,* 5, 431 (1985).
56. Stoker, A. W. and Bissell, M. J., *J. Virol.,* 62, 1008 (1988).
57. Savatier, P., Bagnis, C., Thoraval, P., Poncet, D., Belakebi, M., Mallet, F., Legras, C., Cosset, F.-L., Thomas, J.-L., Chebloune, Y., Faure, C., Verdier, G., Samarut, J., and Nignon, V., *J. Virol.,* 63, 513 (1989).
58. Dougherty, J. P., Wisniewski, R., Yang, S., Rhode, B. W., and Temin, H. M., *J. Virol.,* 63, 3209 (1989).
59. Cosset, F.-L., Legras, C., Chebloune, Y., Savatier, P., Thoraval, P., Thomas, J.-L., Samarut, J., Nignon, V. M., and Verdier, G., *J. Virol.,* 64, 1070 (1990).

Chapter 2

Methods for Studying Genetic Variation of the Human Immunodeficiency Virus (HIV)

*Gerald Schochetman, Shambavi Subbarao, and Marcia L. Kalish**

Contents

I. Introduction

The impact of the extensive genetic heterogeneity of human immunodeficiency virus (HIV) type 1 on the current AIDS pandemic has yet to be defined. The genetic diversity of HIV-1 strains in a geographic region is, in part, a function of the length of time that the virus has been present in the population. Genetic diversity may be further influenced by migration of infected people from other endemic geographic regions. To understand the molecular epidemiology of HIV-1 and to best design a candidate vaccine, it is essential to understand the evolution, geographic distribution, and genetic diversity of HIV-1 strains in different countries. To determine the geographic distribution of HIV-1 strains, specimens from a large number of studies, involving over 24 countries, have been collected by laboratories at the Centers for Disease Control and Prevention (CDC) for genetic analysis. Many of these studies have been conducted since 1988, and include the following countries: the U.S., Haiti, Honduras, Canada, Zaire, Cote d'Ivoire, Uganda, Malawi, Rwanda, Zambia, Kenya, Tanzania, Nigeria, Peru, Brazil, Thailand, Senegal, Cape Verde, Guinea Bissau, China, Taiwan, Myanmar, Malaysia, and India. In the performance of these studies the following protocols were used for DNA extraction and amplification, viral subtype-specific hybridization, and DNA sequencing.

II. Specimen Preparation

A. Extraction of DNA from Peripheral Blood Mononuclear Cells (PBMCs)

Working under a biological safety hood, co-cultured cells or PBMCs from whole blood collected in heparin, citrate, or ethylene diamine tetraacetatic acid (EDTA) and purified using Ficoll-Hypaque are split into two aliquots of 10^6 PBMCs and centrifuged for 5 min at 200 × *g* (see Table 2.1 for the approximate number of PBMCs per given volume of whole blood). The supernatant is removed and the cell pellets washed twice with phosphate-buffered saline to remove potential inhibitors of the polymerase chain reaction (e.g., heparin). DNA is extracted from the cells by a standard detergent/proteinase K lysis procedure.[1] The procedure involves resuspending the washed cells in 250 μl of lysis buffer (50 m*M* KCl, 10 m*M* Tris-HCl, pH 8.3, 2.5 m*M* $MgCl_2$, 25 μ*M* EDTA, 0.45% NP40, 0.45% Tween-20; proteinase K added prior to use to a final concentration of 60 μg/ml), and incubating the tubes in a 56°C water bath for 1 h. The volume of lysis buffer added to the washed cells should be adjusted so that a final concentration of approximately 100,000 to 150,000 PBMCs per 25 μl of lysis buffer is achieved. Following lysis at 56°C, the tubes are clamped to prevent the caps from opening and the proteinase K is heat inactivated at 95°C for 15 min. The resulting lysates are then stored frozen at –20°C until needed for polymerase chain reaction (PCR) amplification. Alternatively, PBMCs can be collected and partially purified in Vacutainer Cell Preparation Tubes (CPT™, Becton Dickinson, Franklin Lanes, NJ).* These tubes incorporate sodium citrate as the anticoagulant, a polyester gel plug, and a density gradient medium for immediate processing of plasma and PBMCs from whole blood in the collection tube. After separation, the PBMCs are pelleted from the plasma, washed, and the DNA is extracted and stored as described above.

TABLE 2.1
Blood Leukocyte Cell Populations

5.6 l of blood in a 70-kg person
500,000 white blood cells per 65 μl whole blood
White Blood Cells
350,000 polymorphonuclear cells (70%)
150,000 mononuclear cells (30%)
Mononuclear Cells
15,000 monocytes (10%)
15,000 large granular lymphocytes (LGL) (10%)
15,000 B-cells (10%)
105,000 T-cells (70%)
T-cells
37,000 CD8 (suppressor) cells (35%)
68,000 CD4 (helper) cells (65%)

* Use of trade names is for identification purposes only and does not imply endorsement by the Public Health Service or the U.S. Department of Health and Human Services.

B. Extraction of DNA from Dried Blood Spots

A convenient way to collect and ship blood specimens is in the form of dried blood samples collected on filter paper. This method has been used for more than 30 years to screen newborns for metabolic disorders. A 0.5 cm^2 piece is cut from filter paper-bound dried blood samples (S&S No. 903 filter paper, Schleicher & Schuell, Inc., Keene, NH) and placed in an Eppendorf tube with 500 µl of lysis buffer (same as used for PBMCs) containing 60 µg/ml proteinase K and 0.5% sodium dodecyl sulfate (SDS). The specimens are incubated overnight at 56°C in an Eppendorf thermomixer (Hamburg, Germany) to facilitate lysis of the cells. The next day, the lysates are extracted twice with phenol/chloroform, the aqueous phase transferred to a fresh tube, and the DNA is precipitated by the addition of 1/10 volume of 3 *M* sodium acetate and 2.5 volumes of 95% ethanol. The addition of 10 µg of carrier RNA (yeast tRNA) to each tube before precipitation may help the recovery of small amounts of DNA. Following precipitation, the DNA pellets are washed once with cold 70% ethanol and desiccated under vacuum in a Speed-Vac (Savant). The dried pellets are reconstituted in 25 to 100 µl deionized water and stored at –20°C.

C. Extraction of DNA from Tissues

Occasions arise when the only specimen available for HIV testing from a specific person is tissue that has been fixed in formalin and embedded in paraffin wax. The type of fixative employed and the length of fixation time are critical considerations for subsequent DNA amplification. In general, tissue that has been fixed in 10% buffered formalin for less than 3 d should yield satisfactory results. The procedures for the extraction of DNA from tissues have been adapted from previously published protocols.[2]

Sections 5 to 10 µm thick are cut, using a new, disposable microtome blade for each specimen, to prevent carry-over contamination from a previous tissue. Tissue sections are deparaffinized before extraction by adding 1 ml of xylene (or solvent of choice, e.g., Propar or Histoclear) to each microcentrifuge tube containing a tissue section. The tissues are allowed to sit at room temperature for about 15 min to dissolve the paraffin completely. The tubes are then centrifuged at 12,000 × *g* for 5 min. The supernatant is decanted and the pellet, which may be difficult to see, is washed by adding 1 ml of 95 to 100% ethanol, and briefly inverting the tube to mix. The material is centrifuged at 12,000 × *g* for 5 min and the ethanol wash is repeated. The supernatant is carefully decanted and the tissue pellets are desiccated under vacuum. The desiccated tissue can be used immediately for DNA extraction or can be stored at –20°C until required.

If frozen or fixed tissues (non-embedded) are being used, the specimens are washed twice in phosphate-buffered saline (PBS) to remove the fixative. The tissue is cut into small pieces (2 to 5 mm^2) in a biological safety hood, and individual pieces are placed in sterile microfuge tubes. The tissues are then digested by the addition of 0.5 ml of lysis buffer-proteinase K solution (50 m*M* KCl, 10 m*M* Tris-HCl, pH 8.3, 2.5 m*M* $MgCl_2$, 25 µ*M* EDTA, 0.45% NP40, and 0.45% Tween 20; proteinase K is added to obtain a final concentration of 400 µg/ml). The smaller deparaffinized tissues are digested by adding 50 to 100 μl of the lysis buffer-proteinase K solution. The digestions are carried out at 56°C for 3 h. For larger pieces of tissue, the incubation time is extended for approximately 1 h. Following digestion, the tubes are vortexed for 30 sec and the proteinase K is heat inactivated at 95°C for 15 min. The residual undigested tissue material is removed by centrifuging the digestions at 12,000 × *g* for 5 min. The supernatants (lysates) are stored frozen at –20°C. Small volumes of the lysates are used in the PCR reaction. More typically, several 10-fold serial dilutions of the lysates are made in order to dilute out inhibitors of the PCR step. Alternatively, nucleic acids in the tissue lysates can be bound to silicon dioxide particles[3] under high salt conditions. These particles act as the solid phase for the capture of nucleic acids. Following several washes the nucleic acids are eluted from the silica particles, free of inhibitors, and used for PCR.

III. HIV DNA Amplification and Detection

A. Diagnostic PCR

We have performed diagnostic PCR on over 25,000 individual PBMC DNA samples from specimens collected worldwide. The most significant application of diagnostic PCR is on pediatric specimens. Due to the presence of maternal antibodies, serological determination of HIV infections in infants is not reliable. Most maternal antibodies are cleared by 18 months of age, at which time HIV negative babies will be seronegative. Many of our specimens come from studies involving large pediatric transmission cohorts. Multiple sequential specimens from each baby are received and PCR test results are compared to serological tests at 18 months of age. Diagnostic PCR involves the amplification of conserved regions of the *gag* gene followed by quantitation of the amount of DNA amplified using a hybridization protection assay (HPA). Approximately 1 μg of DNA from each sample (lysed PBMCs) representing about 150,000 PBMCs is used in a standard 100 μl PCR reaction consisting of 10 m*M* Tris-HCl, pH 8.3, 50 m*M* KCl, 0.025 m*M* EDTA, 2.5 m*M* $MgCl_2$, 0.2 m*M* each dNTP, 2.5 units Taq DNA polymerase, and 0.02 μg of each primer. Duplicate tubes of DNA are amplified with each set of *gag*-specific primer pairs SK38/SK39 and SK145/SK150 (Table 2.2) for 35 cycles using the following cycling profile: thermal ramping to 95°C for 1 min 15 sec, hold at 95°C for 1 min, ramping to 55°C for 1 min 30 sec, hold at 55°C for 30 sec, ramping to 60°C for 30 sec, hold at 60°C for 2 min 30 sec. Amplifications are carried out in a 96-well GeneAmp PCR System 9600 thermocycler (Perkin Elmer Cetus, Norwalk, CT) without the use of oil. After amplification is complete, the tubes are removed from the thermocycler and moved to a different lab for HIV-1 detection. In order to avoid contamination, the tubes containing amplified DNA are never opened in the lab where the amplifications are performed. The HPA used for HIV-1 nucleic acid detection (Gen-Probe Inc., San Diego, CA) uses two sets of acridinium ester (AE) labeled-oligonucleotide probes that are specific for two regions of the *gag* gene that are amplified by the primer pairs

TABLE 2.2
HIV-1 Primer Sequences

Primer	Sequence (5′→3′)	Reference	Gene	Orientation
MK369	TGGAGCCAGTAGATCCTAGACTAGAGCCCT	BRU 5413-5442	tat	→
MK616	AATGGTGAGTATCCCTGCCTAACTCTATT	BRU 7961-7933	*env*/gp41	←
MK603	CAGAAAAATTGTGGGTCACAGTCTATTATGGGGTACCT	BRU 5914-5931	*env*/C1	→
CO602	GCCCATAGTGCTTCCTGCTGCTCCCAAGAACC	BRU 7410-7379	*env*/gp41	←
MK650	AATGTCAGCACAGTACAATGTACAC	BRU 6538-6562	*env*/C2	→
CO601	TTCTCCAATTGTCCCTCATATCTCCTCCTCCA	BRU 7258-7227	*env*/V5	←
CL1028	TCTCTAGCAGTGGCGCCCGAACAGGGAC	BRU 171-199	*gag*/P17	→
AB1033	TCTATCCCATTCTGCAGCTTCCTCATTGAT	BRU 977-948	*gag*/P17	←
CL1029	TTTGACTAGCGGAGGCTAGA	BRU 307-326	*gag*/P17	→
AB1032	GGTGATATGGCCTGATGTACCATTTGCCCCTG	BRU 781-749	*gag*/P17	←
AB1060	ACACTAGAAGAAATGATGACAGCATGTCAGGGAGT	BRU 1359-1394	*gag*/P24	→
AB1061	CAGAGCCAACAGCCCCACCAG	BRU 1729-1749	*gag*	→
SK38	ATAATCCACCTATCCCAGTAGGAGAAAT	BRU 1091-1117	*gag*/P24	→
SK39	TTTGGTCCTTGTCTTATGTCCAGAATGC	BRU 1204-1177	*gag*/P24	←
SK145	AGTGGGGGGACATCAAGCAGCCATGCAAAT	BRU 905-934	*gag*/P24	→

SK150	TGCTATGTCACTTCCCTTGGTTCTCTC	BRU 1046-1019	*gag*/P24	←
AB1071	TGTTGACAGGTGTAGGTCCTACTAATACTGTACCTA	BRU 2049-2084	*pol*	←
Group O				
AB1081	AAAAACAGGCTCTATTCTATGTATCAGAT	ANT 70 6751-6779	*env*/V2	→
AB1083	TGCCGCTGCGCCCATAGTGCTACCTGCTGCACTTAG	ANT 70 7880-7845	*env*/C4	←
AB1080	GTTACTTGTACACATGGCAT	ANT 70 6993-7012	*env*/C2	→
AB1082	AGTTCTCCATATATCTTTCAT	ANT 70 7709-7688	*env*/C3	←
AB1079	TTTACCCATGCATTTAAAGTCCTGGGGGAGATGGC	ANT 70 1305-1271	*gag*/P17	←
AB1078	TGATGTACCATTTGTCCCTG	ANT 70 1269-1249	*gag*/P17	←
AB1075	TGGCCTCCGGGGGGCACGAGGCCAGGCAATT	ANT 70 2167-2198	*pol*	→
AB1077	TTTACTTTTGGTCCATCCATTCCTGGTTT	ANT 70 2669-2641	*pol*/protease	←
AB1074	TTTGCCTCCCTCAAATCCCTCTTTGGGACAGACCAA	ANT 70 2299-2335	*pol*	→
AB1076	TTTACTGGCACTGGGGCTATGGG	ANT 70 2636-2614	*pol*/protease	←

[a]The group M HIV-1s are the major subtypes of HIV-1 responsible for the worldwide AIDS epidemic. The group O HIV-1s (HIV-1s that currently do not fit into any of the defined sequence subtypes of the group M viruses) have sequences that are as distinct from each other as they are from the group M subtypes. See Myers, G. et al., *Human Retroviruses and AIDS, 1994*, Los Alamos National Laboratory, Los Alamos, New Mexico.

SK38/39 and SK145/150 (Table 2.2). Acridinium esters are nonradioactive, highly chemiluminescent compounds. Hybridization of the labeled HIV-specific probes with the target PCR amplified DNA protects the chemiluminescent AE from subsequent alkaline hydrolysis.[4] The following procedure is applied using reagents and controls supplied with the chemiluminescent DNA probe and detection system (Gen-Probe Inc.). Amplified DNA and appropriate negative and positive controls are denatured at 95°C for 5 min, followed by the addition of the AE-labeled probes, and allowed to hybridize at 60°C for 30 min. Alkaline hydrolysis of the acridinium esters of unbound probes is carried out at 60°C for 10 min. Once hydrolyzed, the AE group is no longer chemiluminescent and the chemiluminescence being emitted from the bound probe is read in a luminometer (Leader-1, Gen-Probe Inc.) as relative light units (RLUs). RLU readings of greater than 10,000 are considered as HIV positive. Most HIV negative samples have maximal RLU values of only several hundred. Therefore, there is approximately a 2-log differential between a positive and a negative test.

B. Nested and Anchored PCR

Amplifications of the desired regions of the HIV genome are routinely performed using either an anchored or nested PCR procedure. Approximately 1 μg of PBMC DNA is used in a standard 100 μl reaction consisting of 10 m*M* Tris, pH 8.3, 40 m*M* NaCl, 0.01 m*M* EDTA, 0.2 m*M* each dNTP, 1.5 m*M* $MgCl_2$, 2.5 units of Taq DNA polymerase, and 0.15 μg of each primer. The PCR profile for the primary round of amplification is 96°C for 1 min, 65°C for 3 min 30 sec (for fragments less than 1000 bp in length) or 5 min (fragments greater than 1000 bp in length), and the cycle is repeated 40 times in the automated thermocycler 480 (Perkin-Elmer Cetus). Five microliters of the amplified product are diluted 50-fold and 5 μl of the diluted DNA is used as template DNA for 30 cycles under the same conditions (as the primary amplification) with a set of nested primers (nested PCR) or a single nested primer used in combination with one of the primers from the primary reaction (anchored PCR). A list of primer sequences routinely used in our laboratory for HIV-1 DNA amplification of the *gag*, *pol*, and *env* genes from PBMC samples collected worldwide are included in Table 2.2. The size and approximate concentrations of amplified products are verified by electrophoresis of 2 to 5 μl of DNA on a 1% agarose gel, stained with ethidium bromide, and compared to either the ΦX174/Hae III or 1 kb ladder molecular weight markers (Gibco/BRL, Grand Island, NY). DNA concentrations can also be calculated using the optical density conversions in Table 2.3.

TABLE 2.3
Quantitation of DNA

1 A_{260} unit dsDNA = 50 μg/ml = 0.15 m*M*
1 A_{260} unit ssDNA = 33 μg/ml = 0.10 m*M*
Average weight of a DNA base pair = 650 Da
Mol. wt. of dsDNA = (number of base pairs) × (650 Da/base pair)

C. HIV-1 Subtype-Specific Oligonucleotide Hybridization-Based Subtyping

Our laboratory has developed subtype-specific oligonucleotide probes for subtyping HIV-1 strains from a number of countries. The selection of subtype-specific probes is based on sequence variation between subtypes in conserved regions of the *env* gene of HIV-1. For example, the hybridization-based method has been used successfully to subtype several hundred HIV-1 strains from Thailand. The probes used display 100% specificity and sensitivity for the Thai subtypes B′ and E.[5] Nested primers MK650/CO601 are used to amplify the C2-V5 domain of gp120. A small aliquot (5 to 10 μl) of the amplified DNA is then subjected to alkaline denaturation for 5 min at room temperature by the addition of a tenth volume of denaturing solution (4 *M* NaOH, 100 m*M* EDTA). A small volume (1 to 2 μl) of the denatured DNA samples is spotted onto a Duralon membrane (Stratagene, La Jolla, CA), UV cross-linked, and allowed to air dry. Type-specific oligonucleotide probes are 3′ end-labeled with digoxigenin-dUTP using the DIG DNA Labeling and Detection Kit (Boehringer Mannheim, Indianapolis, IN). The nylon membranes are prehybridized at 37°C for 1 h, followed by hybridization with the DIG-labeled subtype-specific probes at 37°C for 1 h. The unbound probes are removed by washing the membranes following standard protocols (Genius System User's Guide, Boehringer Mannheim, Indianapolis, IN). Colorimetric detection of bound probe is carried out with nitroblue tetrazolium and X-phosphate using the Genius system. The subtyping of HIV strains from Thailand using the hybridization-based approach showed excellent concordance with the results obtained using site-specific immunoassays based on the prototype V3 loop amino acid sequences for each genotype and antisera from the patients.[6]

IV. DNA Cloning and Transformation Techniques

A. Cloning Strategies

Cloning of PCR products is valuable for the long-term maintenance of DNA samples for sequence analysis and the generation of probes. DNA extracted directly from PBMCs, tissues, or co-cultured PBMCs can be amplified by nested or anchored PCR, and then cloned into the appropriate vector. The PCR-amplified DNA is first digested with either a single restriction endonuclease or with two different restriction endonucleases, which allows for directional cloning in a suitably digested vector. This is facilitated by the incorporation of appropriate restriction sites into the 5′ end of the PCR primers. Several vector systems can be used, including M13 (Bethesda Research Laboratories), pBluescript (Stratagene Cloning Systems), pGEM (Promega Corp.), and the M13 TA (M13mp18, M13mp19, M13mp10) vectors developed in our laboratory by C.-C. Luo and N. de la Torre. Our system is based on the TA cloning vector (pCRII) originally developed by Invitrogen (San Diego, CA). The TA cloning system takes advantage of the terminal transferase activity of *Taq* polymerase, which adds a single, 3′ A-overhang to each end of the PCR product. The vector is linearized with a specific restriction enzyme which generates single 3′ T-overhangs on either end of the cut site, enabling direct ligation of PCR products by A–T base pairing. The original pCRII vector and the M13 TA vectors have the *lacZα* complementation fragment allowing for blue-white color screening of recombinants. The M13 TA vectors were constructed by cloning a synthesized fragment of DNA containing two XcmI sites between the *Bam*HI site and *Sal*I sites in the M13 multiple cloning region. The vector is linearized with XcmI, leaving single 3′-T overhangs, which makes it suitable for TA cloning. This procedure permits ligation of the PCR-amplified DNA directly into the vector without restricting endonuclease digestion.

B. Ligation

Ligation of double-stranded PCR-amplified DNA with the TA cloning M13 vectors is achieved by mixing linearized vector and amplified DNA (1:2 molar ratio), 1 unit T4 DNA ligase, and T4 DNA ligase buffer in a total volume of 10 μl. The reactions are incubated overnight at 14°C or at room temperature for 4 h.

C. Transformations

The ligations are transformed into *Escherichia coli* XL2-Blue Ultracompetent Cells (Stratagene, La Jolla, CA). XL2-Blue cells carry the F′ episome encoding for pili that allow infection by filamentous M13 phage. The F′ episome also contains the *lacI*^{q}ZΔM15 mutation, which, together with the M13 *lacZα* gene, allows for blue-white selection of recombinant plaques on media supplemented with the chromogenic substrate X-gal, and the *lac* operon inducer IPTG. The required amount of competent cells are thawed on ice and treated with β-mercaptoethanol (added to a final concentration of 25 m*M*), which has been shown to increase transformation frequencies two- to threefold. For each transformation, 100 μl of treated competent cells are placed in sterile polypropylene tubes on ice. Small volumes (1 μl of a 1:10 dilution) of the ligation mix are added to the cells on ice, gently mixed, and allowed to incubate on ice for 45 min. The transformation mixtures are then heat-shocked at 42°C for 30 to 45 sec and placed on ice for at least 2 min. The transformation mixtures are then transferred to a sterile tube containing 3 ml YT top-agar and 150 μl of XL2-Blue lawn cells (an exponentially growing culture of host cells). The contents of the tube are briefly mixed by gentle vortexing and poured onto a Luria-agar plate. The top agar is allowed to solidify and the plates are inverted and allowed to incubate overnight at 37°C.

D. Identification of Recombinant Phage

Plaques containing recombinant M13 phage can be detected as white plaques if isopropyl-β-D-thiogalactoside (IPTG) and X-gal are added to the top-agar. We apply a different approach to eliminate false positives using plaque hybridizations with two different 3′-digoxigenin-labeled probes, each complementary to one strand of the cloned DNA fragment. This allows for the identification of phage containing opposite strands of the inserted DNA fragment for subsequent use in sequencing. The plaques are lifted onto nylon membranes, UV cross-linked, and hybridized with the probes at 37°C. Positive plaques (those containing probe-specific cloned DNA) are detected by chemiluminescence using the substrate Lumi-Phos 530, or colorimetrically using nitroblue tetrazolium and X-phosphate, as described by the manufacturer (Genius System, Boehringer Mannheim Co., Indianapolis, IN).

E. M13 Template Preparation for Sequencing Reactions

Templates for sequencing are prepared by growing the M13 phage from recombinant plaques and extraction of single-stranded DNA from the phage. A well-isolated positive plaque is inoculated into 5 ml of Luria broth containing a 1:50 dilution of a fresh overnight culture of XL2-Blue host cells. The culture tubes are incubated at 37°C with vigorous shaking for 6 to 8 h. The cultures are centrifuged at room temperature to pellet the bacterial cells. The pellet is used for extraction of M13 RF DNA using a small-scale plasmid prep and is stored as a stock of the clone. The culture supernatant is transferred to a sterile polypropylene tube and the phage is precipitated by adding 1.2 ml of a 20% polyethylene glycol (PEG) and NaCl solution (volume of PEG/NaCl can be scaled according to the volume of supernatant) and incubating the tubes on ice for 1 h. The phage is pelleted by centrifugation at 4°C and all the residual PEG/NaCl solution is carefully removed from the pellet. The phage pellets are resuspended in 500 μl of Tris-EDTA; pH 8.0 (TE) buffer and extracted once with equal volumes of phenol, followed by a phenol/chloroform (1:1) extraction and finally extracted with chloroform. Single-stranded template DNA is precipitated with 1:10 volume of 3 *M* sodium acetate and 2.5 volumes of 95% ethanol. The DNA is washed with 70% ethanol and dissolved in 50 μl of deionized water for storage at –20°C. The concentration of DNA is determined by measuring the optical density using the conversions in Table 2.3. Alternatively, the approximate concentration can be estimated by agarose gel electrophoresis.

V. Sequencing Protocols

A. Direct Sequencing of Amplified DNA

We have used both Taq dye-primer cycle sequencing and the Taq dye terminator cycle sequencing methods (Applied Biosystems Inc., Foster City, CA) to perform direct sequencing of purified PCR products. Automated sequencing using these methods provides a convenient way of determining the HIV-1 DNA sequence profile for a given population consisting of several hundred infected persons. The direct sequencing of DNA amplified from lysed PBMC extracts, without having to first generate individual plasmid/M13 clones from the amplified DNA, does not sacrifice sequence reliability, and generates a "consensus" sequence representing the most common nucleotide at each position for all the quasispecies variants within an HIV-infected person. Purification of nested PCR products prior to sequencing is carried out using either Wizard PCR prep minicolumns (Promega Corp., Madison, WI) or QIA-Quick-Spin PCR Purification columns (Quiagen Inc., Chatsworth, CA) adhering to protocols supplied by the manufacturer. The passage of DNA through the columns removes primer-dimers, amplification primers, and salts. Purification of PCR products is necessary to reduce the intensity of background "noise" on sequencing chromatograms.

Automated sequencing using the dye-labeled primer technology is especially useful for surveys of large populations of infected persons, in which the same region of the HIV genome from each sample is being sequenced using the same primer binding site. Dye-primer cycle sequencing makes use of four identical primers, each labeled to a different fluorescent dye. The DNA primers are synthesized and labeled by Applied Biosystems Inc. (Foster City, CA), using sequences provided by us. The reactions are performed using reagents and protocols supplied with the Taq Dye Primer Cycle Sequencing Kit (Applied Biosystems Inc.). The reactions are carried out in a 96-well plate format in a Perkin Elmer Model 9600 thermocycler using the following profile: (95°C for 30 sec, 55°C for 30 sec, and 70°C for 1 min) for 15 cycles; then (95°C for 30 sec, 70°C for 1 min) for 15 cycles. At completion, the samples are held at 4°C. All four reaction tubes (A, C, G, and T) are combined into an ethanol mixture containing 80 μl of 95 to 100% ethanol with 1.5 μl 3 *M* sodium acetate (pH 5.3). The DNA is precipitated, washed with 70% ethanol, and dried. The DNA pellet is resuspended in 6 μl of deionized formamide/50 m*M* EDTA (pH 8.0) 5:1 (v/v) and stored at –20°C. Prior to electrophoresis, the samples are denatured at 95°C for 2 to 3 min, placed on ice, and immediately loaded onto a pre-electrophoresed 6% acrylamide gel. Sequencing gels are run on the ABI Model 373A Sequencer or 370A Sequencer upgraded to the Model 373A capacity.

The following analysis has been included to demonstrate the accuracy of the direct sequencing approach in determining the major quasispecies present within an HIV-infected person. The C2-V3 sequences from six M13 clones of PCR-amplified DNA from PBMCs isolated from an HIV-1-infected person were obtained using the ABI dye labeled M13 universal –21 primer. The sequences were aligned (DNASIS software) and a consensus sequence of the six clones (CONSENSUS) was derived and is shown in Figure 2.1. In this particular example, there is a high degree of sequence homology among the clones. Where the sequences do differ, as indicated by a missing asterisk below the consensus sequence, the most frequently occurring nucleotide (>50%) was chosen for the consensus sequence. The PCR product used to generate the six M13 clones was directly sequenced using HIV-envelope specific dye-labeled primers. The sequence obtained (DIRECT) is also shown in Figure 2.1. The direct sequence of the PCR product, in this instance, is almost identical to the consensus sequence of the M13 clones. Bases at different positions of the DIRECT sequence correspond to the most frequently occurring base present in the clonal sequences at a given position. For example, at position 56, there are five clones carrying thymine and one with adenine (Figure 2.1). In the DIRECT sequence, base 56 is a thymine, which is consistent with the finding that 83% of the clones sequenced show a thymine at that position. Another example is seen at position 32, where four clones contain cytosine, one adenine, and one thymine (Figure 2.1). The corresponding base in the direct PCR sequence is a cytosine. It should be noted that direct sequencing determines the most common nucleotide at each position in a population of quasispecies.

Although, in the overwhelming majority of positions, the most commonly occurring base is appropriately represented in direct sequencing, exceptions do occur. In the example in Figure 2.1, there are discrepancies between the CONSENSUS and DIRECT sequences at three different positions: 203, 214, and 271. While cytosine occurs with a lower frequency (33%) then thymine at position 203 in the sequences of the clones, the direct sequence assigns a cytosine at this position. Similarly at positions 214 and 271, where adenine and guanine occur with equal frequencies in the sequences of the clones, the direct sequence assigns a guanine at both these positions. We have observed on sequence chromatograms that the signal for adenine tends to be under-represented, possibly due to the exhaustion of adenine-specific labeled primer or terminator. This decrease in signal intensity becomes more apparent about 300 to 350 bases into the sequence and could explain why direct sequencing "calls" a guanine at both positions 214 and 271. A judicious choice of the primer annealing site can ensure that the signal intensity will not affect the accuracy of the sequence. For example, to obtain accurate sequences of the immunodominant V3 loop region (positions 85–189, Figure 2.1) we use primers that anneal approximately 200 bases upstream of the V3 loop coding region. We have found through repeated experiments that direct sequencing of the V3 region is both reproducible and reliable.

Occasionally, an amino acid at a given position in the translation of a direct sequence is either not seen or poorly represented in the amino acid translations of the cloned sequences. Following are possible explanations: (1) direct sequencing does not imply an inherent linkage between nucleotides in a codon. From a direct sequence, the amino acid deduced from the translation of given codon is actually based on a "consensus codon", i.e., each of the three bases represent the most commonly found bases for each position of that codon and the amino acid encoded by the "consensus codon" can be different from amino acids encoded by codons in individual clones. This would lead to an incorrect interpretation of the "consensus codon" in cases where major quasispecies variants exist, each of which has substitutions at differing positions within the same codon. (2) There was a biased selection during the cloning procedure, and the cloned sequences do not truly represent all the genetic variants present in the amplified DNA; or (3) there were an insufficient number of clones sequenced to arrive at a true consensus. Nevertheless, direct sequencing of PCR amplified proviral DNA is easy to perform and the sequences generated are sufficiently reliable for studying phylogenetic relationships and for determining the genetic diversity of HIV-1 in a population of infected individuals.

This dye-terminator cycle sequencing method is useful for obtaining sequences of large regions of DNA. Using dye-labeled terminators one can sequence with the same primers that were used to amplify the DNA. The initial sequence generated is

```
                 1        10        20        30        40        50
Clone1    AAGAGGTAGTAATTAGATCTGAAAATTTCACGGACAATGTTAAAACCAT
Clone2    AAGAGGTAGTAATTAGATCTGAAAATTTCATGGACAATGTTAAAACCAT
Clone3    AAGAGGTAGTAATTAGATCTGAAAATTTCACGGACAATGTTAAAACCAT
Clone4    AAGAGGTAGTAATTAGATCTGAAAATTTCAAGGACAATGTTAAAACCAT
Clone5    AAGAGGTAGTAATTAGATCTGAAAATTTCACGGACAATGTTAAAACCAT
Clone6    AAGAAGTAGTAATTAGATCTGAAAATTTCACGGACAATGTTAAAACCAT
CONSENSUS AAGAGGTAGTAATTAGATCTGAAAATTTCACGGACAATGTTAAAACCAT
          **** ************************* *******************
DIRECT    AAGAGGTAGTAATTAGATCTGAAAATTTCACGGACAATGTTAAAACCAT

                  60        70        80        90       100
Clone1    AATAGTGCAGCCGAATGAATCTGTACAAATTAATTGTACAAGACCCAACA
Clone2    AATAGTGCAGCTGAATGAATCTGTACAAATTAATTGTACAAGACCCAACA
Clone3    AATAGTGCAGCTGAATGAATCTGTGCAAATTAATTGTACAAGACCCAACA
Clone4    AATAGTGCAGCTGAATGAATCTGTACAAATTAATTGTACAAGGCCCAACA
Clone5    AATAGTGCAGCTGAATGAATCTGTACAAATTAATTGTACAAGACCCAACA
Clone6    AATAGAGCAGCTGAATGAATCTGTACAAATTAATTGTACAAGACCCAACA
CONSENSUS AATAGTGCAGCTGAATGAATCTGTACAAATTAATTGTACAAGACCCAACA
          ***** ***** ************ ***************** *******
DIRECT    AATAGTGCAGCTGAATGAATCTGTACAAATTAATTGTACAAGACCCAACA

                 110       120       130       140       150
                         . V3 loop .
Clone1    ACAATACAAGAAAAGGTATACATATAGCACCGGGGAGAGCATTTTATGCA
Clone2    ACAATACAAGAAAAAGTATACATATAGCACCGGGGAGAGCATTTTATGCA
Clone3    ACAATACAAGAAAAAGTATACATATAGCACCGGGGAGAGCATTTTATGCA
Clone4    ACAATACAAGAAAAAGTATACATATAGCACCGGGGAGAGCATTTTATGCA
Clone5    ACAATACAAGAAAAAGTATACATATAGCACCGGGGAGAGCATTTTATGCA
Clone6    ACAATACAAGAAAAAGTATACATATAGCACCGGGGAGAGCATTTTATGCA
CONSENSUS ACAATACAAGAAAAAGTATACATATAGCACCGGGGAGAGCATTTTATGCA
          ************** ***********************************
DIRECT    ACAATACAAGAAAAAGTATACATATAGCACCGGGGAGAGCATTTTATGCA

                 160       170       180       190       200
Clone1    ACAGGAGAAATAATAGGAGATATAAGACAAGCACATTGTAACCTTAGTAG
Clone2    ACAGGAGAAATAATAGGRGATATAAGACAAGCACATTGTAACCTTAGTAG
Clone3    ACAGGAGAAATAATAGGAGATATAAGACGAGCACATTGTAACCTTAGTAG
Clone4    ACAGGAGAAATAATAGGAGATATAAGACAAGCACATTGTAACCTTAGTAG
Clone5    ACAGGAGAAATAATAAGAGATATAAGACAAGCACATCGTAACCTTAGTAG
Clone6    ACAGGAGAAATAATAGGAGATATAAGACAAGCACATTGTAACCTTAGTAG
CONSENSUS ACAGGAGAAATAATAGGAGATATAAGACAAGCACATTGTAACCTTAGTAG
          *************** ************ ******* *************
DIRECT    ACAGGAGAAATAATAGGAGATATAAGACAAGCACATTGTAACCTTAGTAG

                 210       220       230       240       250
Clone1    CATAAAATGGAGGAACACTTTAAGACAGCTAGCTAAAAAATTAAAAGAAC
Clone2    CATAAAATGGAATGACACTTTAAGACAGATAGCTAAAAAATTAAAAGAAC
Clone3    CACAAAATGGAATGACACTTTAAGACAGATAGCTAAAAAATTAAAAGAAC
Clone4    CACAAAATGGAATGACACTTTAAGACAGACAGCTAAAAAGTTAAAAGAAC
Clone5    CATAAAATGGAATAACACTTTAAGACAGATAGCTAAAAAATTAAAAGAAC
Clone6    CATAAAATGGAATAACACTTTAAGACAGATAGCTAAAAAATTAAAAGAAC
CONSENSUS CATAAAATGGAATAACACTTTAAGACAGATAGCTAAAAAATTAGAAGAAC
          ** ********   **************  ********* **********
DIRECT    CACAAAATGGAATGACACTTTAAGACAGATAGCTAAAAAATTAAAAGAAC

                 260       270       280       290       300
Clone1    AATTTGGAAATAAAACAATAGTCTTTAATCAATCCTCAGGAGGGGAACCA
Clone2    AATTTGGAAATAAAACAATAATCTTTAATCAATCCTCAGGAGGGGAACCA
Clone3    AATTTGGAAATAAAACAATAGTCTTTAATCAATCCTCAGGAGGGGAACCA
Clone4    AAATTGGAAATAAAACAATAATCTTTAATCAATCCTCAGGAGGGGAACCA
Clone5    AATTTGGAAATAAAACAATAGTCTTTAATCAATCCTCAGGAGGGGAACCA
Clone6    AATTTGGAAATAAAACAATAATCTTTAATCAATCCTCAGGAGGGGAACCA
CONSENSUS AATTTGGAAATAAAACAATAATCTTTAATCAATCCTCAGGAGGGGAACCA
          ** ***************** *****************************
DIRECT    AATTTGGAAATAAAACAATAGTCTTTAATCAATCCTCAGGAGGGGACCCA
```

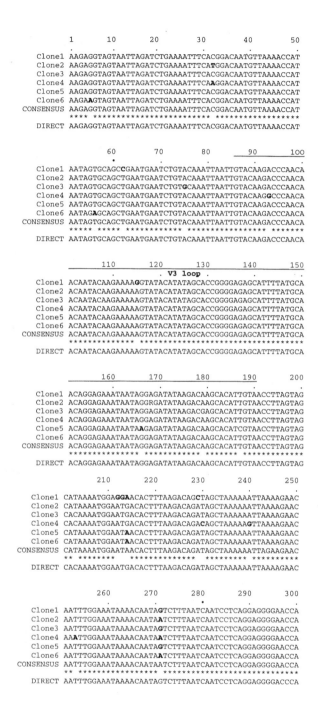

FIGURE 2.1
Comparison of direct sequencing of the C2-V3 region amplified from PBMC DNA lysates (isolated from an HIV-infected patient) with sequences of six clones of the same PCR fragment.

then used to design additional primers in the 3′ region, which are used to sequence the following 300 to 350 nucleotides of the DNA fragment and so on to allow sequencing by "walking" down the HIV genome region of interest. Dye-terminator sequencing is performed using the Taq Dye-Deoxy™ Terminator Cycle Sequencing Kit (Applied Biosystems Inc., Foster City, CA). The method relies on four ABI dye-labeled dideoxy nucleotides: G, A, T, and C DyeDeoxy terminators. When these terminators replace standard dideoxy nucleotides in the sequencing reaction (using AmpliTaq DNA polymerase), a dye-label is incorporated in the DNA along with the terminating base. All four termination reactions are performed in a single tube instead of four individual reactions needed in the dye-primer cycle sequencing. The reactions are carried out in a 96-well plate format in a Perkin Elmer 9600 thermocycler using the following profile: (95°C for 30 sec, 50°C for 15 sec, and 60°C for 4 min) for 25 cycles; then held at 4°C indefinitely. At completion, the sequencing reactions are passed through Centri-Sep columns (Princeton Separations Inc., Adelphia, NJ) to remove the unincorporated dye-labeled terminators by centrifugation. The reactions are vacuum-dried and the dried pellet is resuspended in 3 to 4 μl of deionized formamide/50 m*M* EDTA (pH 8.0) 5:1 (v/v). The sample is denatured at 95°C for 2 to 3 min, placed on ice, and immediately loaded onto a pre-electrophoresed 6% acrylamide gel.

Using the two automated sequencing methods described above we have sequenced both different regions of the *env* gene, as well as the entire gp120 and gp41 genes, and portions of the *pol* gene from different HIV-1 strains.

VI. Data Management and Primary Analysis

A. Data Acquisition and Entry

The DNA sequences are read using: (1) the ABI Automatic Data Collection Software for the Macintosh;[7] (2) the IBI Pustell Sequence Gel Reader for the IBM-compatible personal computer; and (3) the BioImage DNA Film Reader for the Sun Microsystems SPARC 2 computer.

B. Primary Analysis of Genetic Sequence Data

Primary analysis is performed to compile, align, and translate all nucleotide sequence data to identify frame shifts, deletions, insertions, and internal stop codons. To verity the integrity of sequences generated, we compare them with known HIV sequences to ensure that the sequence is HIV-specific and represents the proper gene fragment using available computer software for the IBM PC: DNASIS (Hitachi, San Bruno, CA), PROSIS (Hitachi, San Bruno, CA), PC Gene (IntelliGenetics, Mountainview, CA), IBI Pustell Sequence Analysis (IBI Kodak, New Haven, CT), and CLUSTAL version V Multiple Sequence Alignment package.[8] Available computer software for the Sun Microsystems include IntelliGenetics Suite of DNA Analysis (IG Suite) (IntelliGenetics, Mountainview, CA), BioImage DNA Sequence Assembly Manager (Millipore, Marlboro, MA), Genetic Data Environment (GDE) (developed by the Ribosomal Data Project, University of Illinois, Urbana, IL), Multiple Alignment Sequence Editor (MASE),[9] Viral Epidemiology Signature Pattern Analysis (VESPA),[10] and Pattern Induced Multiple Alignment (PIMA),[11,12] Pustell Sequence Gel Reader, or the BioImage DNA Sequence Film Reader for Sun computers (Millipore Imaging Products, Ann Arbor, MI).

C. Database Search

A sequence data database search is used for comparison between our generated sequences and data reported by other laboratories. The programs Fasta (IBI Pustell), Blast (GDE), and IntelliGenetics Suite are used for database searches. The following databases are available and updated on either a quarterly or monthly basis, as new data are released: GenBank, EMBL, and the Los Alamos National Laboratory HIV database (to access database via worldwide web type: *http://hiv-web.lanl.gov/*or to transfer data from database via internet type: *ftp ncbi.nlm.nih.gov,* type *anonymous* for the login name and your e-mail address for the password, then change directory to *"repository/aids-db"* to download entries).

D. Phylogenetic Analysis

We use a variety of software for inferring phylogenetic relationships. The sequences are routinely aligned using MASE, CLUSTAL, ESEE, or GDE, which provide an interface to subsequent phylogenetic analysis programs. Phylogenetic analysis is done on the basis of the distance matrix (program DNADIST), DNA maximum likelihood (program DNAML), DNA Neighbor-Joining (program DNAN-J), DNA Parsimony and Protein Parsimony (program DNAPARS and PROTPARS, respectively). Final tree topologies are obtained using the programs DRAWGRAM, MCCLADE, and TREETOOL. HIV-1 trees are unrooted using the SIV-CPZ sequence as an outgroup. We maintain other current releases of PHYLIP distributed by J. Felsenstein,[13] FastDNAml distributed by G. J. Olsen (University of Illinois, Urbana, IL), and PAUP distributed by D. L. Swofford (Illinois Natural History Survey, Champaign, IL). These programs are run on Sun Microsystem, PC, and Macintosh computers.

Acknowledgments

The authors wish to thank the following people for their various contributions to the development of these procedures: Chi-Cheng Luo, Nicholas de la Torre, Ken Robbins, Teresa Brown, Jennifer Rapier, Atlena Carter, Theresa Hoenes, and Danuta Pieniazek. We also wish to thank Mark Rayfield for reviewing the manuscript.

References

1. Rogers, M. F., Ou, C. Y., and Rayfield, M., et al., *N. Engl. J. Med.,* 320, 1649 (1989).
2. Shibata, D. K., in *Diagnostic Molecular Pathology – A Practical Approach,* Herrington, C. S. and D'McGee, J. O., Eds., Oxford University Press, Oxford, 1992, 85.
3. Boom, R., Sol, C. J. A., and van der Noordaa, J., et al., *J. Clin. Microbiol.,* 28, 495 (1994).
4. Ou, C. Y., McDonough, S. H., Cabanas, D., et al., *AIDS Res. Hum. Retroviruses,* 6, 1323 (1990).
5. Subbarao, S., Luo, C. C., and Limpakarnjanarat, K., et al., *AIDS,* 10 (1996).
6. Pau, C. P., Kai, M., and Holloman-Candal, D. L., et al., *AIDS Res. Hum. Retroviruses,* 10, 1369 (1994).
7. Staden, R., *Computer Appl. Biosci.,* 6, 387 (1990).
8. Higgins, D. G. and Sharp, P. M., *Computer Appl. Biosci.,* 5, 151 (1989).
9. Faulkner, D. V. and Jurka, J., *Trends Biochem. Sci.,* 13, 321 (1988).
10. Korber, B. and Myers, G., *AIDS Res. Hum. Retroviruses,* 8, 1549 (1992).
11. Smith, R. F. and Smith, T. F., *Proc. Natl. Acad. Sci. U.S.A.,* 87, 118 (1990).
12. Smith, R. F. and Smith, T. F., *Prot. Eng.,* 5, 35 (1992).
13. Felsenstein, J., *Cladistics,* 5, 164 (1989).

Chapter 3

Methods for HIV/SIV Gene Analysis

Akio Adachi, Meiko Kawamura, Kenzo Tokunaga, and Hiroyuki Sakai

Contents

0-8493-4412-3/96/$0.00+$.50

I. Generation of Biologically Active Proviral Clones

The first step toward a better understanding of gene products of human and simian immunodeficiency viruses (HIV/SIV) is molecular cloning of intact HIV/SIV genomes. Once biologically active clones are obtained, it is quite easy to do a systematic genetic study. As the first cloning vehicles, phage vectors have been routinely and successfully used.[1-3] The source of viral DNA for cloning is integrated proviral DNA or unintegrated viral DNA prepared from cells infected with a target virus. Both types of viral DNA can be used as starting materials for molecular cloning.[3-6] However, to facilitate transfection analysis, cloned unintegrated viral DNAs need to be reconstituted to have proviral structure.[4-6] In general, recombinant phage clones are stable and are the source of the next plasmid cloning.

It is important to have plasmid clones to perform a functional analysis of HIV/SIV genomes. Care must be taken in selecting an appropriate plasmid vector.[7] It is sometimes difficult or almost impossible to clone a long HIV/SIV genome into high copy number plasmids (e.g., pUC19). If the insert cloned in this kind of vector is unstable in bacteria (HB101, DH5, etc.), try the following procedure.

1. Transform the cells with ligated plasmid DNAs, and grow cells at 37°C on agar plates in the presence of appropriate antibiotics (e.g., 50 μg/ml of ampicillin).
2. Pick up colonies, and grow cells at 37°C in liquid culture for several hours in the absence of antibiotics.
3. Check DNA prepared by the rapid small-scale alkali method.[7]
4. Grow the bacteria carrying the correct plasmid DNA at 30 to 33°C for large-scale culture. The concentration of the antibiotics in liquid media should be low (e.g., 10 to 20 μg/ml of ampicillin). If possible, grow the cells in the absence of the antibiotics.
5. Prepare DNA by the alkali method[7] as described below and confirm the structure of the DNA carefully by extensive restriction enzyme mapping.
6. Test the DNA by transfection as described below.

If full-length plasmid DNA with high purity cannot be obtained even by this procedure (mixture of full-length and partially deleted DNAs), low copy number plasmids such as pBR322 should be used instead for cloning vehicles.

II. Preparation of Plasmid DNA

Purity of plasmid DNA is essential for efficient transfection. Purification of the DNA by equilibrium centrifugation in CsCl-ethidium bromide gradients[7] always gives good results. Alternatively, easy and rapid purification of plasmid DNA by precipitation with polyethylene glycol (PEG) can be used, as shown here.

1. Grow cells in 100 ml of selective LB medium for 24 h.
2. Pellet cells at 4000 rpm, 4°C, 15 min.

3. Resuspend pellet by vortexing in 6 ml of 50 m*M* glucose, 10 m*M* ethylenediaminetetraacetic acid disodium salt dihydrate (EDTA), 25 m*M* Tris-HCl (pH 8.0).
4. Add powdered lysozyme to a final concentration of 5 mg/ml, and mix thoroughly by inversion.
5. Incubate on ice for 20 min.
6. Add 12 ml of 0.2 *N* NaOH, 1% sodium dodecyl sulfate (SDS), and mix carefully and thoroughly by inversion.
7. Incubate on ice for 10 min.
8. Add 7.5 ml of 3 *M* potassium acetate (pH 4.8 to 5.0), and mix carefully by inversion.
9. Incubate on ice for 20 min.
10. Centrifuge at 8000 to 10,000 rpm for 20 min.
11. Remove supernatant, avoiding white precipitate, and add 100 to 200 μl of boiled RNase A (10 mg/ml).
12. Incubate for 30 to 60 min at 37°C.
13. Extract twice with phenol (as long as possible), and once with chloroform.
14. Extract three times with ether.
15. Add two volumes of ethanol to precipitate DNA.
16. Centrifuge at 8000 rpm for 15 min.
17. Dissolve pellet in 2 ml of H_2O, and add 1 ml of 1.5 *M* NaCl, 20% PEG (No. 6000 or 8000), and mix.
18. Incubate on ice for 1 h.
19. Centrifuge at 8000 to 10,000 rpm for 10 min.
20. Discard supernatant, and wash pellet thoroughly with 70% ethanol.
21. Dissolve DNA in an appropriate volume of H_2O (usually 1 ml).

III. Methods for Transfection

Two types of cells (adherent and nonadherent) have been used for transfection analysis of HIV/SIV genomes. Adherent and nonadherent cells are efficiently transfected by modified calcium-phosphate coprecipitation (CPC) and DEAE-dextran methods, respectively.[8-10]

CPC Method

For transfection, growing monolayer cells with 80 to 90% confluence are generally used. Determine the best condition for each cell line.

Solutions

1. 10X TB (transfection buffer): 1% dextrose, 1.4 *M* NaCl, 50 m*M* KCl, 10 m*M* Na_2HPO_4, 0.2 *M* Hepes, pH 7.05.
2. 2.5 *M* $CaCl_2$
3. Glycerol

Procedure

1. Prepare 1 ml (for 25 cm^2 culture flask) of 1X TB containing plasmid DNA and mix well.
2. Add drop-wise and mix thoroughly 50 µl of 2.5 *M* $CaCl_2$, and let sit for 30 min on ice.
3. Pipet vigorously, and add the mixture to a culture flask.
4. Leave at room temperature for 30 min, with occasional rocking.
5. Add 5 ml of growth medium, and incubate at 37°C for 6 to 10 h.
6. Remove medium, and add 15 to 30% glycerol in serum-free medium (v/v) for 1 to 3 min.
7. Remove glycerol, and wash completely with serum-free medium.
8. Add growth medium, and incubate at 37°C for a few days (usually 2 d).

DEAE-Dextran Method

For transfection, cells in growing phase should be used. Determine the best condition for each cell line. See Table 3.1 as reference.

TABLE 3.1
Conditions for Transfection of Suspension Culture

Cell Line	Cell Number for Each Transfection	Concentration of DMSO (%)	Boosting Time (min)
Sup T1	5×10^6	20	10
A3.01	3×10^6	20	10
H9	3×10^6	20	9
Molt4-clone8	5×10^6	30	10
U937	4×10^6	20	10
CEMx174	3×10^6	30	10
M8166	3×10^6	30	10
MT-4	3×10^6	20	10–12
Jurkat	5×10^6	20	8–9
HL-60	5×10^6	10	10
THP-1	5×10^6	20	10

Solutions

1. 5 mg/ml DEAE-dextran
2. 1 *M* Tris-HCl (pH 7.3)
3. RPMI 1640 medium
4. Dimethyl sulfoxide (DMSO)
5. Hypertonic solution: 1 *M* sucrose, 20% PEG No. 6000, 0.2 *M* NaCl, 70 m*M* Tris-HCl (pH 7.3).

Procedure

1. Prepare 1 ml of transfection solution: e.g., 942 μl of serum-free RPMI 1640, 50 μl of 1 *M* Tris-HCl (pH 7.3), 8 μl of 5 mg/ml of DEAE-dextran. Plasmid DNA in μg × 2 = DEAE-dextran in μg; e.g., 5 mg/ml × 8 μl of DEAE-dextran = 40 μg of DEAE-dextran = 2 × 20 μg of plasmid DNA.
2. Add DNA solution (volume less than 50 μl) drop-wise to transfection solution prewarmed at 37°C, and mix well.
3. Incubate at 37°C for 10 min.
4. Pipet the mixture vigorously several times, and add slowly to cell pellet and mix gently. Before use, cells are washed once with serum-free RPMI 1640, and pelleted by centrifugation at 1000 rpm for 5 min. Cell pellet is tapped enough to suspend in the DNA mixture easily.
5. Incubate at 37°C for 30 min, with tapping at an interval of 10 min.
6. Add several times separately and mix gently 1 ml of 10 to 30% DMSO in hypertonic solution (v/v) prewarmed at 37°C.
7. Incubate at 37°C for 8 to 12 min.
8. Add quickly 12 ml of serum-free RPMI 1640, and mix by inversion.
9. Centrifuge at 1000 rpm for 5 min, and resuspend in 5 ml of serum-free RPMI 1640.
10. Centrifuge at 1000 rpm for 5 min, and resuspend in 5 ml of growth medium.
11. Incubate at 37°C for the desired time.

IV. Construction of Mutant and Recombinant Clones

Methods for constructing mutants and for making chimeric recombinants between biologically distinct clones are now well established.[11-17] In this section, a rapid and reliable method to subclone long DNA fragments (longer than 10 kbp) into plasmid vectors is described.

1. Cleave DNAs with appropriate restriction enzymes, and electrophorese into agarose gel (both the inserts and vectors).
2. Identify the band necessary for construction, and excise gel fragment.
3. Cut into small pieces with a razor blade, and put into Eppendorf tubes.
4. Add 400 to 600 μl of phenol, and mash gel with glass rod.
5. Make 0.1 to 0.3 *M* NaCl, vortex briefly, and freeze.
6. Thaw, vortex briefly, and freeze.
7. Centrifuge at 14,000 rpm for 5 min, remove aqueous phase, and save.
8. Vortex phenol phase briefly, and freeze.
9. Centrifuge at 14,000 rpm for 5 min, remove aqueous phase, and combine with the first collection.
10. Extract twice with phenol and three times with ether, and add two volumes of ethanol to precipitate DNA.

11. Determine DNA concentration by ethidium bromide dot estimation.
12. Ligate the DNAs at concentrations of 50 to 100 μg/ml (molar ratio = insert 3:vector 1), and transform cells. The number of bacterial colonies generated will be small, and therefore use as much DNA as possible for transformation.
13. Pick up colonies, and do rapid screening.[7]

V. Single-Round Replication Assay

To analyze the processes of HIV/SIV replication quantitatively, a system designated single-round replication assay (SRA) has been developed.[10,18,19] SRA has been used to determine the critical stage in the viral replication cycle at which a certain viral protein is required.[20-24] As indicated in Table 3.2, SRA utilizes replication-defective proviral clones carrying a bacterial chloramphenicol acetyltransferase (CAT) gene and is dependent on transcomplementation for production of infectious vector virus. Simply by monitoring CAT (transfection and infection) and reverse transcriptase (RT) activities (transfection), the viral replication process can be estimated quantitatively.[10,20,21] Appropriate mutations are introduced into the genes of interest of these reporter constructs or their equivalents, as described above. By determining the defective replication phase of these mutants, it is possible to know when the viral proteins of interest are necessary (Table 3.3).

TABLE 3.2
Various Reporter Constructs

Construct	Description
pH1-CAT	Reporter for Tat activity. Standard LTR-CAT construct driven by HIV-1 LTR.[27]
pH2-CAT	Reporter for Tat activity. Standard LTR-CAT construct driven by HIV-2 LTR.[27]
pSA-CAT	Reporter for Tat activity. Standard LTR-CAT construct driven by SIV_{AGM} LTR.[27]
pMD-CAT	Reporter for Tat activity. Standard LTR-CAT construct driven by SIV_{MND} LTR.[27]
pNLgCAT-R1	Reporter for Rev activity of HIV-1.[30]
pNLgCAT-R2	Reporter for Rev activity of HIV-2.[30]
pNLgCAT-Ra	Reporter for Rev activity of SIV_{AGM}.[30]
pNLgCAT-Rm	Reporter for Rev activity of SIV_{MND}.[30]
pNLenCAT	Replication-defective HIV-1 carrying CAT in the *vpu-env* region for SRA.[19] Nef mutant has been generated.[26]
pNLnCAT	Replication-defective HIV-1 carrying CAT in the *nef* gene for SRA.[10] Various mutants have been generated.[10,19,20,22-24]
pGHnCAT	Replication-defective HIV-2 carrying CAT in the *nef* gene for SRA.[21] Vpx and Env mutants have been generated.[21]
pGLnCAT	Variant of pGHnCAT with a wide cellular host range carrying CAT in the *nef* gene.[21] Vpx and Env mutants have been generated.[21]
pMAnefCAT	SIV_{MAC} carrying CAT in the *nef* gene.[18] Vif and Env mutants have been generated.[18]

TABLE 3.3
Assay Methods for Single-Round Replication of HIV/SIV

Replication Phase	Methods
Early (from adsorption to integration)	1. SRA (CAT-virus infection) 2. Entry assay (infection:antigen capture) 3. Virus DNA synthesis — unintegrated and integrated (infection:PCR)
Late (from transcription to virion release)	1. SRA (transfection-CAT, RT, antigen capture) 2. Western blotting, immunoprecipitation (transfection) 3. Tat assay (transfection) 4. Rev assay (transfection) 5. Electron microscopy (transfection)

Transfection Analysis in Late Replication Phase (From Transcription to Virion Release)

1. Transfect cells with 10 to 30 μg of plasmid DNAs (reporter plus effector clones), as described above, and incubate for 2 to 3 d.
2. Remove cell supernatants, and save for RT assay, infection experiments, etc.
3. Make cell extract for CAT assay, etc.
4. Determine RT and CAT activities by standard methods (see Reference 10, on assays for Tat and Rev activities), and compare with those of wild-type reporter. Defectiveness of the mutants analyzed can be determined.

Infection Analysis in Early Replication Phase (From Adsorption to DNA Integration)

1. Infect cells (2×10^6) with CAT-virus (10^5 RT units) prepared by transfection,[10,20,21] and incubate at 37°C overnight. Input viral dose may be changeable.[21]
2. Wash infected cells three times with serum-free medium, and incubate at 37°C for 2 more days.
3. Make cell extract, and determine CAT activity by standard methods (see Reference 10 for assays for Tat and Rev activities).
4. Compare the activity with that of wild-type and the defectiveness of the mutants can be determined.

VI. Assays for Viral Entry and Viral DNA Synthesis

If some viral protein is demonstrated to be important for the early phase of the viral replication cycle, it is quite meaningful to determine whether this protein is required for the process of viral entry, viral DNA synthesis, nuclear transport of viral DNA, and integration of viral DNA into cellular DNA. By using regular mutants[4-6,11] or CAT-viruses (Table 3.2), these events can be analyzed quantitatively.[10,24-26]

A. Entry Assay[25]

1. Adsorb cells (10^6) with viruses (10^6 RT units or 100 ng of viral core antigen) in 5 ml medium at 37°C for 1 h.
2. Centrifuge at 1000 rpm for 5 min, and resuspend cells in 500 μl of 0.05% trypsin in phosphate-buffered saline (PBS) containing 0.02% EDTA.
3. Incubate at 37°C for 5 min, and quickly add 6 ml of growth medium.
4. Centrifuge at 1000 rpm for 5 min, and resuspend cell pellet in 6 ml of serum-free medium.
5. Centrifuge at 1000 rpm for 5 min, and resuspend cell pellet in 6 ml of PBS containing 1% serum.
6. Centrifuge at 1000 rpm for 5 min.
7. Remove supernatant completely, and add 200 μl of lysis buffer: 10 m*M* Tris-HCl (pH 7.5), 10 m*M* NaCl, 0.5% NP-40.
8. Vortex thoroughly, and incubate at 4° C overnight.
9. Determine levels of viral core antigen (p24 for HIV-1) by the Core Antigen Assay kit (manufactured by Coulter Immunology, Hialeah, FL).
10. Compare the results with that of wild-type virus. Env(–) mutant virus is a good negative control. Defectiveness of the mutants can be determined by this comparison.

B. Assay for Viral DNA Synthesis by Polymerase Chain Reaction (PCR)[25]

1. Virus stocks

Seed virus is obtained from infected cells or from transfected cells. Virus from transfection should be treated with DNase I (culture fluids are digested with 20 μg of DNase I per ml at 37°C for 30 min in the presence of 10 m*M* $MgCl_2$) before infection to completely remove the plasmid DNA. Virus obtained from transfected T-cell lines gives good results. Env(–) mutants are good negative controls for experiments involving PCR analysis of cells.

2. Infection

1. Infect cells (10^7) with viruses (2×10^6 RT units), and incubate at 37°C for 4 h.
2. Wash three times with serum-free medium, and add 10 ml of growth medium.
3. Harvest cells for PCR analysis at the designated time.

3. Cell lysis for PCR

1. Wash 2.5 × 10^6 cells three times with PBS.
2. Lyse in 100 μl of PCR lysis buffer: 50 m*M* KCl, 10 m*M* Tris-HCl (pH 8.3), 1.5 m*M* $MgCl_2$, 0.01% gelatin, 0.45% NP-40, 0.45% Tween 20, 60 μg/ml proteinase K.
3. Incubate at 55°C overnight.
4. Inactivate proteinase K by boiling for 10 min.

4. Chromosome DNA extraction

1. Wash 5–10 × 10^6 cells three times with PBS.
2. Suspend in 10 ml of digestion buffer consisting of 100 m*M* NaCl, 10 m*M* Tris-HCl (pH 8.0), 25 m*M* EDTA (pH 8.0), 0.5% SDS, 50 to 100 μg/ml RNase A, and 200 units/ml of RNase T_1, and incubate at 37°C for 30 min.
3. Add proteinase K to final concentration of 100 μg/ml, and incubate at 55°C overnight.
4. Extract very gently with phenol twice, and then with ether three times.
5. Add two volumes of ethanol, and spool DNA.
6. Add appropriate volume of H_2O, and let sit at 4°C.

5. PCR analysis

1. Each 100 μl reaction mixture contains: 100 pmol of each primer, 0.2 m*M* of each dNTP, 50 m*M* KCl, 10 m*M* Tris-HCl (pH 8.3), 1.5 m*M* $MgCl_2$, 0.01% gelatin, 5 units of *Taq* DNA polymerase (Stratagene, La Jolla, CA), crude cell lysate of 2.5–50 × 10^4 cells or 1 to 2 μg of chromosome DNA.
2. Each amplification by PCR (Iwaki Thermal Sequencer TSR-300, Tokyo, Japan) consists of 93°C for 120 sec, 25 to 30 cycles of 93°C for 60 sec, 57°C for 90 sec, 72°C for 120 sec, and finally 72°C for 300 sec.
3. Primer pairs: the following are routinely used to detect HIV-1 viral DNA in cells.

   For early reverse transcription product: CACACACAAGGCTACTTCCCT (nt 57-77 of HIV-1 clone pNL432, GenBank data base accession number M19921) and CTGCTAGAGATTTTCCACACTGAC (nt 635-612)

   For intermediate product: TGCTGTTAAATGGCAGTCTAGC (nt 6990-7011) and CCAGACTGTGAGTTGCAACAG (nt 7927-7907)

   For completed product: CACACACAAGGCTACTTCCCT (nt 57-77) and CCGAGTCCTGCGTCGAGAGATC (nt 701-680)
4. Specificity of amplified products is determined by Southern blot hybridization.[10,25]

VII. Assays for Tat and Rev Activities

As shown in Table 3.2, various reporter clones are now available to assess Tat and Rev activities of viruses of the four major HIV/SIV groups.[27-31] CAT activity expressed in cells cotransfected with reporters and various HIV/SIV clones is an indication of these activities essential for viral replication. It is also possible, by this system, to classify some viral clones into known groups of HIV/SIV.[28]

1. Transfect cells with a combination of reporter and effector clones (Table 3.2) as described above. For determination of Tat activity in adherent cells, 1 μg of reporters and 0.1 μg of effectors are generally used. For determination of Rev activity in adherent cells, 10 μg of each clone is generally used for transfection.
2. Incubate cells at 37°C for 48 h.
3. Harvest cells by scraping into culture fluids, and centrifuge at 3000 rpm for 10 min.
4. Discard supernatant, add 100 to 500 μl of 0.25 *M* Tris-HCl (pH 7.8), and vortex well.
5. Freeze then thaw three times, vortexing well each time, and centrifuge at 3000 rpm for 10 min.
6. Remove supernatant and save for CAT assay.
7. Determine CAT activity in the cell extracts (usually 20 μl of extracts is enough) by standard methods.[10]

VIII. Conclusion

Major methods used for the study of HIV/SIV gene function are summarized in Table 3.3. HIV/SIV carry many unique genes in their genome, i.e., *tat, rev, vif, vpr, vpx, vpu,* and *nef,* not found in simple retroviruses. The molecular basis of the functions of these extra gene products is still unclear in most cases. These viral proteins may interact with viral structural proteins, i.e., Gag, Pol, and Env, and also with some cellular factors to maximize viral replication and spread. Genetic analyses based on the techniques shown in Table 3.3 have revealed many aspects of HIV/SIV replication. However, further systemic and molecular genetic studies are critical for complete understanding of complex HIV/SIV virology.

References

1. Sambrook, J., Fritsch, E. F., and Maniatis, T., Eds., *Molecular Cloning: A Laboratory Manual,* 2nd ed., Cold Spring Harbor Laboratory Press, Cold Spring Harbor, New York, 1989, 2.1.
2. Overbaugh, J., Dewhurst, S., and Mullins, J. I., in *Techniques in HIV Research,* Aldovini, A. and Walker, B. D., Eds., Stockton Press, New York, 1990, 131.
3. Adachi, A., Gendelman, H. E., Koenig, S., Folks, T., Willey, R., Rabson, A., and Martin, M. A., *J. Virol.,* 59, 284 (1986).
4. Shibata, R., Miura, T., Hayami, M., Sakai, H., Ogawa, K., Kiyomasu, T., Ishimoto, A., and Adachi, A., *J. Virol.,* 64, 307 (1990).
5. Shibata, R., Miura, T., Hayami, M., Ogawa, K., Sakai, H., Kiyomasu, T., Ishimoto, A., and Adachi, A., *J. Virol.,* 64, 742 (1990).
6. Sakai, H., Sakuragi, J., Sakuragi, S., Shibata, R., Hayami, M., Ishimoto, A., and Adachi, A., *Arch. Virol.,* 125, 1 (1992).

7. Sambrook, J., Fritsch, E. F., and Maniatis, T., Eds., *Molecular Cloning: A Laboratory Manual,* 2nd ed., Cold Spring Harbor Laboratory Press, Cold Spring Harbor, New York, 1989, 1.1.
8. Wigler, M., Pellecer, A., Silverstein, S., Axel, R., Urlaub, G., and Chasin, L., *Proc. Natl. Acad. Sci. U.S.A.,* 76, 1372 (1979).
9. Takai, H. and Ohmori, H., *Biochim. Biophys. Acta,* 1048, 105 (1990).
10. Sakai, H., Kawamura, M., Sakuragi, J., Sakuragi, S., Shibata, R., Ishimoto, A., Ono, N., Ueda, S., and Adachi, A., *J. Virol.,* 67, 1169 (1993).
11. Adachi, A., Ono, N., Sakai, H., Shibata, R., Ogawa, K., Kiyomasu, T., Masuike, H., and Ueda, S., *Arch, Virol.,* 117, 45 (1991).
12. Sakai, H., Sakuragi, J., Sakuragi, S., Shibata, R., and Adachi, A., *Virology,* 189, 161 (1992).
13. Sakai, H., Sakuragi, S., Sakuragi, J., Kawamura, M., Shibata, R., and Adachi, A., *J. Gen. Virol.,* 73, 2989 (1992).
14. Shibata, R., Sakai, H., Kiyomasu, T., Ishimoto, A., Hayami, M., and Adachi, A., *J. Virol.,* 64, 5861 (1990).
15. Shibata, R., Kawamura, M., Sakai, H., Hayami, M., Ishimoto, A., and Adachi, A., *J. Virol.,* 65, 3514 (1991).
16. Shibata, R. and Adachi, A., *AIDS Res. Hum. Retroviruses,* 8, 403 (1992).
17. Sambrook, J., Fritsch, E. F., and Maniatis, T., Eds., in *Molecular Cloning: A Laboratory Manual,* 2nd ed., Cold Spring Harbor Laboratory Press, Cold Spring Harbor, New York, 1989, 15.1.
18. Shibata, R., Sakai, H., Kawamura, M., Tokunaga, K., and Adachi, A., *J. Gen. Virol.,* 76, 2723 (1995).
19. Sakai, H., Furuta, R. A., Tokunaga, K., Kawamura, M., Hatanaka, M., and Adachi, A., *FEBS Letters,* 365, 141 (1995).
20. Sakai, H., Shibata, R., Sakuragi, J., Sakuragi, S., Kawamura, M., and Adachi, A., *J. Virol.,* 67, 1663 (1993).
21. Kawamura, M., Sakai, H., and Adachi, A., *Microbiol. Immunol.,* 38, 871 (1994).
22. Sakai, H., Tokunaga, K., Kawamura, M., and Adachi, A., *J. Gen. Virol.,* 76, 2712 (1995).
23. Sakuragi, J., Sakai, H., Kawamura, M., Tokunaga, K., Ueda, S., and Adachi, A., *Virology,* 212, 251 (1995).
24. Kawamura, M., Tokunaga, K., Sakai, H., and Adachi, A., unpublished results.
25. von Schwedler, U., Song, J., Aiken, C., and Trono, D., *J. Virol.,* 67, 4945 (1993).
26. Tokunaga, K. and Adachi, A., submitted.
27. Sakuragi, J., Fukasawa, M., Shibata R., Sakai, H., Kawamura, M., Akari, H., Kiyomasu, T., Ishimoto, A., Hayami, M., and Adachi, A., *Virology,* 185, 455 (1991).
28. Sakuragi, J., Sakai, H., Sakuragi, S., Shibata, R., Wain-Hobson, S., Hayami, M., and Adachi, A., *Virology,* 189, 354 (1992).
29. Sakai, H., Siomi, H., Shida, H., Shibata, R., Kiyomasu, T., and Adachi, A., *J. Virol.,* 64, 5833 (1990).
30. Sakai, H., Shibata, R., Sakuragi, J., Kiyomasu, T., Kawamura, M., Hayami, M., Ishimoto, A., and Adachi, A., *Virology,* 184, 513 (1991).
31. Sakai, H., Sakuragi, J., Sakuragi, S., Kawamura, M., and Adachi, A., *Arch. Virol.,* 129, 1 (1993).

Chapter 4

Analysis of Human Immunodeficiency Virus Reverse Transcriptase Function

Vinayaka R. Prasad, William C. Drosopoulos, and Yvonne Kew

Contents

0-8493-4412-3/96/$0.00+$.50

I. Introduction

Antiviral strategies against human immunodeficiency virus (HIV) are a focal point of research directed at combating the acquired immunodeficiency syndrome (AIDS). A pressing need to find a "chink in the armor" of HIV has led to a thorough study of key viral functions that can be blocked effectively. Inhibitors of HIV reverse transcriptase (RT), the preferred target from the start, are the most studied, both biochemically and clinically.[1,2] This is reflected in the fact that all four of the currently approved drugs for anti-AIDS treatment, 2′,3′-deoxy, 3′-azidothymidine (AZT), 2′,3′-dideoxyinosine (DDI), 2′,3′-dideoxycytosine (DDC), and 2′,3′-dideoxy, 2′,3′-didehydrothymidine (D4T), are anti-reverse transcriptase drugs.[1,2] However, the development of viral resistance in response to long-term treatment of infected individuals is a major problem.[3,4] Since a vaccine is not forthcoming in the foreseeable future, a major thrust in this area is toward developing drugs that are less likely to lead to resistance, as well as combination therapies targeting one or more viral functions. Towards this goal, one must acquire the requisite knowledge of structure-function of HIV RT.

RT consists of RNA-dependent DNA polymerase (RDDP) and ribonuclease H (RNAse H) activities. The assays described in this chapter are aimed primarily at a dissection of the RDDP function of HIV RT employing recombinant enzyme. Several expression constructs that permit the expression of HIV-1 RT in *Escherichia coli* are currently available.[5-10] Most assays described here assume the availability of purified HIV-1 RT that is devoid of bacterial or cellular DNAses and RNAses. These assays will help the study of various aspects of RDDP activity and are divided into five sections. In these sections, methods employed to measure the extent of polymerization by HIV RT, template switching, fidelity, and processivity are discussed. Also included, at the end, is a section that describes a method used for genetic screening for functional variants of HIV RT.

II. Measurement of Polymerase Activity

A. RNA-Dependent DNA Polymerase Activity

The RDDP activity of HIV-1 RT can be measured on both homopolymeric and heteropolymeric RNA templates with appropriate oligodeoxynucleotide primers. Historically, homopolymeric RNA templates were used in the determination of RT activity in cell culture media as a means of detecting retrovirus particles. The popular use of homopolymers reflects the relative ease of use and the commercial availability of such polymers. More recently, with a view to accurately recapitulate the *in vivo* reaction, many investigators have begun to use heteropolymeric templates.

1. Homopolymeric assays

HIV RT can employ both poly(rC) and poly(rA) templates in RDDP reactions. However, if poly(rC)·oligo(dG) is used as template-primer with magnesium as the divalent cation, the RDDP activity of recombinant HIV-1 RT produced in *E. coli* can be measured in crude lysates with no interference from bacterial polymerases. This is due to the fact that *E. coli* DNA polymerase I, which displays RDDP activity, is inactive under these conditions.[11] Mutational analyses of HIV-1 and HIV-2 RT involving several dozens of mutants employed crude lysates to determine the effect of substitution, insertion, and deletion mutations on RDDP activity and has yielded important information on the structure-function of HIV-1 RT.[11-15]

We routinely carry out RDDP assays in a 50 μl reaction volume, containing 50 m*M* Tris-HCl, pH 8.0, 10 m*M* dithiothreitol, 80 m*M* KCl, 6 m*M* $MgCl_2$, 1 μ*M* poly(rC) (3′-OH ends), oligo(dG) (3′-OH ends), 10 μ*M* [α-^{32}P]dGTP (specific radioactivity, 2.3 Ci/mmol). The reactions, in duplicate, are initiated with the addition of HIV RT (crude lysate from bacteria expressing the recombinant RT, solubilized virion preparation, or purified enzyme) and incubated for 60 min at 37°C. The reactions are stopped by spotting on to DE-81 filter squares and the filter squares washed with 2X SSC (300 m*M* sodium chloride and 30 m*M* sodium citrate, pH 7.0) for 30 min with two changes. The filters are placed in 5 ml of scintillation fluid (Ecoscint H, National Diagnostics, USA) and the radioactivity determined in a liquid scintillation counter (1218 Rack Beta, LKB-Wallac, Sweden). The radioactivity detected as counts per minute (cpm) are used to calculate the picomoles of deoxynucleoside monophosphate (dNMP) incorporated into the nascent DNA.

2. Heteropolymeric assays

In vivo, HIV reverse transcriptase, like all retroviral RTs, copies a heteropolymeric RNA of about 8 to 10 kb using the host-derived $tRNA^{lys,3}$ as primer.[17,18] Yet, most investigators employ homopolymers for routine assaying of RT since most biochemical studies have detected no differences in the gross enzymatic properties as measured in the homopolymeric and heteropolymeric template-primer systems. Heteropolymeric templates have facilitated the use of sophisticated methods to study fidelity, template switching, processivity, and other aspects of HIV RT function that were poorly understood previously. Heteropolymeric assays can be carried out with (1) a viral RNA template:$tRNA^{lys,3}$ primer;[19,20] (2) a nonviral heteropolymeric RNA such as ribosomal RNA;[21] or (3) a synthetic oligoribonucleotide.[22] In the latter two cases, an appropriate oligodeoxynucleotide primer is utilized.

Templates resembling viral RNA — It has been shown that, *in vitro,* HIV-1 RT binds specifically to natural $tRNA^{lys,3}$. However, when viral RNA template is provided along with natural $tRNA^{lys,3}$ primer, HIV RT can efficiently form binary complexes with the tRNA but cannot initiate DNA synthesis in the absence of nucleocapsid (NC) protein.[20] Thus, it appears that NC may be required to promote the annealing of natural $tRNA^{lys,3}$. This is supported by the recent finding that NC promotes selective annealing of complementary DNA strands.[23] However, with short DNA oligonucleotides containing a primer binding site (PBS) as template, HIV-1 RT can efficiently initiate DNA synthesis from *in vitro* synthesized human $tRNA^{lys,3}$ or native *E. coli* $tRNA^{gln,2}$ primers.[19]

Ribosomal RNA as template — White et al.[21] employed 16S *E. coli* ribosomal RNA as template for RT assay with a complementary 15-mer oligodeoxynucleotide as primer. In their study, comparative measurements of sensitivity of HIV-1 RT to inhibition by nucleoside and non-nucleoside analogs on ribosomal RNA, gapped duplex DNA, and poly(rA)·oligo(dT) templates showed widely different ID_{50} values. While the reasons for such differences are unknown, it is important to consider these template-specific effects when determining the potency of a given inhibitor.

Defined oligoribonucleotides as templates —Several groups have also employed short oligoribonucleotides as template for the analysis of HIV-1 RT function. For example, Reardon and Miller[22] used a 44-mer RNA fragment with a 21-mer oligodeoxynucleotide for their kinetic studies on the HIV-1 RT. Such templates are ideal, since the small size decreases the likelihood of secondary structure formation.

B. DNA-Dependent DNA Polymerase Activity

The DNA-dependent DNA polymerase (DDDP) activity of HIV-1 RT is a measure of the function required for second strand or plus strand DNA synthesis during viral replication. As with RDDP activity, the DDDP activity can also be measured on homopolymeric templates such as poly(dA) with oligo(dT) primer.[24,25]

C. Polymerase Assays that Do Not Require Purified RT Preparations

While solution assays for RDDP activity can easily be performed in crude lysates without interference from the *E. coli* DNA polymerase I, assaying for DNA-dependent DNA polymerase activity of HIV-1 RT poses additional problems due to the involvement of multiple polymerases. This problem has been circumvented by an initial separation of the crude proteins by electrophoresis on a sodium dodecyl sulfate (SDS) polyacrylamide gel, renaturation of the proteins, followed by incubation with template-primer and radioactively labeled substrate deoxynucleoside triphosphates (dNTPs).[26] When studying a large number of mutants, this procedure helps obtain a quick measure of the polymerase activity and can be used for measuring both RDDP and DDDP activities.

The assay we routinely employ in the laboratory is a modification of the *in situ* gel assay[8] in that, following the electrophoresis of crude lysates containing HIV-1 RT, the proteins are transferred to nitrocellulose membrane before renaturation and assaying for activity. First, a bacterial lysate is prepared from 1 ml of induced bacterial culture, as described previously.[7,27] The insoluble pellet is dissolved in SDS-urea sample buffer (1% $NaDodSO_4$, 1% β-mercaptoethanol, 0.01 *M* sodium phosphate, pH 7.2, 6 *M* urea) and applied to standard SDS polyacrylamide gel.[28] After the electrophoresis, the proteins are electroblotted to nitrocellulose filter (Schleicher and Schuell) at room temperature by standard procedures employed in immunoblot techniques (20 m*M* Tris base, 200 m*M* glycine). The nonspecific binding sites on the paper are blocked by incubating the nitrocellulose paper in 0.02% bovine serum albumin (BSA) (Fraction V, Sigma) solution prepared in 30 m*M* Hepes, pH 7.5, for

1 h at room temperature. To facilitate the renaturation of the proteins bound to the membrane, the blot is incubated in buffer (50 m*M* Tris-HCl, pH 7.5, 1 m*M* EDTA, 2 m*M* dithiothreitol, 0.2 *M* NaCl, 10% glycerol, and 0.1% NP-40) for 12 h at 4°C with several changes of buffer. The nitrocellulose paper is then preincubated at 37°C in a heat-sealed pouch containing reverse transcriptase reaction buffer, as described above, but without nucleotides, for 30 min. The reaction buffer is then replaced with a cocktail complete with [α-^{32}P] dNTP. At the end of the reaction, the nitrocellulose filter is washed and fixed in ice-cold 10% trichloroacetic acid for 10 min with three changes of the solution. This step also helps remove unincorporated dNTPs. The filter is then air-dried an autoradiographed. A schematic representation of the procedure is given in Figure 4.1.

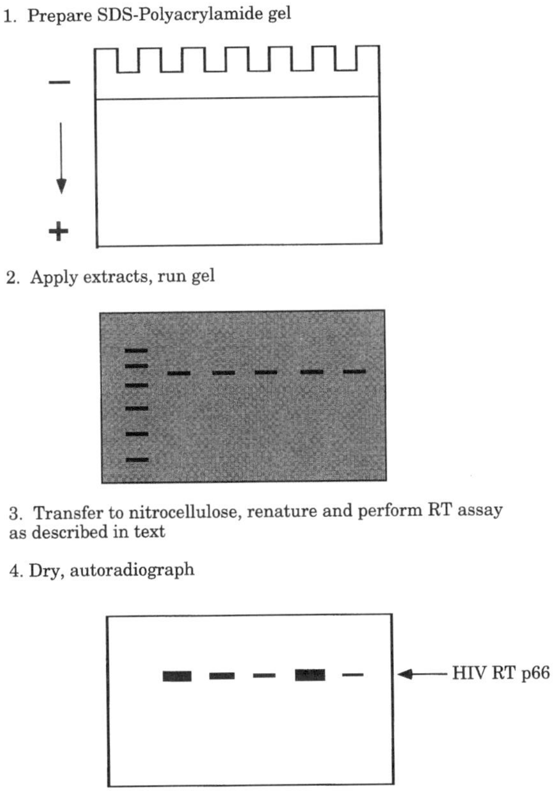
1. Prepare SDS-Polyacrylamide gel
–
+
2. Apply extracts, run gel
3. Transfer to nitrocellulose, renature and perform RT assay
as described in text
4. Dry, autoradiograph
HIV RT p66

FIGURE 4.1
Schematic of the *in situ* gel assay for DNA polymerase activity. Crude lysates of bacteria expressing the HIV-1 RT are applied to an SDS polyacrylamide gel. Following electrophoresis, the proteins are electrotransferred to nitrocellulose paper. The transferred proteins are allowed to renature. The HIV-1 RT reaction is carried out as described in text. The intensity of the bands on the autoradiogram correspond to the level of activity displayed by each mutant. The single band depicted in each case corresponds to the 66-kDa subunit of the HIV-1 RT heterodimers.

One can also cast the template-primer into the SDS polyacrylamide gel and, following electrophoresis, allow the proteins to renature in the gel support itself before initiating the polymerase reaction.[29,30] One caveat to both of these approaches is that the level of activity measured can often be a reflection of the ability of the protein to renature into an enzymatically active form. The method has been employed to assess the effect of mutations on both RDDP and DDDP activities of HIV-1 RT.[29,30]

III. Template Switching *In Vitro*

The *in vitro* assays described above provide a gross measure of the polymerase activity of HIV RT in various template-primer settings. *In vivo,* however, the process of reverse transcription is far more complex. Following viral entry, the synthesis of viral DNA on the viral RNA template catalyzed by the reverse transcriptase involves a series of DNA synthesis reactions involving initiation, elongation, and strong stop DNA formation alternating with DNA strand transfers that result in the creation of terminal sequence redundance in the resulting double-stranded proviral DNA.[17,18]

The two strands of proviral DNA are synthesized sequentially — the minus strand DNA is synthesized first, followed by the plus strand DNA synthesis. First, the host cell-derived $tRNA^{lys,3}$ anneals to the primer binding site (PBS) located downstream of the U5 element and primes the minus DNA synthesis. In the subsequent elongation step, moving toward the 5′ end of the RNA genome and copying U5 and R regions, RT reaches the 5′ end, resulting in a short DNA fragment containing a minus strand copy of the U5 and R, the minus strand strong stop DNA. This strong stop DNA is subsequently translocated to the 3′ end of the RNA genome, where the annealing is accomplished via homology between the R elements. This event, variously called strand transfer, translocation, template switching, etc., facilitates the resumption of minus DNA synthesis. The RNAse H activity of HIV RT, presumably in a concerted fashion[31] with the minus DNA polymerization, degrades the genomic RNA bound to the newly made DNA. As a result, the RNA genome is nearly completely digested, except for a short oligonucleotide consisting of a stretch of purines, located upstream of the U3 sequence. The RNAse H-resistant oligonucleotide serves as the primer for the synthesis of plus strand DNA employing the minus strand DNA as template. The synthesis proceeds to the end of the minus DNA extending up to the first modified base of tRNA primer, resulting in a plus strand strong stop DNA, which essentially is a complete copy of LTR with U3, R, and U5 elements. The second template switch involves the annealing of the PBS sequences of the plus strand DNA, derived by copying the tRNA primer, to the minus strand copy of PBS located at the other end. This second strand transfer event allows for the completion of both strands of proviral DNA.

Template switching is a key function of HIV RT *in vivo* that is likely to influence rates of mutation and recombination by HIV RT.[32] Therefore, it will be important to examine this process at a biochemical level to help dissect this function as well as to facilitate delineating the structural elements of RT that govern this function through mutational analysis.

Template switching involves translocation of nascent DNA from one position of the template to another position of the same or different template strand. Such events can be simulated *in vitro* and have been studied by reconstituting purified RT with synthetic oligonucleotides denoted as donor or acceptor templates.[33-36] The sequences selected as templates for reverse transcription can be from a retroviral genome or sequences from a plasmid. RNA donor template is used to study minus strand strong stop DNA transfer.[33] Similarly, one can utilize a DNA donor template to examine plus strand strong stop DNA transfer. Studies designed to investigate the first transfer process have utilized both RNA[33-35] and DNA acceptor templates.[34]

In experiments designed to study strand transfer reactions, the following conditions must be provided. First, the donor and acceptor templates have to contain a common nucleotide stretch or a homologous region to facilitate the transfer reaction. Furthermore, primers should be specific for the donor template and not anneal to the acceptor template. Finally, the templates should be constructed to allow the distinction of their products. For example, donor template-directed synthesis can be designed to yield smaller-length products than strand transfer products, such that they are distinguishable on a denaturing polyacrylamide sequencing gel.

A standard assay for studying strand transfer activity is described.[36] First the primer, a 20-mer deoxyoligonucleotide is 5′-[^{32}P] end-labeled using T4 polynucleotide kinase. The primer is combined with the donor template at a 2:1 (primer/donor) molar ratio in 10 m*M* Tris-HCl, pH 7.5, 1 m*M* EDTA, 50 m*M* KCl solution, and the mixture is heated to 65°C for 10 min and then slowly cooled to room temperature. A preincubation mixture is made in 20 μl reaction volume containing various amounts of HIV RT, 2 n*M* primer-donor template, and 200 n*M* acceptor template in buffer A (50 m*M* Tris-HCl, pH 8.0, 80 m*M* KCl, 1 m*M* dithiothreitol, 0.1 m*M* EDTA, 2% glycerol) for 3 min. The reaction is initiated with the addition of $MgCl_2$ and dNTPs in 5 μl of buffer A to give a final concentration of 6 m*M* $MgCl_2$ and 1 m*M* each of dATP, dCTP, dGTP, and dTTP. The reactions are incubated at 37°C for 1 h. At the end of the incubation, the reactions are stopped with 20 μl of gel loading buffer (90% formamide, 10 m*M* EDTA, pH 8.0, 0.1% xylene cyanol, 0.1% bromophenol blue).

The reactions are then loaded along with size markers onto a denaturing gel of 8% polyacrylamide and the reverse transcription products are separated by electrophoresis.[49] The gel is dried and analyzed by autoradiography. For quantitation, the primer-extended radioactive products are measured by densitometry. The lane from a reaction with no acceptor template should only display a product corresponding to the size of donor-template directed DNA synthesis, while a lane with no donor template should have no product. Finally, the lane containing both donor and acceptor templates may show, depending on the extent of strand transfer, products derived from strand transfer as well as products that finished on the donor template (Figure 4.2).

The efficiency of strand transfer depends on several factors. Previous studies showed that various parameters affect the strand transfer event. Luo and Taylor[35] determined that the frequency of template switching increased with (1) increasing reaction time, (2) increasing acceptor/donor template ratio, and (3) increasing lengths of overlapping or homologous region, with a minimum of greater than 10 bases required. Additionally, conditions that promote polymerase pausing, such as the

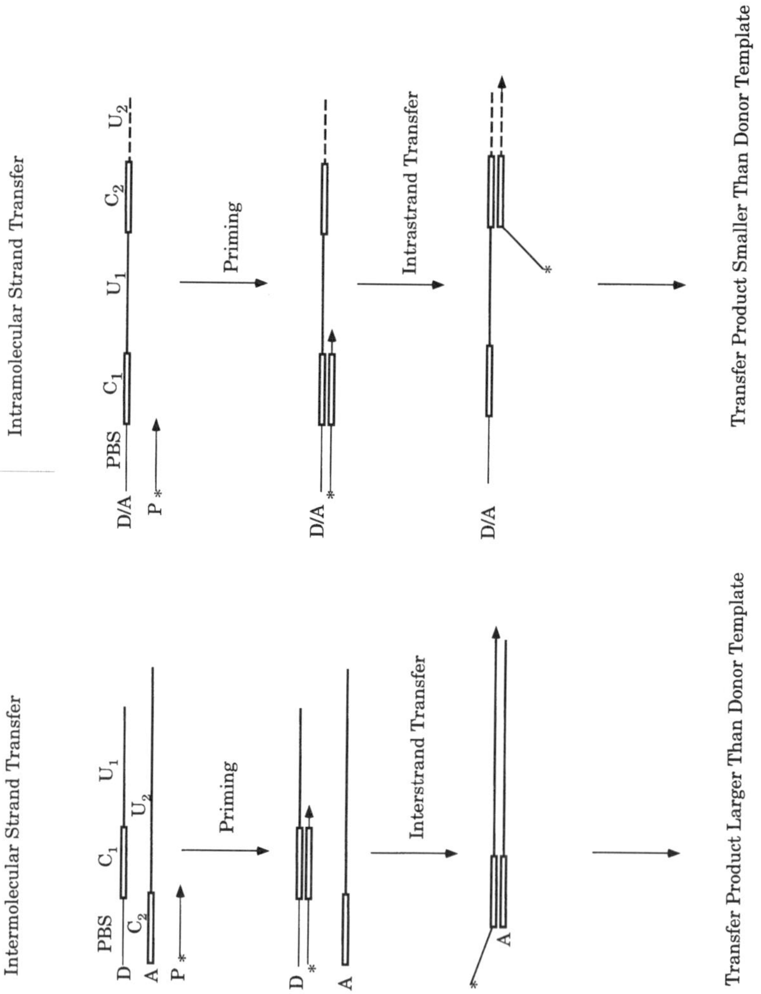
Intramolecular Strand Transfer
D/A
PBS
C1
U1
C2
U2
P
Priming
Intrastrand Transfer
Transfer Product Smaller Than Donor Template
Intermolecular Strand Transfer
D
A
P
Priming
Interstrand Transfer
Transfer Product Larger Than Donor Template

FIGURE 4.2

introduction of a hairpin structure, which serves as a polymerase pause site on the donor template, or reducing the dNTP concentration (from 50 to 0.2 m*M*), increase the rate of strand transfer events.[36] Further, template switching is more frequent when the donor template is RNA.[35] Although not absolutely required, when RNA donor template is used, RNase H activity substantially enhanced strand transfer.[35,36]

IV. Fidelity of Synthesis

The primary *in vivo* function of HIV RT, the replication of viral genome, should be error-free, to ensure the generation of viable progeny virions. However, it is common knowledge that HIV displays a high degree of genetic variability.[37,38] This high mutation rate has been attributed to two causes: the low fidelity of replication by HIV RT and selective forces such as the immune system and other biological forces that exist *in vivo*.[39,40] In order to assess the contribution of HIV RT to viral mutation rates, several systems have been developed to facilitate the measurement of fidelity of HIV RT.[41-48] These include *in vitro* kinetic assays using synthetic or natural templates,[47] as well as forward and reverse mutation assays using biologically active templates.[41,42] The kinetic assays, in general, are preferred since they allow one to examine specific catalytic parameters that could affect polymerase fidelity, such as nucleotide misinsertion efficiency and specificity, misextension efficiency, polymerase processivity, and template sequence composition. Methods for evaluating HIV RT fidelity in terms of misinsertion efficiency are presented below.

A. Nucleotide Misinsertion

Nucleotide misinsertion, the addition of noncomplementary nucleotides onto an extending primer terminus, represents the initial event in the enzymatic induction of certain mutations by HIV RT. As such, the contribution of nucleotide misinsertion

FIGURE 4.2
A schematic of the *in vitro* strand transfer reaction. Both inter- and intramolecular transfers are depicted. The templates are shown with solid lines and boxes representing unique and common sequences, respectively. The primers are shown with an arrow to indicate the direction of synthesis, with the asterisk indicating the 5′ location of the ^{32}P label. D, donor template; A, acceptor template; D/A, template can act as a donor as well as an acceptor; P, 5′ end labeled primer containing sequences complementary to the primer binding site; PBS, primer binding site; C_1 and C_2, common sequences shared between donor and acceptor templates; U_1 and U_2, unique sequences. Intermolecular strand transfer: the D and A templates are engineered to allow distinction of the products made from D alone and those that originated from D but completed on A. The primer anneals to PBS of D and becomes extended by reverse transcriptase. When the synthesis involves template jump, from C_1 of D to C_2 of A, the size of the product will be greater than D. When no translocation or jump occurs during the elongation step, the product size will be equal to that of D. Intramolecular strand transfer: in contrast to the above example, synthesis involving intramolecular strand transfer produces a product shorter than the one made by synthesis of the complete D/A template. When no jump occurs during RT elongation, the resulting product is the same size as D/A. However, a product that is derived from template jumping from C_1 to C_2 will lack U_1 sequences and thus will be smaller than D/A.

to overall polymerase fidelity of HIV RT has been examined extensively, particularly with regard to misinsertion kinetics.[42,46-48] A versatile system, based on gel mobility shifts, that has been effectively used for measuring kinetic parameters of misinsertion was described by Yu and Goodman.[47] This system, described in detail below, allows for misinsertion events to be characterized using a variety of homo- and heteropolymeric RNA/DNA and DNA/DNA template-primer substrates in a primer extension assay. Briefly, a 5′-[^{32}P] end-labeled oligonucleotide primer is annealed to an RNA or DNA template. The annealed primer is positioned on the template such that the insertion event to be examined is targeted either directly to the 3′ terminus of the primer, i.e., "standing start" conditions, or to a position two or more nucleotides downstream of the 3′ terminus, i.e., "running start" conditions. The primer extension reaction is carried out either in the presence of one dNTP (correct or incorrect) for standing start templates or in the presence of a saturating concentration of the correct running start dNTP plus varying concentrations of the same or one other dNTP for running start templates (Figure 4.3). Reactions are terminated following a short extension period where less than 20% of the annealed primers are extended, at which point the reactions are at a steady state and the probability of polymerase reassociation with previously extended primers is low. Extension products are then analyzed via polyacrylamide gel electrophoresis (PAGE) followed by autoradiography. The relative band intensities for extension events are measured by densitometry of the autoradiogram.

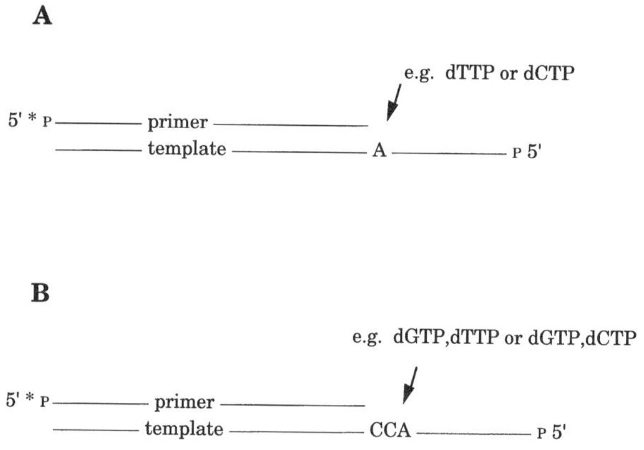

FIGURE 4.3
Template-primers employed in the measurement of misinsertion frequency. (A) Standing start; (B) running start. Insertion is measured as extension from the 3′-OH terminus of 5′-[α^{32}P] end-labeled primer. In the examples illustrated here, the insertion events being examined, a comparison of correct (dTTP) versus incorrect (dCTP) insertion, are targeted opposite a template A nucleotide. In the case of running start template-primer (B), the running start nucleotide (i.e., dGTP) is also included in the assay reaction.

B. Measuring Misinsertion

The assay presented here, with the exception of template and primer selection and preparation, is essentially as developed by Yu and Goodman.[47]

Template-primers — Selection of templates and primers, as well as procedures for preparing 5′-[^{32}P] end-labeled oligonucleotide primers and heteropolymeric RNA or DNA templates, are as described below (see next section) for processivity assay, except that the templates may have a minimum length of 100 nucleotides. Labeled primers are annealed to templates by mixing a 74 n*M* solution of 5′-[^{32}P] end-labeled primer and a 81 n*M* solution of template RNA or DNA in hybridization buffer (50 m*M* Tris-HCl, pH 8.0, 2 m*M* β-mercaptoethanol, 50 μg/ml acetylated BSA), heating to 65°C (for DNA templates) or 55°C (for RNA templates), then slow cooling to room temperature over 2 h. Hybridized template-primers are then stored at –20°C.

Assay — An RT-template primer mixture (solution A) is made by combining 10 μl enzyme with 85 μl hybridized template-primer at an enzyme to template-primer molar ratio of 1 to 7 (dilute RT in hybridization buffer if necessary). Reactions are initiated as follows: for standing start reactions, combine equal volumes (3 μl) of solution A and solution B (100 m*M* Tris-HCl, pH 8.0, 12 m*M* $MgCl_2$, 20 m*M* dithiothreitol, and only one dNTP [correct or incorrect] at varying [0.2–12000 μ*M*] concentrations); for running start reactions equal volumes (3 μl) of solution A and solution C (solution B plus running start dNTP at an empirically determined saturating concentration). Reaction mixes are incubated at 37°C for 2 (standing start) or 3 (running start) min, then terminated by adding 12 μl of 20 m*M* EDTA in 95% formamide. A 3 μl aliquot of terminated reaction mix is removed, heated to 100°C for 5 min, loaded onto a 16% polyacrylamide/8 *M* urea gel, and electrophoresed to separate the reaction products. As a size standard, 5′-[^{32}P] end-labeled primer is co-electrophoresed with reaction products. The gel is then dried and autoradiographed according to standard protocols.[49] Autoradiographs are then scanned on a densitometer to determine integrated band intensities of each extension product.[50]

Interpretation of data — Interpretation of the gel-based assays for DNA polymerase fidelity and the kinetic equations to derive the misinsertion efficiency were first defined by Goodman and coworkers and are summarized below (for a detailed account, see Reference 51). For a single extension event at a given target site, T, the integrated intensity of the band corresponding to this extension product, I_T, is related to reaction velocity, V, in a standing start reaction by the equation:

$$V = \left(100\ \ I_T\right) / \left(I_{T\pm1} + 0.5\ \ I_T\right) t \tag{1}$$

where t equals reaction time and I_{T-1} equals the intensity of the band corresponding to the primer prior to extension at target site T, or if $I_T \ll I_{T-1}$ then simply by the equation:

$$V = \left(I_T\right) / t \tag{2}$$

or in a running start reaction by the equation:

$$V = \left(I_T / I_{T\pm1}\right)\left(I_T + I_{T\pm1}\right) / t \tag{3}$$

Reaction velocities for insertion of correct and incorrect nucleotides are determined over a range of dNTP concentrations and then used to calculate kinetic parameters K_m and V_{max}. From these parameters the fraction of misinserted nucleotides, f_{ins}, is obtained, defined by the equation:

$$f_{ins} = \frac{\left(V_{max}/K_m\right)\text{ incorrect}}{\left(V_{max}/K_m\right)\text{ correct}} \tag{4}$$

The f_{ins} value provides a direct measure of misinsertion efficiency for a given misinsertion event.

V. Processivity

Polymerization is thought to be "processive" if the polymerase, upon the addition of a dNMP to the primer, translocates to the new primer position with a greater probability than dissociating from the template-primer. In contrast, polymerization is believed to be "distributive" if, after the dNMP addition, the dissociation of the polymerase from the template-primer complex occurs with a greater probability than its translocation to the new primer position. Polymerase processivity is measured by the number of nucleotides polymerized onto a nascent chain during a single binding event by a polymerase before dissociation from a template-primer, and is an important enzymatic parameter that can influence the fidelity of HIV RT, possibly by affecting misextension kinetics. The processivity also plays a key role in the strand transfer events that characterize retroviral DNA synthesis. Most methods employed to study the processivity of HIV RT[46,52-57] measure the incorporation of nucleotide triphosphates into defined template-primers (primer extension) under conditions that allow for only one processive cycle. Such conditions are achieved by including, in the reaction mixture, an "enzyme trap" (challenger), which sequesters the enzyme upon dissociation from the primer-template complex. To monitor the incorporation of nucleotides, polymerization reactions are performed with 5′-[^{32}P] end-labeled oligonucleotide primers or α-[^{32}P] dNTP, depending on choice of template. Reaction products are analyzed via PAGE followed by autoradiography (labeled primer) or via liquid scintillation counting (labeled dNTP) and time courses for product formation determined to establish nucleotide incorporation rates.

A. Measuring Processivity

The assay presented here provides a basic framework for studying HIV RT processivity on heteropolymeric RNA or DNA templates. The selection of template sequence is left to the investigator; however, template length should be at least 500 nucleotides. Primer length should be kept between 18 and 22 nucleotides and should be designed to allow for a minimum extension of >80% of template length. To avoid multiple priming sites, primers should not have >50% complementarity to any template sequence other than the targeted hybridization site. Primers should also be free of sequences with potential for forming secondary structures (e.g., hairpins). The assay described here is mainly based on those described by Dudding et al.[53] and Yu and Goodman.[47]

Templates — In a versatile approach that allows one to generate either RNA or DNA templates of the same sequence, the DNA fragment of the investigator's choice is inserted into a plasmid vector downstream of a T7 promoter. For RNA template preparation, the vector is first linearized with the appropriate restriction endonuclease and run-off transcription carried out with T7 polymerase using commercially available transcription kits (e.g., Ambion or Promega). The RNA transcripts are purified via denaturing PAGE (8% polyacrylamide/8 *M* urea/TBE [Tris. Borate/EDTA]) followed by electroelution using a Bio-Rad Electroeluter (BioRad) pretreated for 1.5 h with 1X TBE/5% diethyl pyrocarbonate (DEPC). The eluted RNA is ethanol precipitated and resuspended in DEPC-treated H_2O and spectrophotometrically quantitated. To prepare DNA templates, asymmetric polymerase chain reaction (PCR) with Vent Polymerase (New England Biolabs, Beverly, MA) is employed.[63] A suitable primer pair is selected and PCR is carried out with the T7 plasmid vector template at a primer-to-primer ratio of 10:1, with the primer that yields the desired template DNA strand in excess. Following PCR, the PCR products are separated from unextended primers and unincorporated dNTPs by a spun-column procedure,[49] or by using a commercial purification column, such as Wizard PCR Preps purification columns (Promega, Madison, WI), and the recovered material further purified via agarose gel electrophoresis (3% low-melt) followed by gel slice elution using Wizard PCR kit (Promega). The purified template DNA is then quantitated by spectrophotometry and stored at –20°C.

Primers — Chemically synthesized oligodeoxynucleotide primers are gel purified via denaturing PAGE.[49] The purified primers are then 5′ end-labeled by standard protocol[49] using T4 polynucleotide kinase and γ-[^{32}P] ATP. Radiolabeled primers are then purified from unincorporated ATP using commercially available kits (e.g., Promega) according to manufacturer's recommendations. Labeled primers are stored at –20°C.

Hybridization of template-primers — Labeled primers are annealed to templates by heating a mixture of 300 n*M* 5′-[^{32}P] end-labeled primer and 600 n*M* template RNA or DNA in hybridization buffer (50 m*M* Tris-HCl, pH 7.5, 50 m*M* KCl) to 65°C then slow cooling to room temperature over 2 h. Hybridized template-primers are then stored at –20°C.

Assay — Standard processivity assay is performed as follows: 4 to 200 n*M* RT enzyme is mixed with 4 n*M* template-primer in 1X assay buffer (50 m*M* Tris-HCl, pH 8.0, 80 m*M* KCl, 1 m*M* EDTA, 1 m*M* dithiothreitol, 0.1% Triton X-100) in a total volume of 45 μl, and the enzyme allowed to bind to template-primer by preincubation at 37°C for 10 min. The polymerization reaction is then initiated by adding 5 μl of prewarmed (37°C) 1X assay buffer containing 2 m*M* dNTPs, 60 m*M*

$MgCl_2$, and, in challenged assays, 1 mg/ml heparin as enzyme trap. Reaction mixes are incubated at 37°C for 10 min then terminated via the addition of 0.5 *M* EDTA, pH 8.0, to a final concentration of 50 m*M*. To analyze the reaction products, a 14 μl aliquot of terminated reaction mix is removed, heated to 75°C for 4 min, then electrophoresed in an 8% polyacrylamide/8 *M* urea gel to resolve the polymerization products. As a size standard, 5′-[^{32}P] end-labeled *Hae*III digested φX174 RF DNA is electrophoresed alongside the reaction products. In a separate lane, 5′-[^{32}P] end-labeled primer is also electrophoresed. Following electrophoresis, the gel is dried and autoradiographed according to standard methods.[49]

Interpretation of data — Over long reaction times, nucleotide incorporation reaches a plateau and, at this limiting level, processivity can be accurately evaluated.[57] The reaction period (10 min) selected for this assay allows for this incorporation plateau to be reached. As such, a simple visual inspection of the autoradiographic band pattern produced from a challenged reaction provides a direct measurement of the average number of nucleotides incorporated into a labeled primer per initial RT-template-primer complex.

VI. A Screen for Genetic Variants

For rapid isolation of mutants that display polymerase activity, a method that allows the detection of HIV-1 RT activity in bacterial colonies has been developed.[58] The method, to be used in combination with an HIV-1 RT expression vector, can be employed to isolate revertants of an inactive parent[58] or to isolate drug-resistant mutants.[59] Isolation of pseudorevertants will help obtain further insights into the spatial arrangements of binding sites for various substrates, viz., dNTPs, primer, and template RNA and DNA. The technique can also be used to predict the possible generation, *in vivo,* of variant viruses that are resistant to novel inhibitors. To isolate such genetic variants, one needs to examine a large number of clones arising *in vivo* or from random mutagenesis *in vitro,* selecting or screening for an active variant. Since there is no biological selection available for HIV RT, the *in situ* screening technique is of great utility.

The screening is based on the detection of the enzymatic activity of HIV RT expressed in bacteria, *in situ,* in a replica of the colonies. One begins with an HIV RT expression plasmid in bacteria. The assay works best when trp E-HIV-1 RT fusion constructs are used, although non-fusion constructs that encode authentic RTs have also been used successfully.[60,64,65] Colonies carrying the trp E-HIV-1 RT expression plasmid are grown on ordinary petri plates and a lift of the colonies onto a nitrocellulose filter is prepared. The expression of the fusion protein is induced *in situ* in the colonies on the filter, followed by lysis of the bacteria in the colonies, and immobilization of the released proteins. The filter is then directly incubated in RT reaction cocktail yielding an assay for the synthesis, on RNA templates, of radioactively labeled complementary DNA. The labeled reaction product is retained on the filter until fixation, and is detected by autoradiography (Figure 4.4).

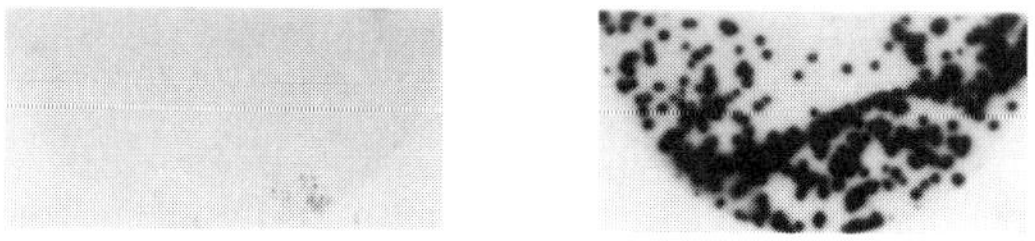

pATH2 pHRTRX2

FIGURE 4.4
Autoradiogram showing a typical *in situ* colony screening assay. The half circles represent portions of the colony lifts onto nitrocellulose filter that were processed as described in the text for detection of RNA-dependent DNA polymerase activity of HIV-1 RT. Left, colonies carrying the expression vector alone without the HIV-1 RT insert (pATH2); right, colonies containing the HIV-1 RT expression vector, pHRTRX2. (From Prasad, V. R. and Goff, S. P., *J. Biol. Chem.*, 264, 16689, 1989. With permission.)

A. Assay

Bacteria carrying the trp E-HIV RT expression plasmid[8] are plated on M9 plates with casamino acids (5 mg/ml), tryptophan (20 μg/ml), and the appropriate antibiotic selection to form about 10^2 to 10^3 colonies per 100-mm dish. After incubation at 37°C for 16 to 18 h, at a time when the colonies are still very small (diameter <1 mm), the colonies are lifted onto nitrocellulose filter circles and the filters placed, colony side up, on M9 plates containing casamino acids but lacking tryptophan. The colonies are incubated for 2 h at 37°C (the remnant of the colonies on the original plate are also incubated to maintain a master stock of the library being screened). The filters are transferred to M9 plates containing casamino acids plus 10 μg/ml 3β-indole acrylic acid, and the colonies incubated an additional 6 h at 37°C to induce the trpE-RT fusion protein.

Following induction for the expression of trp E-HIV-1 RT, the filters are placed, colony side up, on 0.4-ml drops of lysozyme solution (20 μg/ml in TEND buffer [50 m*M* Tris-HCl, pH 7.5, 0.5 m*M* EDTA, 0.3 *M* NaCl, and 1 m*M* DTT] for 30 min at 4°C. The filters are then placed on 1-ml drops of a solution of 0.4% NP-40 in TEND buffer for 15 min at room temperature to lyse the bacteria. The proteins are immobilized to the filters by placing the filters on Whatman 3M paper under a hand-held UV torch set to long wavelength (Mineralight Lamp Model UVGL-25 [multi-band UV, wavelengths 254/366 nm]) for 10 min. The filters are washed in a solution of 0.08% BSA (Sigma) in buffer (30 m*M* HEPES, pH 7.5) for 3 h to saturate nonspecific binding sites. The excess debris is removed during an overnight soaking in a solution containing 50 m*M* Tris-HCl, pH 7.5, 1 m*M* EDTA, 2 m*M* dithiothreitol, 0.2 m*M* NaCl, 10% glycerol, and 0.1% NP-40. The filters are removed from the wash gently, allowing excess glycerol to drain, and soaked for 30 min in RT cocktail containing template and primer, but lacking triphosphates [50 m*M* Tris-HCl, pH 8.0, 10 m*M* $MgCl_2$, 20 m*M* DTT, 60 m*M* NaCl, 0.05% NP-40, 10 mg/ml poly r(C) and 5 mg/ml oligo d(G)]. The reaction is initiated by transfer of the filter to RT cocktail containing triphosphates (10 μ*M* [α-^{32}P] dGTP; 1 Ci/mmol) and allowed to proceed for 30 min at room temperature. Finally, the filters are fixed in cold 10%

trichloroacetic acid (TCA) with two changes at 10 min intervals, air dried, and exposed to X-ray film for 3 to 4 h. An autoradiogram of a typical screen is shown in Figure 4.4.

The success of this protocol is due partly to the fact that, under the conditions in which the HIV RT activity is assayed, the endogenous DNA polymerase I enzyme is not active. The DNA polymerase I of *Escherichia coli* displays RNA-dependent DNA polymerase activity on poly (rA)·oligo (dT) template primers with manganese as the divalent cation. In this procedure, however, the reaction cocktails utilize poly (rC)·oligo (dG) with magnesium as the source of divalent cation.

Several applications of this technique are possible and, as indicated below, some of these have already been proven: (1) the isolation of second-site revertants of linker insertions of HIV RT,[58] (2) the isolation of drug-resistant variant reverse transcriptase molecules,[59] (3) isolation of enzymatically active hybrid RT molecules, starting with overlapping, inactive fragments of RT genes from different retroviruses, forcing homologous recombination in bacteria to form active recombinants. One application, the isolation of drug resistant variants, is described below.

B. Isolation of Drug Resistant RTs *In Vitro*

The colony screening method has been used both to isolate drug resistant mutant RTs expressed in bacteria, and to detect the presence of drug resistant RTs in patient-derived viruses. In the former, one begins with a random pool of mutations created in an HIV-1 RT expression construct, transforms the library into *E. coli,* carrying out screening, as described above, in the presence of inhibitory concentrations of a drug (Figure 4.5). The screen would display rare mutants with decreased sensitivity to the drug, if these were present in the library. This approach was successfully used to isolate a dideoxyguanosine triphosphate-resistant variant of HIV-1 RT. This mutant, with a Glu89Gly alteration, also displayed resistance to ddTTP, ddCTP, ddATP, AZTTP, as well as to phosphonoformic acid (PFA), and a loss of magnesium preference.[59] The X-ray crystal structure of HIV-1 RT[61] has shown that the 89th residue makes contact with the template DNA strand and is located very close to the dNTP addition site.

With the increased realization that other viral targets are as likely to evolve into drug-resistant forms during therapy as RT, our hopes for a chemotherapeutic intervention for AIDS are dependent on the development of a new generation of drugs that not only inhibit RT effectively but are designed to preclude the development of resistance. Such attempts are fueled by the recent description of the X-ray crystal structure of HIV-1 RT[61,62] and the potential to generate high-resolution structures of mutant RTs. The availability of such structural information, the mutants generated *in vivo* and *in vitro,* and the biochemical tests exemplified by those described above should be major steps in the right direction.

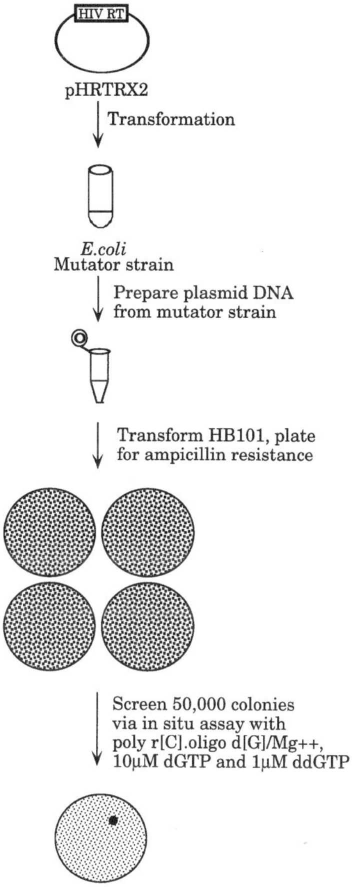

FIGURE 4.5
Schematic of *in situ* colony screening for drug resistant RT variants. The HIV-1 RT expression plasmid is subjected to random mutagenesis by passage in the mutator strain of *E. coli* mutD5. The pool of mutants is transformed into the *E. coli* strain HB101 and up to 50,000 colonies are plated. Following induction for expression as described in the text, the colonies are lysed *in situ* and RT reactions performed in the presence of an RT-specific inhibitor. Rare colonies that show activity are then isolated and retested.

References

1. Mitsuya, H., Yarchoan, R., and Broder, S., *Science,* 249, 1533 (1990).
2. Mitsuya, H. and Broder, S., *Nature,* 325, 773 (1987).
3. Larder, B. A., Darby, G., and Richman, D. D., *Science,* 243, 1731 (1989).
4. Larder, B. A. and Kemp, S. D., *Science,* 246, 1155 (1989).
5. Le Grice, S. F. J., Beuck, V., and Mous, J., *Gene,* 55, 95 (1987).
6. Hizi, A., McGill, C., and Hughes, S. H., *Proc. Natl. Acad. Sci. U.S.A.,* 85, 1218 (1988).
7. Tanese, N., Sodroski, J., Haseltine, W., and Goff, S. P., *J. Virol.,* 59, 743 (1986).
8. Tanese, N., Prasad, V. R., and Goff, S. P., *DNA,* 7, 407 (1988).

9. di Marzo Veronese, F., Copeland, T. D., Oroszlan, S., Gallo, R. C., and Sarngadharan, M. G., *J. Virol.,* 62, 795 (1988).
10. Larder, B., Purifoy, D., Powell, K., and Darby, G., *EMBO J.,* 6, 3133 (1987).
11. Prasad, V. R. and Goff, S. P., *Proc. Natl. Acad. Sci. U.S.A.,* 86, 3104 (1989).
12. Prasad, V. R. and Goff, S. P., *Ann. N.Y. Acad. Sci.,* 616, 11 (1990).
13. Hizi, A., Barber, A., and Hughes, S. H., *Virology,* 170, 326 (1989).
14. Hizi, A., Hughes, S. H., and Shaharabany, M., *Virology,* 175, 575 (1990).
15. Larder, B. A., Purifoy, D. J. M., Powell, K. L., and Darby, G., *Nature,* 327, 716 (1987).
16. Kew, Y., Song, Q., and Prasad, V., *J. Biol. Chem.,* 269, 15331 (1994).
17. Weiss, R., Teich, N., Varmus, H., Coffin, J., Eds., *RNA Tumor Viruses,* Vol. 1, Cold Spring Harbor Laboratory Press, Cold Spring Harbor, New York, 1984.
18. Goff, S. P., *J. AIDS,* 3, 817 (1990).
19. Kohlstaedt, L. A. and Steitz, T. A., *Proc. Natl. Acad. Sci. U.S.A.,* 89, 9652 (1992).
20. Barat, C., Lullien, V., Schatz, O., Keith, G., Nugeyre, M. T., LeGrice, S. F. J., and Darlix, J. L., *EMBO J.,* 8, 3279 (1989).
21. White, E. L., Buckheit, R. W. J., Ross, L. J., Germany, J. M., Andries, K., Pauwels, R., Janssen, P. A. J., Shannon, W. M., and Chirigos, M., *Antivir. Res.,* 16, 257 (1991).
22. Reardon, J. E. and Miller, W., *J. Biol. Chem.,* 265, 20302 (1990).
23. Tsuchihashi, Z. and Brown, P. O., *J. Virol.,* 68, 5863 (1994).
24. Hoffman, A. D., Banapour, B., and Levy, J. A., *Virology,* 147, 326 (1985).
25. Rey, M. A., Spire, B., Dormont, D., Barre-Sinoussi, F., Montagnier, L., and Chermann, J. C., *Biochem. Biophys. Res., Commun.,* 121, 126 (1984).
26. Spanos, A., Sedgwick, S. G., Yarranton, G. T., Hubscher, U., and Banks, G. R., *Nucleic Acids Res.,* 9, 1825 (1981).
27. Roth, M., Tanese, N., and Goff, S. P., *J. Biol. Chem.,* 260, 9326 (1985).
28. Laemmli, U. K., *Nature,* 227, 680 (1970).
29. Starnes, M. C., Gao, W., Ting, R. Y. C., and Cheng, Y., *J. Biol. Chem.,* 263, 5132 (1988).
30. Boyer, P. L., Ferris, A. L., and Hughes, S. H., *J. Virol.,* 66, 1031 (1992).
31. Furfine, E. S. and Reardon, J. E., *J. Biol. Chem.,* 266, 406 (1991).
32. Temin, H. M., *Proc. Natl. Acad. Sci. U.S.A.,* 90, 6900 (1993).
33. Peliska, J. A. and Benkovic, S. J., *Science,* 258, 1112 (1992).
34. DeStefano, J. J., Bambara, R. A., and Fay, P. J., *J. Biol. Chem.,* 269, 161 (1994).
35. Luo, G. and Taylor, J. M., *J. Virol.,* 64, 4321 (1990).
36. DeStefano, J. J., Mallaber, L. M., Rodriguez-Rodriguez, L., and Bambara, R. A., *J. Virol.,* 66, 6370 (1992).
37. Saag, M. S., Hahn, B. H., Gibbons, J., Li, Y., Parks, E. S., Parks, W. P., and Shaw, G. M., *Nature,* 334, 440 (1988).
38. Meyerhans, A., Cheynier, R., Albert, J., Seth, M., Kwok, S., Sninsky, J., Morfeldt-Manson, L., Asjo, B., Wain-Hobson, S., *Cell,* 58, 901 (1989).
39. Preston, B. D. and Garvey, N., *Pharm. Tech.,* 16, 34 (1992).
40. Cheynier, R., Henrichwark, S., Hadida, F., Pelletier, E., Oksenhendler, E., and Wain-Hobson, A. B. S., *Cell,* 78, 373 (1994).
41. Roberts, J. D., Bebenek, K., and Kunkel, T. A., *Science,* 242, 1171 (1988).
42. Preston, B. D., Poisez, B. J., and Loeb, L. A., *Science,* 242, 1168 (1988).
43. Takeuchi, Y., Nagumo, T., and Hoshino, H., *J. Virol.,* 62, 3900 (1988).

44. Perrino, F. W., Preston, B. D., Sandell, L. L., and Loeb, L. A., *Proc. Natl. Acad. Sci. U.S.A.,* 86, 8343 (1989).
45. Weber, J. and Grosse, F., *Nucleic Acids Res.,* 17, 1379 (1989).
46. Kati, W., Johnson, K. A., Jerva, L. F., and Anderson, K. S., *J. Biol. Chem.,* 267, 25988 (1992).
47. Yu, H. and Goodman, M. F., *J. Biol. Chem.,* 267, 10888 (1992).
48. Ricchetti, M. and Buc, H., *EMBO J.,* 12, 387 (1993).
49. Sambrook, J., Fritsch, E. F., and Maniatis, T., *Molecular Cloning: A Laboratory Manual,* 2nd ed., Cold Spring Harbor Laboratory Press, Cold Spring Harbor, New York, 1989.
50. Boosalis, M. S., Petruska, J., and Goodman, M. F., *J. Biol. Chem.,* 262, 14689 (1987).
51. Goodman, M. F., Creighton, S., Bloom, L. B., and Petruska, J., *Crit. Rev. Biochem. Mol. Biol.,* 28, 83 (1993).
52. Huber, H. E., McCoy, J. M., Seehra, J. S., and Richardson, C. C., *J. Biol. Chem.,* 264, 4669 (1989).
53. Dudding, L. R., Nkabinde, N. C., and Mizrahi, V., *Biochemistry,* 30, 10498 (1991).
54. Kopp, E. B., Miglietta, J. J., Shrutkowski, A. G., Shih, C.-K., Grob, P. M., and Skoog, M. T., *Nucleic Acids Res.,* 19, 3035 (1991).
55. Dirani-Diab, R. E., Andreola, M.-L., Nevinsky, G., Tharaud, D., Barr, P. J., Litvak, S., and Tarrago-Litvak, L., *FEBS Lett.,* 301, 23 (1992).
56. Beard, W. A. and Wilson, S. H., *Biochemistry,* 32, 9745 (1993).
57. Bavand, M. R., Wagner, R., and Richmond, T. J., *Biochemistry,* 32, 10543 (1993).
58. Prasad, V. R. and Goff, S. P., *J. Biol. Chem.,* 264, 16689 (1989).
59. Prasad, V. R., Lowy, I., De Los Santos, T., Chiang, L., and Goff, S. P., *Proc. Natl. Acad. Sci. U.S.A.,* 88, 11363 (1991).
60. Saag, M. S., Emini, E. A., Laskin, O. L., Douglas, J., Lapidus, W. I., Schleif, W. A., Whitley, R. J., Hildebrand, C., Byrnes, V. W., Kappes, J. C., Anderson, K. W., Massari, F. E., and Shaw, G. M., *N. Engl. J. Med.,* 329, 1065 (1993).
61. Jacobo-Molina, A., Ding, J., Nanni, R. G., Clark, A. D. J., Lu, X., Tantillo, C., Williams, R. L., Kamer, G., Ferris, A. L., Clark, P., Hizi, A., Hughes, S., and Arnold, E., *Proc. Natl. Acad. Sci. U.S.A.,* 90, 6320 (1993).
62. Kohlstaedt, L. A., Wang, J., Friedman, J. M., Rice, P. A., and Steitz, T. A., *Science,* 256, 1783 (1992).
63. Levin, J., personal communication.
64. Emini, E., personal communication.
65. Hughes, S., personal communication.

Chapter 5

Tat Activation Studies: Use of a Reporter Cell Line

Estuardo Aguilar-Cordova, Javier Chinen, and John W. Belmont

Contents

0-8493-4412-3/96/$0.00+$.50

I. Background

HIV-1, like other lentiviruses, encodes genes for regulatory proteins that control expression of its structural genes and of themselves.[1] Tat is one of these regulatory proteins and is essential for viral replication.[2,3] Tat is a potent positive regulator, which acts on all viral transcripts and has been proposed to increase transcription, stability of the transcripts, and efficiency of transcript translation.[1,4-6]

A. The *tat* Gene and Protein

The *tat* gene consists of a 5′ non-coding exon and two coding exons, which encode a small 86-amino acid nuclear protein.[7] The first 67 amino acids encoded by the second exon are sufficient for trans-activation.[8] Tat contains a conserved cysteine-rich region postulated to function by binding metal ions and interacting with other proteins, and a highly basic region required for nuclear localization and nucleic acid binding.[9,10] Tat trans-activates transcription of itself, causing a positive feedback loop for a burst of viral production.[11] Cellular factors and other viral proteins such as Rev may also affect *tat* transcription.[4]

B. Mechanisms of Trans-Activation by Tat

Tat mediates its activity through a trans-activation response region (TAR) located downstream of the transcription initiation site (+19 to +42 relative to the HIV-1 cap site).[12-15] It interacts with the transcribed RNA structure of TAR rather than the DNA sequence[16,17] and it affects multiple levels in the pathway of viral protein expression.[1,4-6] The Tat–TAR interaction has been studied extensively and reviewed.[18] *In vitro* and chimeric protein studies have demonstrated Tat binds directly to TAR

RNA.[17,19] The TAR RNA sequence folds into a highly nuclease-resistant stem loop structure with a U-rich "bulge", 6 bp below the loop.[20] Mutations that destabilize the hairpin structure or disrupt the bulge or apex loop decrease trans-activation by Tat.[3,20-23]

Tat has been described as an anti-terminator and as a transcriptional activator.[1,11,20,24,25] If Tat is not present, viral mRNA transcription pauses after the TAR sequence is transcribed and results in the accumulation of short RNA sequences.[11] TAR is not, however, a negative regulatory element, since its absence does not increase the basal level expression from the HIV long terminal repeats (LTR).[13] Tat has other functions in addition to transcription elongation. For example, Laspia et al. demonstrated that both E1A and Tat could synergistically increase the promoter proximal transcription, but only Tat decreased short transcript accumulation and led to a corresponding increase in the number of elongated transcripts.[5,24] In the absence of E1A, Tat was sufficient to increase initiation and elongation. Therefore, two mechanisms for Tat-induced increase of viral transcripts were shown: an increase in the rate of transcription initiation and the production of long transcripts.

C. Cellular Factors in Tat-Mediated HIV-1 Transcription

Many lines of evidence support the involvement of cellular factors in the multiplicity of Tat-mediated effects. Attempts to reproduce Tat trans-activation in nonprimate cells (hamster and mouse models) have failed, but Tat function could be restored in murine/human somatic cell hybrids bearing the human chromosome 12.[26] *In vitro,* Tat binds to the bulge of the TAR sequence,[27] cellular factors bind the TAR loop sequence.[8,28-31] One of these cellular factors, TRP-2, preferentially binds the bulge and may, under certain conditions, compete with Tat for TAR binding.[8] TRP-2 highlights the possibility of identifying cellular factors that may function as trans-dominant negative regulators of Tat. The important role of cellular factors for Tat activity stresses the need to use natural HIV target cells for functional analysis of Tat (i.e., human $CD4^+$ cells).

D. Tat as a Therapeutic Target

The control of viral transcription by Tat protein is a logical point of attack for antiviral therapeutics. The importance of Tat in the viral life cycle and its distinctiveness make it an attractive target for study and therapeutic intervention. Although it might be presumed that inhibition of Tat function would not be very useful during the asymptomatic stage of AIDS, recent reports indicate viral replication continues throughout the latency period. Therefore, curtailment of Tat activity would be beneficial at all stages of the disease.[7,32] Drug and biological treatments have been explored and show possible efficacy.[33,34]

Gene therapy is also a potential anti-HIV therapeutic route. Tat function has been a major target in the gene therapy strategies proposed to date. One approach has been the expression of antiviral or toxin genes (e.g., diphtheria toxin) regulated by the HIV LTR (see below).[35] In the diphtheria toxin example, the goal is to achieve Tat-dependent toxin expression. Thus, cells carrying the HIV LTR–toxin construct would express the toxin and die when infected by HIV, thereby not supporting a productive viral infection. Another approach has been to use the expression of TAR decoys to sequester Tat and thus limit its effects.[36] A third approach is to express Tat trans-dominant negative proteins.[37] Trans-dominants have been generated by designed mutations in *tat*.

One limiting factor in the development of novel therapeutics has been the lack of reliable, safe, and high-throughput assay systems. Most systems described to test anti-Tat therapeutics have been limited to direct HIV infection analysis or cytopathic measurements.[38,39] These methods do not isolate Tat from other viral effects and most utilize live HIV.

E. Models for Tat Analysis

Studies on HIV Tat protein trans-activation activity are mainly performed using modified cell lines with a reporter gene. Ideally, such cell lines should (1) be of human origin from naturally infected tissues, (2) have a quantitative response to Tat trans-activation, (3) be responsive to cellular factors associated with increased viral expression, (4) have low background expression of the reporter gene, (5) be widely disseminated so analyses from different laboratories can be compared, (6) be easy and inexpensive to analyze. In addition, for safety considerations and ease of large-scale screenings, such cell lines should permit assays to be performed in the absence of infectious HIV.

F. The MT-4 Cell Line

Described by Pauwels et al.,[38,39] the MT-4 cell line fulfills several of the criteria mentioned above. MT-4 is a human T cell leukemia virus type 1 (HTLV-1) transformed human T cell sensitive to the cytopathic effects of HIV-1. One drawback to the use of MT-4 in assays for antiviral activity is that it requires the use of live HIV. Results are measured by the number of cells that survive viral infection. MT-4 assays are not specific for Tat activity, but rather general for HIV production/inhibition.

G. CAT Indicator Cell Lines

Many indicator cell lines have been generated by transduction of lymphocytic and monocytic cell lines with vectors containing an HIV LTR promoter linked to the chloramphenicol acetyl transferase (CAT) gene.[40] Tat activity in these cells is measured indirectly by the relative amount of CAT enzyme activity. Most CAT enzyme assays require radioactivity or relatively expensive immunochemical reagents for detection and are not easily accommodated in large-scale screenings.

H. β-galactosidase Reporter Cell Line

Kimpton and Emerman reported a CD4+ HeLa cell line containing a single integrated copy of a β-galactosidase (β-gal) gene under the control of a truncated HIV LTR.[41] This cell line allows titering infective virus particles of HIV samples, but can also be used for Tat-activation studies. β-gal activity is most often analyzed by a colorimetric assay in fixed cells; however, it may also be analyzed from cell lysate samples. HeLa cells are of epithelial origin and not a natural target for HIV infection.

I. Secreted Alkaline Phosphatase Reporter Cell Line

Another reporter gene is the secreted alkaline phosphatase gene (SeAP).[34,42] It has the advantages of being secreted to the media and that it can be easily quantitated by a colorimetric assay. Several assays can be performed from the same culture. Its sensitivity is lower than that of CAT and β-gal, but suitable for many circumstances.

J. Cell Lines that Express Tat Constitutively

Many such cell lines have been described; these cells are used primarily to study effects of Tat on cellular functions and association with cellular proteins. Two examples are the HeLa-tat cells and the Jurkat-tat cells.[43,44]

K. Luciferase as a Reporter Gene

The firefly luciferase gene is a reporter gene that can be rapidly analyzed with relatively inexpensive reagents. Analysis of its expression is quantitative and easy to perform, and it is very sensitive and utilizes nonradioactive substrates. Unlike CAT, SeAP and β-gal, luciferase has a short half-life in Jurkat cells (estimated $t_{1/2}$ 12 h). Thus, background activity is very low and the level of luciferase expression closely parallels *tat* transcription or Tat activity. Luciferase meets all other necessary criteria as a reporter gene.

Jurkat cells are transformed human T cells that retain many normal properties of cell activation, cytokine secretion, and sensitivity to HIV. A tat-sensitive reporter gene expressed in Jurkat cells should provide a system to model most major aspects of HIV gene expression.

L. 1G5 Cells

We constructed an HIV LTR expression plasmid that contains a 450 bp EcoRV-HindIII LTR fragment from HIV-SF2, cloned into SmaI-HindIII sites of pXP-1[45] (Figure 5.1). This plasmid was stably integrated into Jurkat cells by co-transfection with pSV2-neo and selection in G418. 1G5 is a clone derived from this transfection and selected for: (1) low basal level activity, (2) susceptibility to HIV infection, (3) high responsiveness to tat expression, and (4) high responsiveness to T cell activation signals (Figures 5.2 and 5.3). When these cells are transfected with a Tat-expressing plasmid, luciferase activity is increased by 100- to 1000-fold. The same effect can be obtained if the cells are infected with a retroviral vector carrying the *tat* gene. T cell stimulating factors, which activate the HIV promoter via NFkB and other cellular factors, also increased luciferase activity from 1G5 cells.

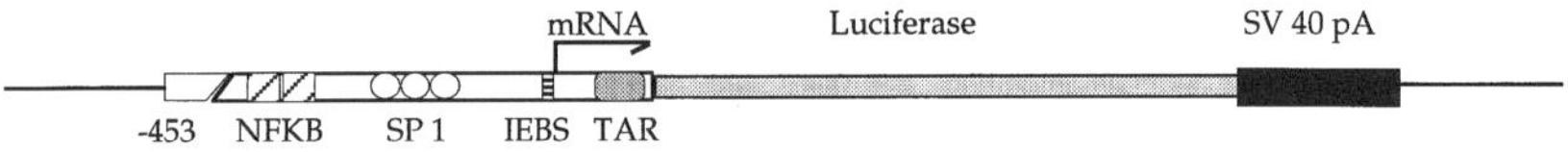

FIGURE 5.1
Diagram of pWTHluc. The expression cassette pWTHluc is cloned in a pUC12 backbone. Expression characteristics of this vector are described in the text.

M. DC10 Cells

The 1G5 cells were stably transfected with a Tat expressing vector to generate DC10 cells, a clone with constitutive Tat and luciferase activity (Figure 5.2). DC10 cells model chronic HIV infected cells, with constitutive expression of Tat.

The 1G5 and DC10 cells can be used very effectively in quantitative analyses of biologic and drug inhibitors of Tat activation.

II. Examples of Applications for 1G5 and DC10 Cell Lines

A. Gene Therapy: Effects of a "Suicide" Gene, DT176

Towards the development of HIV gene therapy strategies, we constructed a Tat-dependent diphtheria toxin vector. This vector has an attenuated diphtheria toxin gene (DT176) driven by a shortened HIV LTR promoter. The presence of Tat activates the expression of the toxin; therefore, HIV infected cells are killed and do not support further HIV replication. The effectiveness of this vector was assayed by transfections into 1G5 and DC10 cells. 1G5 cells were co-transfected with pSV2tat and varying amounts of control or toxin vectors (Figure 5.4). In DC10 cells, the toxin vector was singly transfected and its effects were compared to control construct effects (Figure 5.4). Concomitantly, survival of cells was measured by thiazolyl blue (MTT) cytotoxicity assays. Addition of the Tat-expressing vector to DT176-transfected cells resulted in decreased luciferase activity and increased cell death.

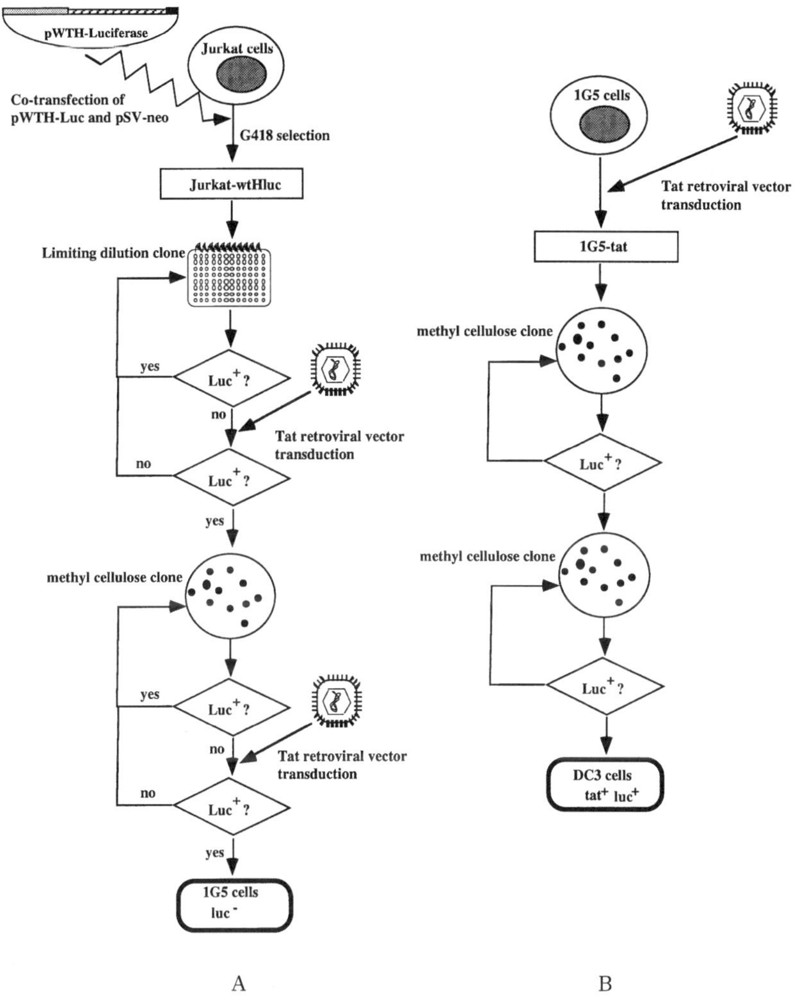

FIGURE 5.2
Strategy for derivation of 1G5 and DC10 cells (A) were derived from Jurkat cells transfected with pWTHluc, single cell cloned by limiting dilution and methyl-cellulose, and selected as described in text. DC10 cells (B) were derived from 1G5 cells transduced with a *tat* retroviral vector, single cell cloned two times by methyl-cellulose colony formation and selected as described.

B. Anti-Tat Drugs

The 1G5 and DC10 cells have also been used for screening of potential Tat-inhibitor drugs. For analyses on 1G5, the cells were infected with a *tat* retroviral vector at the time of drug exposure. This protocol detected inhibition of Tat function as well as events upstream of Tat synthesis (e.g., reverse transcription inhibition). For analyses in DC10, the cells were directly exposed to varying drug concentrations and

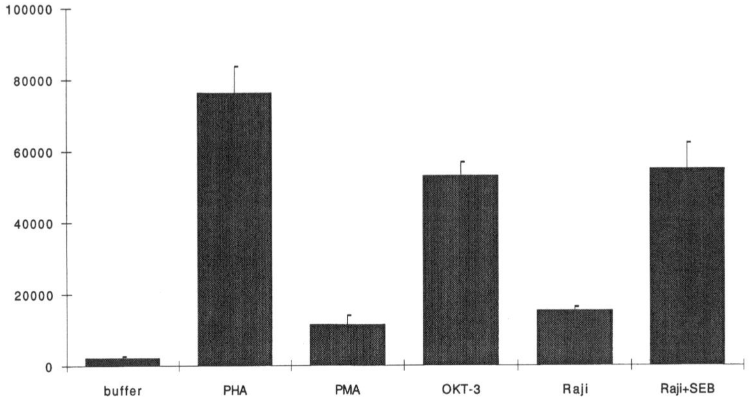

A

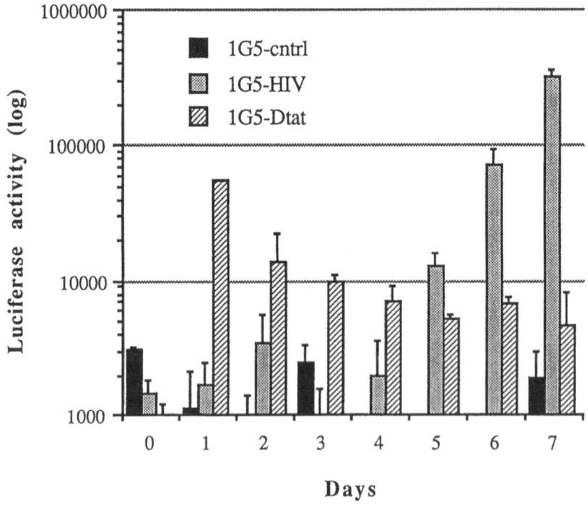

B

FIGURE 5.3
Expression characteristics of 1G5 cells after stimulation and infection. (A) Luciferase expression after stimulation with various mitogens: PHA, phytohemaglutinin at 0.5 μg/ml; PMA, phorbol-12 myristate-13 acetate at 50 ng/ml; OKT 3, anti CD3 antibody; Raji, antigen presenting B cells; Raji+SEB, raji cells plus *Staphylococcus* enterotoxin B superantigen. (B) Time course of 1G5 cells infected with HIV-1 (stippled bars) or transduced with a *tat* retroviral vector (striped bars) or mock infected (black bars). (From Aguilar-Cordova, C. E. et al., *AIDS Res. Hum. Retroviruses,* 10, 293 (1994). With permission.)

the luciferase activity measured after 3 d. The DC10 protocol was designed to detect only Tat-specific inhibition. The viability of the cells was analyzed in parallel to luciferase activity to distinguish cytotoxic from Tat-inhibitory effects. The use of 1G5 and DC10 as reporter cells in HIV/tat inhibition studies was validated by comparing results of known inhibitory drugs on luciferase expression with effects of the drugs on chronic and acute HIV infection of cultured T cell lines measured by reverse transcriptase analysis (Figure 5.5).

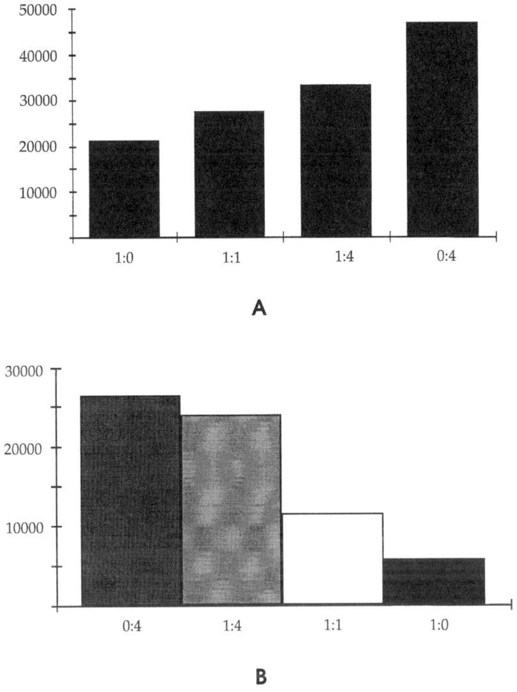

FIGURE 5.4
Effects of anti-tat vectors on luciferase expression by 1G5 and DC10 cells. Dilution effects of toxin vectors on (A) 1G5 cells, and (B) DC10 cells. Controls for co-transfection included pSV-neo (–), pH176 (+), and no transfection. All transfections were standardized to 20 μg of DNA. Abscissa, the ratio of toxin vector (pH176) to carrier plasmid. Ordinate, luciferase activity.

III. Protocols

A. Cell Culture

Cells are maintained in RPMI 1640 medium supplemented with 10% calf serum (Hyclone) in 5% CO_2, at 37°C. Selection for stable transfections using the hygromycin resistance (hyg) gene is done using a concentration of 200 μg/ml of hygromycin.

B. Transfection

1. Transient

1. Grow cells to a density of 10^6 cells/ml.
2. Harvest and pellet cells at 400 *g*.
3. Resuspend in RPMI 1640 10% serum to 2×10^7 cells/ml.

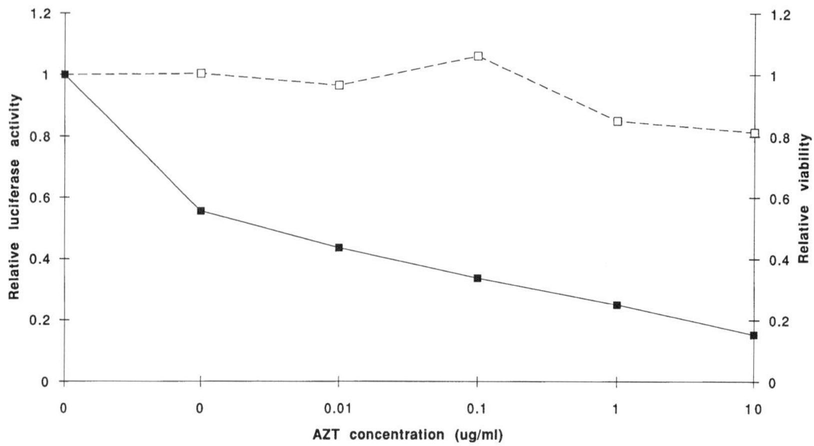

FIGURE 5.5
Antiviral drug effects on luciferase expression from 1G5 cells. 1G5 cells were transduced with a *tat* retroviral vector or infected with HIV-1 in the presence of varying concentrations of 2′,3′-deoxy, 3′-azidothymidine (AZT). The effect of AZT was similar for both viruses; the results shown represent the average of various experiments with each virus. Luciferase activity was measured 3 d after transduction and 6 d after infection. EC_{50} (effective concentration at which 50% of control luciferase activity is inhibited) = 0.0028 μg/ml. Solid line, luciferase activity; dashed line, viability. (From Aguilar-Cordova, C. E. et al., *AIDS Res. Hum. Retroviruses,* 10, 293 (1994). With permission.)

4. Mix 125 μl of cell suspension with the plasmids to transfect. Total amount of DNA per transfection should be 20 μg in 25 μl. DNA concentration can be adjusted with a carrier plasmid (e.g., pUC-12).
5. Electroporate the preparation in 0.2 cm-gap cuvettes using the following conditions: 125 V, 3000 μF, and 13 ohms for resistance (ECM 600BTX electroporator).
6. Transfer electroporated cells with a Pasteur pipette from cuvettes to 1.5 ml growth media in 24-well dishes.
7. Harvest after 18 h and analyze luciferase activity.

2. Stable

1. Follow steps 1 and 2 above.
2. Resuspend cells at 1 × 10^7 cells/ml.
3. Mix 300 μl of cell suspension with the plasmids to transfect. Total amount of DNA per transfection should be 20 μg and use of carrier DNA is not recommended. Selection plasmids have to be in small proportion relative to test plasmid (1:10) to assure presence of the latter in the selected cells. For example: if selecting with hygromycin, mix 2 μg of *hyg*^{r} plasmid with 18 μg of the desired plasmid to transfect.
4. Electroporate the cells in 0.4 cm-gap cuvettes using the following conditions: 230 V, 950 μF, and 13 ohms for resistance (ECM 600BTX electroporator).
5. Transfer electroporated cells with a Pasteur pipette from cuvettes to 1.5 ml growth media, 24-well dishes.

6. Change media adding the selector drug 24 h after transfection. Use electroplated but non-transfected cells as a control for drug selection. Change selection media every 3 to 5 d until all control cells are dead (approximately 14 d).

C. Gene Transfer by Retroviral Vector Transduction

1. Plate 200,000 cells in 1 ml of RPMI 1640, serum 10% media.
2. Add 1 ml of filtered retrovirus supernatant, 10^6 infectious units/ml.
3. Add Polybrene to a final concentration of 4 µg/ml. (Stock: 4 mg/ml Polybrene in water; 1000X)
4. Change to growth media after overnight incubation.
5. Analyze transduced cells after 3 to 5 d.

D. Luciferase Assay

1. Harvest cells in 1.5 ml microcentrifuge tubes.
2. Pellet cells by pulsing in a microcentrifuge for 10 sec and discard supernatant.
3. Resuspend cells in 500 µl of PBS by vortex.
4. Repeat step 2.
5. Add 50 µl of lysis buffer (Luciferase Assay System, Promega) and vortex.
6. Incubate at room temperature for 20 min.
7. Vortex and centrifuge at top speed for 30 sec.
8. Add 10 µl of cell lysate to 50 µl of luciferase substrate (Luciferase Assay System, Promega).
9. Read light emission in a scintillation counter or a luminometer (scintillation counter needs to have the correction coincidence circuit turned off or be set in single photon counting mode).

E. Cytotoxicity Effect

1. MTT assay

1. Grow cells in a 96-well tissue culture plate; use 100 µl per well.
2. After sufficient time for the test agent to have effect (typically 3 to 5 d for drug assays), add 15 µl of dye solution to each well (Promega, MTT assay).
3. Incubate at 37°C in a humidified atmosphere, 5% CO_2, for 4 h.
4. Add 100 µl of the solubilization solution (Promega, MTT assay).
5. Incubate 1 h.
6. Mix the contents of the wells and record the absorbance at 570 nm with a reference wavelength of 630 nm.

2. Alamar Blue assay (Alamar Biosciences, Inc.)

1. Grow cells in a 96-well tissue culture plate, use 100 μl per well.
2. After sufficient time for the test agent to have effect, add 25 μl of Alamar Blue reagent per well.
3. Incubate at 4 h at 37°C.
4. Record absorbance at 570 nm with a reference wavelength of 600 nm, or measure fluorescence with excitation wavelength at 560 nm and emission at 590 nm.

3. Reverse transcriptase assay

1. Pellet supernatant at 70,000 rpm, 1 h in a TL 100 Beckman centrifuge rotor.
2. Solubilize pellet in 15 ml of solubilization mix (50 m*M* Trizma, pH 8.2, 0.2 m*M* leupeptine, 50 m*M* EACA, 0.5% Triton X-100, 5% glycerol, 0.8 *M* NaCl, 2.0 m*M* DTT).
3. Incubate at room temperature for 10 min.
4. Add 25 μl of RT cocktail (50 m*M* Trizma, pH 8.2, 10 m*M* $MgCl_2$, 2 m*M* DTT, 20 μCi ^{3}H-TTP, 0.05 units poly(rA):p(dT)12-18) to each sample.
5. Incubate at 37°C for 1.5 h.
6. Absorb the RT mix to 1 inch square of DEAE cellulose.
7. Incubate at room temperature for 15 to 30 min.
8. Wash the square in 0.25 *M* Na_2HOP_4.
9. Repeat three more times.
10. Wash once in ddH_2O.
11. Wash once in 95% ethanol.
12. Dry the square under vacuum at 80°C for 30 min.
13. Add 1 ml of scintillation fluid to the samples and count in scintillation counter.

IV. Comments

The analysis of Tat activation by using reporter cell lines has been pivotal in understanding Tat function and HIV biology. These cell lines are now being used to evaluate methods to fight HIV infection. Tests of classical and novel anti-HIV drugs can be performed studying the anti-Tat effect alone or the anti-viral effect. Luciferase activity is easy to measure, very sensitive, and can be quantitated accurately. A possible disadvantage is that measuring requires a luminometer, or a scintillation counter with single photon counting capacity. In addition to HIV analyses, the cell lines described above are also useful for studies of T cell activation, since many elements of the HIV promoter respond to T cell activation signals (e.g., NFkB, SP1 sites). These cell lines are available through the AIDS Research and Reference Reagent Program (Repository, (301) 496-8378, Cat. No. 1819).

Acknowledgments

We greatly appreciate the scientific support from our colleagues Drs. Janet Butel, Dorothy Lewis, Andrew Rice, and Wade Harper, as well as technical support from Wendy Fang and Felipe Amaya. The work presented here was supported in part by NIAID, NIH Grant U01 A130243 and ACS Grant #IRG-199. J.C. is supported by a Howard Hughes Medical Institute Predoctoral Fellowship.

References

1. Cullen, B. R. and Greene, W. C., *Cell,* 58, 423 (1989).
2. Dayton, A. I., Sodrowski, J. G., Rosen, C. A., Goh, W. C., and Haseltine, W. A., *Cell,* 44, 941 (1986).
3. Feng, S. and Holland, E. C., *Nature,* 334, 165 (1988).
4. Jones, K. A., *New Biol.,* 1, 127 (1989).
5. Laspia, M. F., Rice, A. P., and Mathews, M. B., *Genes Dev.,* 4, 2397 (1990).
6. Braddock, M., Thorburn, A. M., Kingsman, A. J., and Kingsman, S. M., *Nature,* 350, 439 (1991).
7. Arya, A. K., Guo, C., Josephs, S. F., and Wong-Staal, F., *Science,* 229, 69 (1985).
8. Sheline, C. T., Milocco, L. H., and Jones, K. A., *Genes Dev.,* 5, 2508 (1991).
9. Siomi, H., Shida, H., Maki, M., and Hatanka, M., *J. Virol.,* 64, 1803 (1990).
10. Frankel, A. D., Bredt, D. S., and Pabo, C. O., *Science,* 240, 70 (1988).
11. Kao, S. Y., Cohen, A., Luciw, P. A., and Peterlin, M. B., *Nature,* 330, 489 (1987).
12. Rosen, C. A., Sodroski, J. G., and Haseltine, W. A., *Cell,* 41, 813 (1985).
13. Muesing, M. A., Smith, D. H., and Capon, D. J., *Cell,* 48, 691 (1987).
14. Hauber, J. and Cullen, B. R., *J. Virol.,* 62, 673 (1988).
15. Fisher, A. G., Feinberg, M. B., Josephs, S. F., Harper, M. E., Marselle, L. M., Reyes, G., Gonda, M. A., Aldovini, A., Debouk, C., Gallo, R. C., and Wong-Staal, F., *Nature,* 320, 367 (1986).
16. Jackovitz, A., Smith, D. H., Jalobovitz, E. B., and Capon, D. J., *Mol. Cell Biol.,* 8, 2556 (1988).
17. Southgate, C., Zapp, M. L., and Green, M. R., *Nature,* 345, 640 (1990).
18. Sharp, P. A. and Marciniak, R. A., *Cell,* 59, 229 (1989).
19. Weeks, K. M. and Crothers, D. M., *Cell,* 66, 577 (1991).
20. Selby, M. J., Bain, E. S., Luciw, P. A., and Peterlin, B. M., *Genes Dev.,* 3, 547 (1989).
21. Dingwall, C., Ernber, I., Gait, M. J., Green, S. M., Heaphy, S., Karn, J., Lowe, A. D., Singh, M., and Skinner, M. A., *EMBO J.,* 12, 4145 (1990).
22. Mueburg, M. A., Smith, D. H., and Capon, D. J., *Cell,* 48, 691 (1987).
23. Okamoto, T. and Wong-Staal, F., *Cell,* 47, 29 (1986).
24. Laspia, M. F., Rice, A. P., and Mathews, M. B., *Cell,* 59, 282 (1989).
25. Marciniak, R. A., Cainan, B. J., Frankel, A. D., and Sharp, P. A., *Cell,* 63, 791 (1990).
26. Newstein, M., Stanbridge, E. G., Casey, G., and Shank, P. R., *J. Virol.,* 64, 4565 (1990).
27. Roy, S., Parkin, N. T., Rosen, C., Horvitch, J., and Sovengerg, N., *J. Virol.,* 64, 1402 (1990).

28. Marciniak, R. A., Garcia-Blanco, M. A., and Sharp, P. A., *Proc. Natl. Acad. Sci. U.S.A.,* 87, 3624 (1990).
29. Jones, K. A., Luciw, P. A., and Duchange, N., *Genes Dev.,* 2, 1101 (1988).
30. Gatignol, A., Kumar, A., Rabson, A., and Jeang, K.-T., *Proc. Natl. Acad. Sci. U.S.A.,* 86, 7828 (1989).
31. Gaynor, R., Soultanakis, E., Kuwabara, E., Garcia, J., and Sigman, D. S., *Proc. Natl. Acad. Sci. U.S.A.,* 86, 4858 (1989).
32. Embretson, J., Zupancic, M., Ribas, J. L., Burke, A., Racz, P., Tenner-Racz, K., and Haase, A. T., *Nature,* 362, 359 (1993).
33. Hsu, M.-C., Schutt, A. D., Holly, M., Slice, L. W., Sherman, M. I., Richman, D. D., Potash, M. J., and Volsky, D. J., *Science,* 254, 1799 (1991).
34. Pauwels, R., Andries, K., Desmyter, J., Schols, D., Kukla, M. J., Breslin, J. K., Raeymaeckers, A., Van Gelder, J., Woestenborghs, R., Heykants, J., Schellekens, K., Janssen, M. A. C., De Clercq, E., and Janssen, P. A. J., *Nature,* 343, 470 (1990).
35. Harrison, G. S., Long, C. J., Curiel, T. J., Maxwell, F., and Maxwell, I. H., *Hum. Gene Ther.,* 3, 461 (1992).
36. Sullenger, B. A., Gallardo, H. F., Ungers, G. E., and Gilboa, E., *J. Virol.,* 65, 6811 (1991).
37. Pearson, L., Garcia, J., Wu, F., Modesti, N., Nelson, J., and Gaynor, R., *Proc. Natl. Acad. Sci. U.S.A.,* 87, 5879 (1990).
38. Pauwels, R., Balzarini, J., Baba, M., Snoeck, R., Schols, D., Herdewijn, P., Desmyter, J., and De Clercq, E., *J. Virol. Meth.,* 20, 309 (1988).
39. Pauwels, R., De Clercq, E., Desmyter, J., Balzarini, J., Goubau, P., Herdewijn, P., Vanderhaeghe, H., and Vandeputte, M., *J. Virol. Meth.,* 16, 171 (1987).
40. Schwartz, S., Felber, B. K., Fenyo, E. M., and Pavlakis, G. N., *Proc. Natl. Acad. Sci. U.S.A.,* 86, 7200 (1994).
41. Kimpton, J. and Emerman, M., *J. Virol.,* 66, 2232 (1992).
42. Cullen, B. R. and Malim, M. H., *Methods Enzymol.,* 216, 362 (1994).
43. Terwilliger, E., Proulx, J., Sodroski, J., and Haseltine, W. A., *AIDS,* 1, 317 (1994).
44. Kaputo, A., Sodroski, J. G., and Haseltine, W. A., *AIDS,* 3, 372 (1994).
45. Nordeen, S. K., *BioTech.,* 6, 454 (1988).
46. Aguilar-Cordova, C. E., Chinen, J., Donehower, L., Lewis, D. E., and Belmont, J. W., *AIDS Res. Hum. Retroviruses,* 10, 293 (1994).

Chapter 6

Characterization of HIV Nef Cell-Specific Effects

*Jon W. Marsh and Swapan K. De**

Contents

I. Introduction

The ability to study the role of a viral protein in infectivity is facilitated when its omission results in the failure to achieve a particular step in the infection process, and thus its function can be defined. The HIV protein Nef is nonessential for viral replication *in vitro;* indeed, many of the early HIV-1 isolates lacked Nef expression. Although various activities have been attributed to it, the role of this protein remains an enigma. The study of the capacity of a viral protein to alter a biochemical pathway or cellular process is of greatest interest when the viral-mediated perturbation explains some pathogenic feature seen in the infection. When the protein has no known function, however, one may be limited to exploring the molecular basis for phenotypic changes in cells expressing the protein. Typically, one examines the cell type that is infected by the virus, but the relevance of any acquired phenotype towards pathogenesis remains in question. Alternatively, one can select a highly biochemically characterized cell and, if perturbations are forthcoming, determine the mechanism of the protein-mediated perturbation. In this latter situation, the experimentalist is addressing the molecular interactions of the viral protein with cellular components and is not directly concerned with pathogenesis. The biochemical perturbation generated by the introduced viral protein cannot be lethal, but the perturbation should be quantifiable and reproducible. Ideally, it is a nonessential, inducible biochemical pathway that leads to an easily measurable cell activity.

Our approach to the study of HIV Nef has involved both courses mentioned above: we have looked at the protein's ability (1) to affect CD4 T cell phenotype and T cell activation through the T cell receptor, as measured by interleukin 2 (IL-2) production, and (2) to alter the proliferative response of NIH-3T3 cells to growth factors. CD4 T cells are the prototypic host cell for the HIV-1 viral infection, whereas fibroblasts (NIH-3T3 cells) are not. This chapter describes the methodologies that are being used to characterize the cellular and biochemical interactions of Nef.

II. Expression of the Viral Protein

The expression of a foreign protein by a cell is typically achieved by transfection of a DNA expression vector, which incorporates the structural gene for the protein of interest. A full discussion of vectors for expression in mammalian cells exists.[1,2] One alternative to transfection with DNA is the use of recombinant retroviruses. Retroviral expression vectors can be directly transfected into target cells or can be incorporated into recombinant retroviruses by packaging cell lines. The work described here makes use of the L*nef*SN recombinant retrovirus expression vector as produced by the PA317 packaging cell line. L*nef*SN expresses both the Nef protein and a selection enzyme, neomycin phosphotransferase *(neo)* which imparts resistance to the antibiotic G418. A control vector LN expresses only the *neo* gene. The vector DNA sequences, such as L*nef*SN, are incorporated as RNA into the virions by the *trans*-expression of the viral structural proteins in the packaging line. With PA317, the generated recombinant virus is amphotropic, and thus can infect (and integrate a *nef* gene into) most mammalian cells. These constructs were originally described by Garcia and Miller[3] and have been used to generate cells with stable expression of the *nef* gene. Following incubation of the targeted cell with the PA317-generated recombinant virus, the viral RNA genome is reverse transcribed to DNA and then integrated into the chromosomal DNA of the host cell. Here the proviral DNA behaves as a cellular gene, resulting in stable expression of the protein. Expression of *nef* message from this vector is promoted from the retroviral long terminal repeat (LTR). Following infection by the recombinant retrovirus, the targeted cell is permitted to go through replication, then placed on selection media. With the LN and L*nef*SN vectors (both expressing neomycin phosphotransferase) the selection is for resistance to the drug G418.

A. Method

1. Transduction of *nef* into targeted cell

The supernatants of the packaging lines that contain the recombinant viral particle with the genes of interest (*neo* from PA317/LN cells or *neo* plus *nef* from PA317/L*nef*SN cells) are used to transduce these genes into susceptible cells. The most efficient transfer of virus from packaging line to target cell is through co-culture of the two respective cell lines, but the target cell cannot be adherent like the PA317 cell. The infection process is promoted by the inclusion of 3 to 8 μg/ml polybrene (made as a 1000X stock solution, filter sterilized) to diminish charge repulsion between the cell surface and viral particle. Following co-culture, contaminating PA317 cells are removed by repetitive culture into flasks to adhere (and remove) the packaging line. If the targeted cell is adherent, then the supernatant from the PA317 packaging line is filtered through a 0.45-μm low-binding filter prior to being added to the target cell. This size pore removes all PA317 cells but permits limited passage of the recombinant viral particles. Cells that are to be transduced with the recombinant retrovirus should be in exponential growth for efficient infection.

2. G418 selection

Prior to selection in G418-containing medium, one must titer the non-neophosphotransferase expressing *(neo⁻)* cell line with varied antibiotic G418 concentrations to determine the level at which 100% of *neo⁻* cells will be killed. Typically this range will be 50 to 1000 μg/ml and the incubation will take 5 to 15 d, but some cells, such as B and T cells, require higher levels. Each batch of G418 has a given percent of active species, typically around 50%. In most discussions the concentration of G418 means active weight, so these corrections must be applied in making stock G418 solutions. We make a 20 mg/ml active weight stock solution in culture medium and then filter through a 0.22-μm filter to sterilize. This G418 stock is added directly to the growth medium to reach the appropriate level. Control *neo⁻* cells should be placed in G418 in parallel to assure that non-*neo*-transduced cells are all killed. Following selection in G418, the resistant *neo⁺* populations can then be examined to confirm functional gene expression of the introduced vector.

The co-transduction of *neo* and *nef* genes, followed by selection for G418 resistance, requires only that *neo* be expressed. To demonstrate co-expression of *nef* gene, the cells are assayed for *nef* mRNA and protein expression.

III. Analysis of Steady-State *nef* mRNA and Protein in *nef*-Transduced Cells

A demonstration of specific mRNA expression by Northern blot analysis should be first attempted with total RNA from the cells. If this level of RNA is insufficient, then preparation of polyA⁺ RNA may be done. Detection is achieved with a synthetic oligonucleotide probe, typically 21 to 50 bases long and complementary to a segment of the message. This probe is ^{32}P-labeled with [γ-^{32}P]ATP by phosphorylation with bacteriophage T4 polynucleotide kinase. The protocol described below details our examination of *nef* message expression in NIH-3T3 cells, but is also applicable to other cells.

A. Method: Analysis of Steady-State *nef* mRNA

RNA can be successfully isolated by numerous previously described techniques.[4] In brief, extraction of frozen cells with sodium dodecyl sulfate (SDS)-phenol, followed by chloroform, isoamyl alcohol, and phenol, fully described elsewhere,[5] yielded 80 to 120 μg RNA from 2 × 10^7 cells. To detect *nef* mRNA, 20 μg of the isolated RNA is denatured for 5 min at 65°C in a solution containing 1X MOPS buffer (20 m*M* 3-morpholino-propanesulfonic acid [MOPS], 5 m*M* sodium acetate, 1 m*M* ethylene diamine tetraacetate [EDTA], pH 7.0), 2.2 *M* formaldehyde, and 50% formamide. Denatured RNA (6 μg in 15 μl) is separated by electrophoresis in a 1% agarose gel (5 × 7.5 cm) containing 1X MOPS buffer and 2.2 *M* formaldehyde.[6] During electrophoresis, gels are submerged in running buffer (1X MOPS buffer and 2.2 *M* formaldehyde) that is recirculated constantly to prevent marked changes in pH.

Electrophoresis is carried out by applying a constant voltage (50 V) across the gels. Following electrophoresis the gels are soaked for 40 min in 20 m*M* sodium phosphate (pH 7.0) and transferred to nylon membrane in the presence of 20X SSC (3 *M* NaCl and 0.3 *M* sodium citrate, pH 7.2).[7] Following transfer, the filters are UV cross-linked.[8]

Northern blots are incubated in sealed bags at 55°C in a water bath in 3X SET (450 m*M* NaCl, 6 m*M* EDTA, 90 m*M* Tris, pH 8.0) containing 0.1% SDS. The filters are then incubated in 3X SET containing 10X Denhardt's solution (0.2% each of bovine serum albumin [BSA], Ficoll, and polyvinylpyrolidone)[9] for 2 h and finally 3X SET containing 10X Denhardt's solution and yeast tRNA (250 μg/ml) for 2 h. Filters are hybridized with the labeled probe for 18 h at 55°C in a solution containing 3X SET, 0.1% SDS, 10X Denhardt's solution, and yeast tRNA (250 μg/ml). An end-labeled complementary oligonucleotide probe from a stretch of 27 nucleotides in HIV-1 *nef* gene with the following sequences (nucleotide sequence 8958 to 8984) was synthesized and used as a probe.

5′-CGTCGATGATTACGACTAACACGGACC-3′

This probe is [^{32}P]-5′ end-labeled using T4 polynucleotide kinase. The synthesis reaction is carried out with 25 pmol of oligonucleotide and 25 pmol of [γ-^{32}P]ATP (specific activity 3000 Ci/mmol; 5 mCi/ml in aqueous solution) in a buffer containing 25 m*M* Tris-HCl (pH 7.8), 10 m*M* $MgCl_2$, 5 m*M* dithiothreitol, 0.2 m*M* spermidine, and 100 μ*M* EDTA. 10 units (1 μl) of bacteriophage T4 polynucleotide kinase are added to the mixture and incubated at 37°C for 30 min. Enzyme (10 units) is added once more and, following another 30 min incubation, the reaction is terminated by adding 50 m*M* EDTA.

The probe is purified using a Nuctrap™ push column (Stratagene, La Jolla, CA). In brief, the small column is prewetted by 70 μl of 1X SET (150 m*M* NaCl, 30 m*M* Tris, pH 8.0, and 2 m*M* EDTA). The sample is loaded to the column by adjusting the volume to 70 μl with 1X SET, followed by washing the column with 1X SET to get a final elution volume of 120 μl. The probe has a specific activity of about 2×10^8 dpm/μg.

Following hybridization, the filters are washed at 55°C for 1 h each in 1X SSC (150 m*M* NaCl and 15 m*M* sodium citrate) containing 0.1% SDS, followed by 0.3X SSC containing 0.1% SDS. Autoradiographs of filters are obtained by exposing X-ray film (XAR-05, Kodak, Rochester, NY) at –70°C with an intensifying screen. Following these methods we are able to detect the expression level of mRNA of *nef* in the *nef*-transduced NIH-3T3 cells, as shown in Figure 6.1a. It is also possible to compare the mRNA levels among the different populations of transduced cells by these techniques.

B. Method: Analysis of Steady-State Nef Protein in the *nef*-Transduced Cells

To analyze Nef protein, the cells are labeled with ^{35}S-cysteine and -methionine (100 μCi/ml; ICN Biomedicals, Costa Mesa, CA) for 6 h. Cells are lysed in buffer

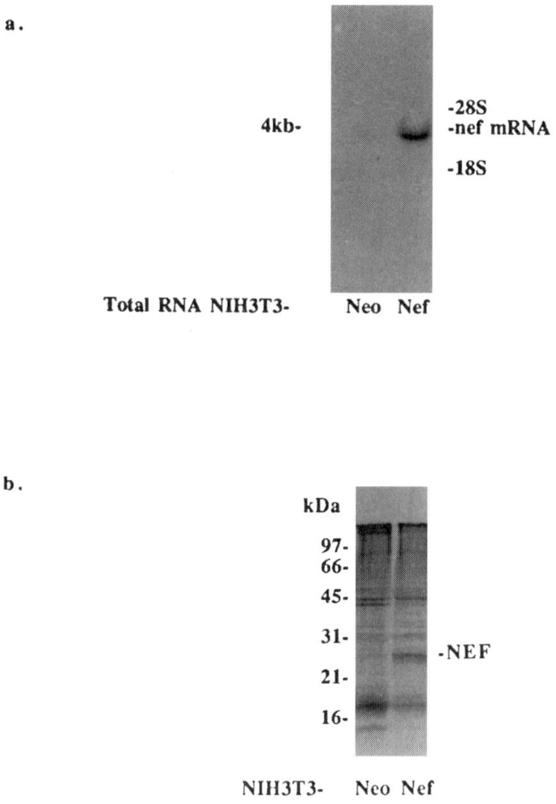

FIGURE 6.1

Analysis of steady-state *nef* mRNA and protein levels in the *nef*-transduced NIH-3T3 cell. (a) The same amount (6 μg) of total RNA from control (Neo) or *nef*-transduced (Nef) cells was run on an agarose gel, transferred to a nylon membrane and hybridized overnight using an oligonucleotide probe at 55°C. The size of the hybridized species was estimated using a single-stranded RNA ladder. (b) Cells were labeled biosynthetically with [^{35}S]-cysteine and -methionine. The cells were collected, solubilized in extraction buffer, and protein was immunoprecipitated with anti-Nef antibody. Immunoprecipitated proteins were analyzed by 10% SDS-polyacrylamide gel electrophoresis. (From De, S. K. and Marsh, J. W., *J. Biol. Chem.*, 269, 6656 (1994).)

containing 50 m*M* Tris-HCl (pH 7.5), 0.1% SDS, 300 m*M* NaCl, 0.5% deoxycholate, 1.0% Triton X-100, 1 mg/ml leupeptin, 1 mg/ml aprotinin, and 50 mg/ml phenylmethylsulfonyl fluoride (PMSF). Cellular extracts are incubated overnight at 4°C with a polyclonal rabbit anti-Nef antisera (AIDS Research and Reference Reagent Program, Rockville, MD, reagent No. 331) and precipitated with formalin-fixed Staph A (Gibco-BRL, Gaithersburg, MD). The precipitate is solubilized and fractionated on a reducing 10% SDS-polyacrylamide gel. The gels are dried and exposed to X-ray film. Specific expression of Nef in the *nef*-transduced line is shown on the autoradiogram in Figure 6.1b.

IV. Effect of Nef on T Cell Activity

Demonstration of stable expression of the Nef protein in G418 selected cells then permits detailed studies of Nef-induced cellular phenotypes and biochemical changes. Below, we detail experimental approaches to characterize the Nef-mediated cellular alterations.

To examine the effect of Nef on cellular activation pathways, we have utilized a T cell hybridoma which can be activated through its T cell receptor. The T cell receptor is associated with at least three tyrosine kinases, as shown in Figure 6.2. Phosphorylation of phospholipase $C_{\gamma 1}$ ($PLC_{\gamma 1}$) leads to its activation, resulting in hydrolysis of phosphatidylinositol 4,5-bisphosphate to yield the secondary messengers, diacylglycerol (DAG) and inositol 1,4,5-trisphosphate (IP_3). DAG activates protein kinase C, while IP_3 binds to an IP_3-specific endoplasmic reticulum receptor, causing release into the cytoplasm of calcium from the vesicular pool. The activation of T cells *in vivo* would normally involve the recognition of presented antigen in the context of an MHC molecule. The recognition by the T cell would be through its T cell receptor and associated co-receptors, which include CD4, CD8, CD28, and numerous adhesion molecules. The binding of the T cell receptor, itself a complex of two chains and the associated multichain CD3 complex, to the appropriate antigen-presenting cell results in the recruitment of tyrosine kinase activity. As mentioned above, this results in the activation of a pathway that includes phospholipase C activity, protein kinase C induction, and rapid, transient elevations in cytosolic free calcium. T cell activation *in vivo* leads to proliferation of antigen-specific T cells and secretion of lymphokines, such as IL-2, to serve in a positive feedback loop to the relevant cells. *In vitro* studies using culturable T cells usually involve transformed T cell lines or T cell hybridomas. Proliferation is not inducible in these cells. With these cell lines, however, activation through the T cell receptor results in release of IL-2. The study of IL-2 promoter activity is relevant to the function of the HIV LTR, as the IL-2 and HIV LTR promoter regions share a number of transcriptional protein binding sites. The study of the influence of Nef on IL-2 expression should correlate with LTR expression.

One of the simplest means to activate T cells through their receptor is with antibodies directed at components of the receptor. Typically, solubilized antibodies are inadequate and must be cross-linked. Murine T cell hybridomas can be activated with antibodies alone. Activation of human cell lines like Jurkat require, in addition to anti-receptor antibodies, less specific reagents, such as phytohemagglutinin (PHA) or phorbol ester.

A. Method

1. Stimulation of T cell hybridomas

For murine T cell hybridomas, exposure of the cells to immobilized anti-CD3 antibody, 2C11, yields sufficient IL-2 generation.[11] Purified and filter-sterilized 2C11

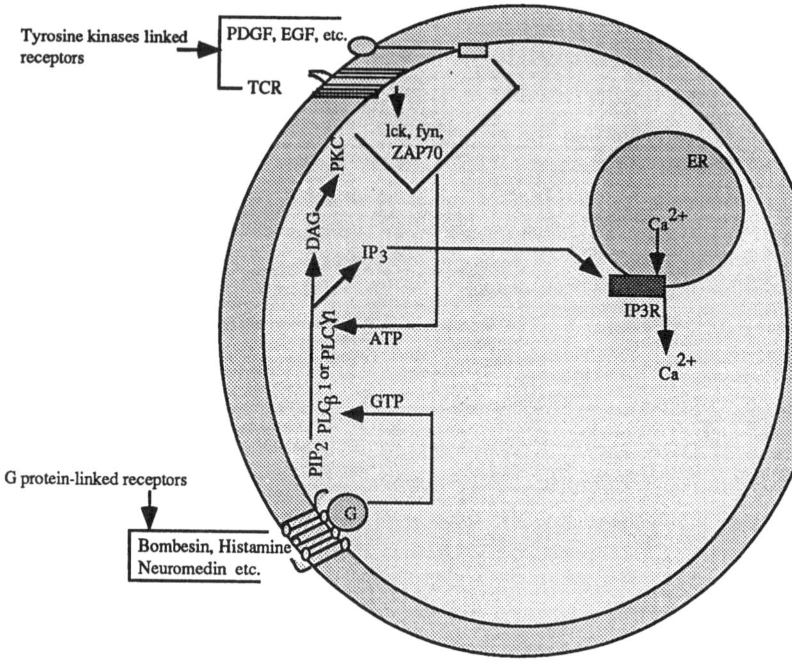

FIGURE 6.2
Receptor-mediated activation pathways for representative tyrosine kinase-linked and G protein-linked receptors. Activation of either phospholipase C ($PLC_{\beta 1}$ or $PLC_{\gamma 1}$) results in the hydrolysis of the membrane-associated phosphatidylinositol 4,5-bisphosphate (PIP_2) to yield the secondary messengers, diacylglycerol (DAG) and inositol 1,4,5-trisphosphate (IP_3). DAG activates protein kinase C (PKC), while IP_3 binds to an IP_3-specific endoplasmic reticulum receptor (IP_3R), causing release into the cytoplasm of calcium from the vesicular endoplasmic reticulum (ER) pool.

is diluted into sterile phosphate-buffered saline (PBS) to a final concentration of 2 to 20 μg/ml. The minimal effective concentration of antibody required to result in optimal IL-2 generation must be determined empirically. With the following conditions we see maximal stimulation at 0.1 μg/well. At 2 μg/ml, 50 μl of antibody solution is aliquoted into each well of a flat-bottom 96-well plate (tissue culture treated, e.g., Costar 3596). For best adherence the antibody solution should contain no carrier protein. The plate is stored at 4°C overnight, and the next day is rinsed with PBS and medium.

The murine T cell hybridoma 2Q23-34.7.9,[12] expressing either the neophosphotransferase *(neo)* gene alone or with the *nef* gene, is grown in RPMI 1640 supplemented with 25 m*M* HEPES, 2 m*M* sodium bicarbonate, 50 μg/ml gentamycin, 1 m*M* sodium pyruvate, 50 μ*M* mercaptoethanol, 100 μ*M* MEM nonessential amino acids, and 10% fetal calf serum. Exponentially growing cells are diluted into fresh media to a final density of 5×10^5 cells/ml. 200 μl are aliquoted into the rinsed antibody-containing wells. The cells are incubated in a 37°C, 5% CO_2 incubator overnight. Although IL-2 secretion is measurable in a few hours, we have found it to be a convenient and reproducible practice to start this antibody-cell incubation at the end of the day, with the IL-2 enzyme-linked immunosorbent assay (ELISA) performed the next morning. An alternative means to activate murine hybridomas with an antibody is to use the anti-Thy1 antibody G7.[13] This monoclonal is unusual in that pre-adsorption to the culture dish is unnecessary. That is, the soluble, free antibody is capable of activating T cells. As before, the optimal concentration must be determined. We use 1 to 10 μg/ml (0.2 to 2 μg/200-μl well). Following antibody stimulation, the cells and media are mixed and then centrifuged; 100 μl is then examined by ELISA for IL-2.

2. IL-2 measurement

Enzyme-linked immunosorbent assays (ELISA) are used to quantitate soluble molecules such as IL-2. They are typically composed of an immobilized capture antibody, which in effect immobilizes the IL-2, and a secondary enzyme-linked antibody that recognizes a different epitope on the immobilized IL-2, in effect immobilizing the enzyme. These assays are efficiently done in 96-well plates. The secondary antibody is typically cross-linked with alkaline phosphatase or horseradish peroxidase. Following the addition of color-forming substrates, the plate is read on a 96-well plate reader, which measures optical density in each well.

ELISA kits are available for many cytokines; however, one may also use paired antibodies and prepare the assay plate from scratch. Below, we will detail the use of an IL-2 kit, plus procedures that can be used in the laboratory-generated version of the assay.

The typical murine IL-2 ELISA kit (e.g., Collaborative Biomedical, Bedford, MA, No. 30032) contains a 96-well flat-bottom plate to which anti-IL-2 antibodies have been immobilized. Immobilization of an antibody can also be achieved as described in the previous section. One hundred microliters of the experimental cellular supernatants, as well as murine IL-2 standards, are incubated in these wells at room temperature for 2 h. The wells are then rinsed with wash buffer (PBS plus 0.05% Tween 20). This can be achieved by a commercial plate washer, or by gently flooding the wells with rinse buffer and tapping the plate dry on paper towels. After a rapid rinse, buffer (250 μl) is added to the wells and emptied after 5 min. After repeating the wash two more times, the plate is drained and 100 μl of the secondary anti-IL-2 antibody peroxidase conjugate is added. With the commercial kit the appropriate concentration of secondary antibody is predetermined; otherwise, it must be determined with standards. Dilution is into a buffer similar to the wash buffer, but containing a carrier protein, such as albumin (0.1 to 1%). The plate is incubated again at room temperature for 30 to 60 min, and the washing protocol is repeated as above. Variations in the required incubation temperature and time can occur with different antibodies. Prior to addition of substrate it is important to remove all bubbles that can remain in the washed wells and can interfere with optical density readings. For peroxidase conjugates we typically use the two component tetramethylbenzidine (TMB; one component is the TMB solution, the other is H_2O_2 solution: e.g., Kirkegaard and Perry, Gaithersburg, MD, No. 50-76-00). These solutions are mixed and 100 μl is immediately added to each well. The optical density at 650 nm is followed kinetically on a kinetic microplate reader (e.g., Molecular Devices, Sunnyvale, CA, V_{max} reader). The value of kinetic readings has been dealt with elsewhere;[14] however, one can also do end point readings at 650 nm, or by adding 100 μl 1 *M* phosphoric acid and measuring optical density at 450 nm. Acidification and reading at this second wavelength more than doubles the sensitivity of the TMB assay. Development of the standard curve from known IL-2 concentrations permits an estimate of unknown values. Figure 6.3 shows a typical T cell IL-2 response to varied anti-Thy G7 antibody concentrations, comparing the presence or absence of Nef expression.

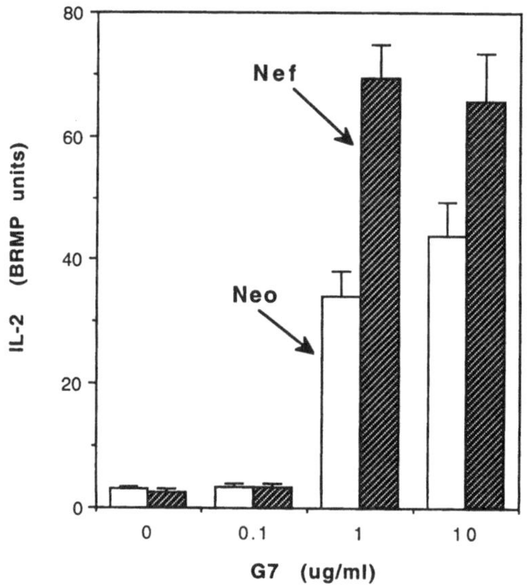

FIGURE 6.3
IL-2 induction in anti-Thy antibody-stimulated T cell hybridomas. Control (Neo) or Nef-expressing cells were exposed to varied G7 antibody concentrations, as detailed in Reference 15, and after 18 h the IL-2 levels were determined by ELISA.

V. NIH-3T3 Activation

The finding that T cell activation pathways could be modified by Nef is complicated, unfortunately, by the lack of resolution of the biochemical pathways used by T cells, or, perhaps more accurately, by the multiplicity of different surface receptors and activation pathways. An alternative approach is to examine the effects of Nef on a highly defined activation pathway, that of growth factor-dependent proliferation in the NIH-3T3 cell.

To examine the effect of Nef on NIH-3T3 activation pathways, we have utilized two growth factors: bombesin, which works through a G protein-linked receptor, and platelet-derived growth factor (PDGF), which delivers its response through the receptor's tyrosine kinase moiety (see Figure 6.2). The bombesin and PDGF receptor systems are coupled to energy-requiring (GTP or ATP, respectively) mechanisms that lead to activation of phospholipase $C_{\beta 1}$ ($PLC_{\beta 1}$) or phospholipase $C_{\gamma 1}$ ($PLC_{\gamma 1}$), respectively. The PDGF pathway closely corresponds to the activation pathway in T cells, where tyrosine kinase activation of $PLC_{\gamma 1}$ occurs. Both phospholipases hydrolyze phosphatidylinositol 4,5-bisphosphate to yield the secondary messengers, DAG and IP_3. The generation of these two metabolites is a common (intercept) point when comparing bombesin and PDGF activation pathways. DAG activates protein kinase C, while IP_3 binds to an IP_3-specific endoplasmic reticulum receptor, causing release into the cytoplasm of calcium from the vesicular pool. These events in turn promote cellular growth and proliferation.

VI. Growth Factor-Mediated Cell Proliferation

The growth of NIH-3T3 cells in media supplemented with fetal calf sera is largely unaffected by the addition of various growth factors, presumably because the sera alone contains a multitude of factors. When the cells are cultured in media with low level sera, proliferation can be mediated by singular growth factors, whose pathways can be interrupted at singular, defined points along an activation pathway. Proliferation in sera-starved cells, as mediated by PDGF and bombesin, can be measured by the incorporation of radiolabeled thymidine into cellular DNA.

A. Method: Measurement of Proliferation of Cells in Culture

Two thousand cells are plated in Dulbecco's modified Eagle medium (DMEM) plus 10% fetal bovine serum (FBS) in each well of a 96-well plate. After 24 h, control and *nef*-expressing cells are sera-starved overnight. The level of sera in this overnight incubation depends on the growth factor to be added. Prior to bombesin addition, cells are cultured in media with 0.5% sera, and the preincubation of PDGF is done in sera-free media. Bombesin or PDGF is added at different concentrations. Following different periods of incubation, 0.1 m*M* (2.5 μCi/ml) [^{3}H]thymidine is added and the cells are harvested after an additional 6 h. The cells are dissociated in an enzyme-free EDTA solution (Specialty Media Inc., Lavallette, NJ) before harvesting in a filtermat (Skatron Inc., Sterling, VA). Filters are then counted in a Beckman LS 5800 liquid scintillation counter. As shown in Figure 6.4, a profound effect of HIV-1 Nef on the NIH-3T3 proliferation is evident in those cells expressing Nef.

VII. Analysis of Signal-Transducing Metabolites in Activated Cells

The loss of growth factor signaling could be due to a loss of the factor receptor or receptor function, or to a lesion in the activation pathway. As shown in Figure 6.2,

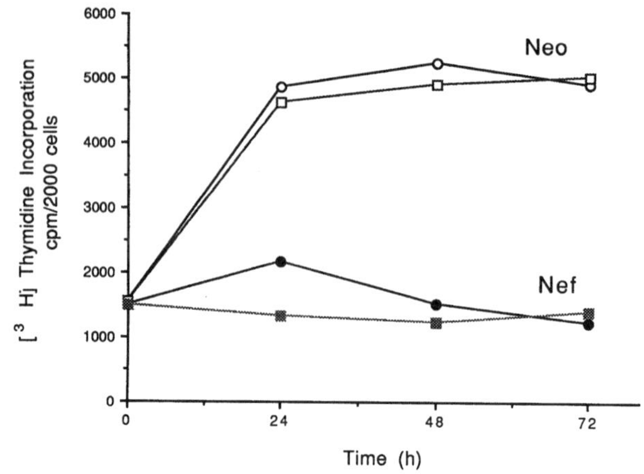

FIGURE 6.4
Bombesin- and PDGF-induced DNA synthesis in NIH-3T3 control (Neo) or Nef-expressing cells. Cells were sera-starved overnight, then stimulated by the addition of bombesin (20 m*M*; squares) or PDGF (1 ng/ml; circles). The cells were harvested at the indicated time following incubation with 0.1 m*M* (2.5 μCi/ml) [^{3}H]thymidine for 6 h. (From De, S. K. and Marsh, J. W., *J. Biol. Chem.*, 269, 6656 (1994).)

the early pathways for PDGF and bombesin are not identical, but following differential activation of phospholipase C activities, the two pathways converge, with generation of IP_3 and DAG. The choice of these two growth factors would permit a differential examination of Nef effects on the early events in activation, if the Nef effect occurred early. Finding that both are altered suggests that Nef inhibition of proliferation is mediated at a common event in the pathways, the earliest being generation of IP_3 and DAG.

A. Method: Analysis of IP_3 Levels in Response to Bombesin and PDGF

An IP_3 radioreceptor assay is used to measure the IP_3 levels in the cells following bombesin (20 m*M*) or PDGF (1 ng/ml) treatment. Cells (1 × 10^6/well) are plated in DMEM containing 10% FBS in the 6-well tissue culture plate overnight. The cells are then changed to 0.8 ml media without serum, and following a 15-h incubation, bombesin and PDGF are added. One hundred percent ice cold trichloroacetic acid (TCA) (0.2 ml/well) is added at the end of the timed incubation, and cellular extracts are prepared and examined according to the manufacturer's protocol (Du Pont NEN, Boston, MA). Briefly, the same amount of cell lysate from different samples and a fixed amount of tracer ([^{3}H]IP_3) are allowed to compete for a constant and limiting amount of receptor. Decreasing amounts of tracer are bound to the receptor as the amount of introduced unlabeled ligand is increased. The amount of [^{3}H]IP_3 bound to the receptor in the presence of standard or test samples is determined by separating the bound from the free [^{3}H]IP_3 by rapid filtration. The bound radioactivity is quantitated by liquid scintillation counting of the filter. If the IP_3 response is normal, as was found with Nef-expressing cells, then an examination of IP_3 receptor function is required. This receptor, when bound by IP_3, releases calcium into the cytosol from the endoplasmic reticulum.

VIII. Determination of Free Cytosolic Calcium Concentration

Release of calcium can be measured by fluorescence spectroscopy, flow cytometry (fluorescence-activated cell sorting, FACS), and digital imaging fluorescence microscopy (digital imaging ratio).[16] With digital imaging fluorescence microscopy, changes in the intracytosolic free calcium can be followed kinetically in single cells. In contrast, fluorescence spectroscopy measures the average change in the population being examined, and FACS analysis examines single cells at one instant from a population with a defined number of cells over a defined period of time.

All of these procedures make use of a calcium-binding molecule whose fluorescence characteristics change upon binding free calcium.[17] Among many of these fluorescent dyes, indo-1, quin-2, quin-3, and fura-2 are the most commonly used for intracellular calcium determination. It is common practice to use the cell-permeant acetoxymethyl (AM) esters of fluorescent calcium indicators. The AM ester of fura-2, the indicator described here, passively crosses the plasma membrane and, once inside the cell, is cleaved to a cell-impermeant product by intracellular esterases. Fura-2 exhibits a shift in spectrum on binding calcium. In the absence of Ca^{2+} the excitation peak of fura-2 is at 380 to 385 nm, and at saturating concentrations of Ca^{2+}, it shifts to 340–345 nm with emission at 510 nm, as shown in Figure 6.5. The ratio of the emission fluorescence at 510 nm with excitations at 340 and 380 nm gives a relative measure of calcium.

A. Method: Measurement of Calcium Release in NIH-3T3 Cells by Digital Imaging Fluorescence Microscopy

Thirty to forty thousand cells are plated overnight on a cover slip (22 × 60 mm) in DMEM containing 10% FBS. The cells are then changed to DMEM without any FBS for 8 h. Following sera starvation, the cells are washed in buffer (pH 7.4) containing 25 m*M* HEPES, 1.8 m*M* $CaCl_2$, 110 m*M* NaCl, 4.56 m*M* KCl, 0.8 m*M* $MgSO_4$, 1 m*M* NaH_2PO_4, 2.4 m*M* $NaHCO_3$, and 0.1% BSA with essential and nonessential amino acids. Cells are then incubated in the same buffer containing 1 μ*M* of fura-2 AM. Following a 20-min incubation, the cells are washed again with the same buffer in the absence of fura-2 AM. The cover slip containing the cells is then placed in an open perfusion chamber and fixed to the objective of a digital imaging microscope. Continuous perfusion of buffers at 40 ml/h through the chamber is maintained during the experiment. Following 5 min of control buffer perfusion

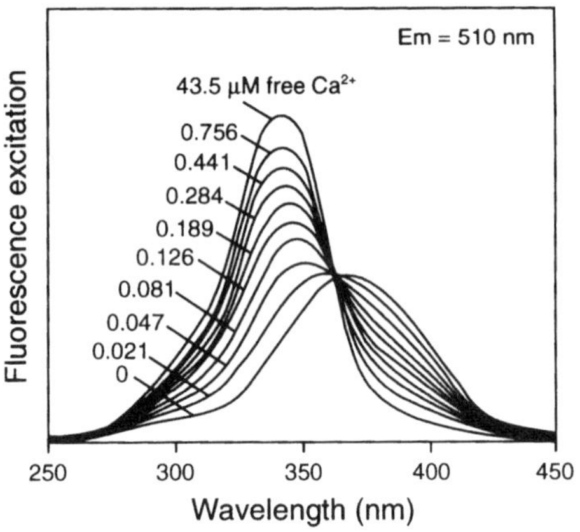

FIGURE 6.5
Excitation spectra of fura-2 in increasing free calcium concentrations (courtesy of Iain Johnson, Molecular Probes, Inc.).

through the chamber, 5 n*M* bombesin or 50 ng/ml PDGF is perfused for about 6 min, and then perfusion is shifted to either control buffer or buffer containing 20 μ*M* thapsigargin (Calbiochem, San Diego, CA) for 3 to 5 min. Alternate images of 340 and 380 nm excitation are captured at 15-sec intervals. The ratio of the 510-nm emission patterns from the two excitation wavelengths (340 and 380 nm) of each cell is calculated and plotted. The calcium levels in the different cells are then compared following bombesin treatment.

The ratio of fura-2 510 nm fluorescence at 340 vs. 380 nm excitation is calibrated against intracellular pH by perfusing fura-2 loaded NIH-3T3 cells with 2 μ*M* nigericin, 2 μ*M* FCCP (carbonyl cyanide 4-[trifluoromethoxy]phenylhydrazone) in the above-mentioned buffer, which contains 130 m*M* K^+ and 5 m*M* Na^+. The cells are then perfused with the same buffer at pH values from 6.0 to 7.8 from which a calibration curve is generated. Intracellular Ca^{2+} concentration is determined from the fluorescence ratio ($R = F_{340}/F_{380}$) using the following equation formulated by Grynkiewicz et al.[18]

$$\left[Ca^{2+}\right]i = K_d\left[\left(R \pm R_{min}\right)/\left(R_{max} \pm R\right)\right]b$$

where K_d = 135 n*M*,[19] $R_{min} = F_{340}/F_{380}$ at 10^{-11} *M* Ca^{2+}, $R_{max} = F_{340}/F_{380}$ at 10^{-3} *M* Ca^{2+}, $b = F_{380}$ at 10^{-11} *M* Ca^{2+}/F_{380} at 10^{-3} *M* Ca^{2+}, and F_{340} and F_{380} equal fluorescence intensities at the 340 and 380 nm excitation frequencies, respectively. R_{min} and R_{max} are measured by perfusing fura-2 loaded NIH-3T3 cells with the above-mentioned buffer, containing 10 m*M* ionomycin and either 1.8 m*M* Ca^{2+} (R_{max}) or 10 m*M* EGTA (ethyleneglycol-bis-[b-aminoethyl ether] *N,N′*-tetraacetic acid) plus 1.8 m*M* Ca^{2+} (R_{min}).

A typical perfusion experiment is demonstrated in Figure 6.6, where the initial exposure of either the control (filled circles) or Nef-expressing (open circles) cells

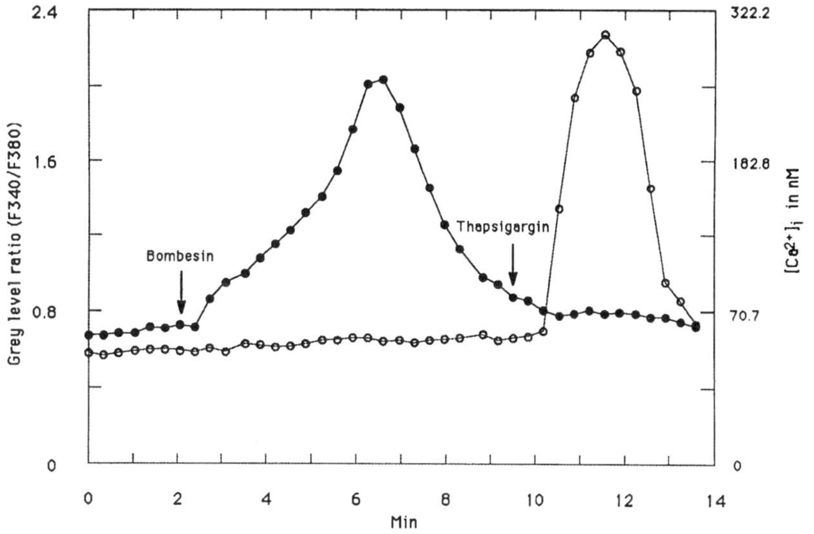

FIGURE 6.6
Free cytosolic calcium levels following sequential addition of bombesin and thapsigargin. NIH-3T3 control *neo* (●) and *nef* (○) expressing cells were grown on coverslip, labeled with fura-2, and washed with control buffer. The cells were perfused with control buffer, switched to a perfusate containing bombesin (5 n*M*) and then to thapsigargin-containing buffer (20 μ*M*) at the indicated arrows. Emitted light was measured at 510 nm (excitation at alternating 340 and 380 nm). The data represent F_{340}/F_{380} ratios among pixels within a single cell. Intracellular Ca^{2+} concentration was determined from a calibration curve generated in a separate experiment using the fluorescence ratio. (From De, S. K. and Marsh, J. W., *J. Biol. Chem.*, 269, 6656 (1994).)

to bombesin results in an immediate increase or no change in intracellular calcium, respectively (as shown by the F_{340}/F_{380} ratio). Control cells, which have now released their IP_3-sensitive calcium stores, do not respond to the addition of the Ca^{2+} ATPase inhibitor, thapsigargin; however, the IP_3-insensitive Nef-expressing cells are responsive to the thapsigargin addition, implying that their vesicular Ca^{2+} ATPase-generated calcium pool is intact.

IX. Nef-Mediated Surface Antigen Modulation

Cell surface antigen modulation is a common event with retroviral infection, but it is usually limited to down regulation of the viral receptor. The mechanism can be attributed to newly synthesized viral *env* product that complexes with the cellular receptor, to which the viral particle normally binds and initiates the infective process. The binding of the cellular and viral moieties can occur at the plasma membrane or an earlier step in the synthetic pathway, and the resultant loss of functional viral receptors from the cell surface has been called "interference".[20] These modified cells are now resistant to superinfection by the same virus or by other viruses that share the same cellular receptor molecule. The HIV receptor, CD4, is also downmodulated by the HIV *env* product, gp160,[21,22] and it is of interest that there are two other HIV gene products, Vpu[23] and Nef, that down-modulate CD4 surface expression. While the CD4 molecule is the HIV receptor and is down-modulated by the HIV *env* product, presumably with an interference-like mechanism, the down-modulation by Nef is not known to be due to a direct interaction with CD4.

The decreased surface expression of CD4 was one of the first recognized cellular phenotypes in Nef-expressing T cells.[3,24] Analysis of CD4 expression is achieved by flow cytometry, commonly specified by the acronym FACS, regardless of whether there is sorting or not. This instrumentation measures the fluorescence intensity of individual cells as they pass through the laser beam. The use of antigen-specific antibodies chemically coupled to fluorescent moieties, such as fluorescein, or fluorescent proteins like phycoerythrin, permits comparisons of the levels of expressed surface antigens. Cell populations are incubated with the fluorescent antibody. If one population has altered antigen levels, the number of bound fluorescent antibodies also changes, resulting in proportional changes in cellular fluorescence intensity. The nuances and complications of this methodology are not easily covered in a chapter, but to appreciate this method as a means of characterizing Nef-induced phenotypes, abbreviated protocols and typical results are discussed.

A. Method: Flow Cytometric Measurement of Surface Antigens

One million cells are suspended in 100 μl modified Hank's buffer (Hank's plus) containing 0.1% bovine serum albumin, 0.1% sodium azide, 25 m*M* HEPES (pH 7.2), but lacking phenol red, and placed on ice. Ten microliters of a 10 mg/ml solution of nonfluorescent murine IgG is added to minimize nonspecific binding. Then 1 μg of the monoclonal fluorescein-conjugated CD4 antibody is added to the cell suspension. The incubation is carried out on ice for 30 min, followed by a 2 ml wash with cold Hank's plus buffer. As a control, 1 μg of a similar isotype monoclonal antibody-fluorescein conjugate that does not recognize any cellular antigen is added to a second aliquot of cells. The washed cell pellets are then resuspended into 0.5 ml Hank's plus buffer, and applied to the FACS instrument.

Graphic representation for a single antigen is commonly a histogram. Fluorescence intensity is on the X-axis and the number of cells (incidents) with that intensity are on the Y-axis. Fluorescence intensity is directly proportional to the number of fluorescent moieties bound; therefore, in order to compare the levels of surface antigen in two or more cell populations, the fluorescent antibody conjugate concentration must be high enough to assure saturation of binding. This level must be determined empirically. A typical run includes examination of background fluorescence. Background fluorescence is due to the inherent fluorescence of a cell (various molecules such as NADH and riboflavin have their own fluorescence spectra that are similar to the commonly used dyes) and to the nonspecifically bound antibody conjugates. While one can do little to minimize the former, the latter can be diminished by including nonspecific, nonconjugated IgG to compete for FcR binding, the addition of competitive anti-FcR antibodies, and by removing aggregates of the fluorescent conjugate antibody by filtration or centrifugation prior to its addition to the cell suspension.

Background fluorescence (dashed line) and anti-CD4 binding to *neo* control cells (solid line) and to Nef-expressing cells (dotted line) are shown in Figure 6.7. Comparison of the mean value of intensity for these populations implies that there is a 90% reduction in CD4 on the surface with Nef expression. The nature of the curves also tells us that there is variation in CD4 surface expression within each population, and that within the selected *nef*-transduced population, essentially all cells have had their CD4 levels decreased. The power of FACS analysis is that thousands of cells are examined one at a time; thus, one gets a true distribution of what is being measured.

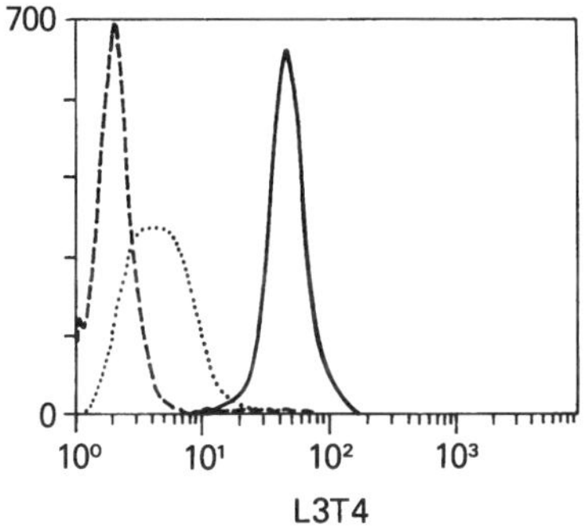

FIGURE 6.7
FACS analysis of *neo* (solid line) or *nef* (dotted line) transduced cells, stained with anti-CD4–fluorescein conjugate. The dashed line represents fluorescence from an irrelevant IgG-fluorescein conjugate, representing background fluorescence. (From Rhee, S. S. and Marsh, J. W., *J. Immunol.,* 152, 5128 (1994).)

X. Biochemical Characterization of Antigen Modulation

Reduction of a surface antigen can be due to decreased protein synthesis, at either the transcriptional or translational level, or to a post-translational phenomenon. When the protein being modulated from the surface of a cell is not totally eliminated, it is first useful to establish whether the viral product alters the steady-state level of the protein, and then to determine if there are any changes in the turnover rate by pulse-chase experiments.

A. Method: Evaluation of Protein Stability by Pulse-Chase Experiment

Ten million exponentially growing cells are starved for 30 min in cysteine- and methionine-free RPMI 1640 and then pulse-labeled for 30 min in labeling medium containing ^{35}S-labeled methionine and cysteine (0.1 mCi/ml; ICN Flow, Costa Mesa, CA). Pulse-labeled cells are washed in complete growth medium and aliquoted for the chase. At the designated chase times individual cell aliquots are washed in cold PBS (pH 7.4) and then lysed in PBS containing 0.1% SDS, 0.5% deoxycholate, 1.0% Triton X-100, 1 μg/ml leupeptin, 1 μg/ml aprotinin, and 1 m*M* PMSF.

Radiolabeled CD4 is immunoprecipitated with biotinylated anti-L3T4 antibody (Pharmingen, San Diego, CA) coupled to the streptavidin-linked magnetic beads (Dynal, Great Neck, NY), followed by SDS-PAGE and autoradiography (Figure 6.8). The radioactivity incorporated into protein is quantified by exposing the dried gel to a storage phosphor screen and scanning the screen with a phosphor imaging instrument (e.g., Fuji Bas 2000). Although it is possible to do so, quantification of radioactivity by fluorography is more difficult. The loss of CD4 radioactivity follows an exponential decline, made linear on the semilog plot (Figure 6.9). The curves are fit with an exponential function $y = A \times 10^{bx}$, where y = radioactivity and x = time in hours. The derived b value is then placed into the formula $t_{1/2} = \ln 2/[-\ln(10^{-b})]$ to obtain the half-life of the labeled protein in the unit of hours. The Nef-expressing cell data fit an exponential function: $y = 1900 \times 10^{-0.0063}$, yielding a half-life of 5 h, from $t_{1/2} = \ln 2/[-\ln(10^{-0.00063})]$. By comparison, the control murine cell line CD4 has a half-life of greater than 2 d. Additionally, the increase in apparent molecular weight of the immunoprecipitated CD4 protein as seen in the gel suggests that the newly synthesized protein becomes glycosylated prior to the Nef-mediated accelerated loss.

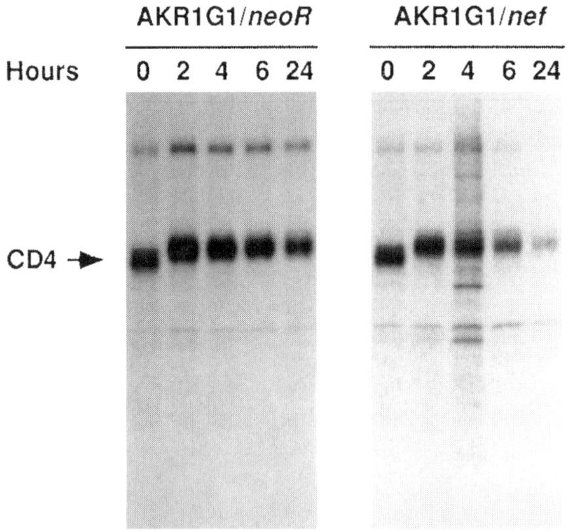

FIGURE 6.8
Pulse-chase of labeled CD4 in *neo*- and *nef*-transduced AKR T cells. Cells were pulsed for 30 sec then chased in non-labeling media for the times indicated. The immunoprecipitated CD4 protein was run on a 10% acrylamide gel, then subjected to autoradiography. (From Rhee, S. S. and Marsh, J. W., *J. Immunol.*, 152, 5128 (1994).)

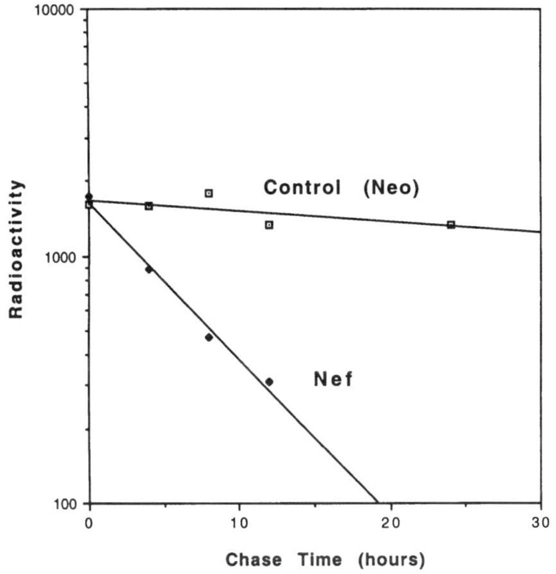

FIGURE 6.9
Semilog plot of CD4 protein radioactivity vs. time of chase. The CD4 band radioactivity was determined by phosphor imaging, and the resultant data for the control or Nef-expressing cells were fit with an exponential function.

XI. Post-Endoplasmic Reticulum Processing

Surface proteins are invariably glycosylated, with carbohydrate maturation of glycoproteins occurring in the Golgi apparatus. This process can be demonstrated in part by the increased resistance of the glycoprotein to endo-β-*N*-acetylglucosaminidase H (endo H).[24] Resistance to this deglycosylating enzyme, which cleaves high-mannose oligosaccharides, occurs with the conversion of the carbohydrate moiety in the medial Golgi to its complex form. Glycoprotein conversion to endo H resistance implies that the protein is being processed normally and is a direct demonstration that it has moved to the Golgi apparatus. To further characterize the Nef-mediated turnover of the CD4 protein, one can examine post-translational cellular processing of the newly synthesized protein in the presence and absence of Nef expression.

A. Method: Endo H Analysis and Post-Golgi Processing

The CD4 immune complex bound to the streptavidin-linked magnetic beads is suspended in 100 μl of 25 m*M* acetate buffer (pH 5.2) containing 0.05% SDS and 1 m*M* PMSF. Twenty milliunits of endo-β-*N*-acetylglucosaminidase H (Boehringer

Mannheim, Indianapolis, IN) are added, and the mixture is incubated for 20 h at 37°C. Samples are washed with 62.5 m*M* Tris-HCl (pH 6.8) and then solubilized in the sample buffer (2% SDS, 2.5% 2-mercaptoethanol, 10% glycerol, 62.5 m*M* tris-HCl [pH 6.8]) for SDS-PAGE.

Treatment of cell lysates with endo H demonstrates that the CD4 protein from both control and Nef-expressing cells gains endo H resistance within 2 h (Figure 6.10a), implying that the CD4 protein movement into the medial Golgi is not inhibited by the presence of Nef.

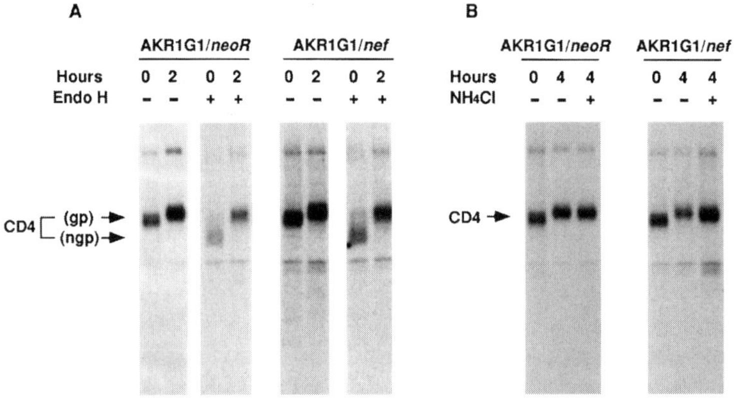

FIGURE 6.10
Examination of post-translation processing of CD4 in control and Nef-expressing cells. (A) Effect of endoglycosidase H on labeled CD4 from cell lysates prior to, or following, a 2-h chase. In both the *neo*- and *nef*-transduced cells, treatment of cell lysates immediately following the 30-min labeling period (0 h) resulted in an approximate 7-kDa loss in apparent molecular weight (from 55 to 48 kDa). These positions are indicated as gp (glycosylated protein) and ngp (non-glycosylated protein). This susceptibility to endo H was lost in both cell lines following the 2-h chase. (B) Effect of addition of 25 m*M* ammonium chloride to chase media. Cells were radio-labeled for 30 min followed by immediate lysis (0 h) or by a 4-h chase in the presence (+) or absence (–) of ammonium chloride, then followed by lysis and immunoprecipitation of CD4 protein. (From Rhee, S. S. and Marsh, J. W., *J. Immunol.*, 152, 5128 (1994).)

One mechanism for the decreased half-life of a protein is intracellular degradation. For newly synthesized proteins, this can occur by acid-independent endoplasmic reticulum degradation, or can be routed to acid-dependent lysosomal degradation. Inclusion of 25 m*M* ammonium chloride to a pulse-chase experiment, as shown in Figure 6.10b, can demonstrate the requirement for acidification. Recovery of a normal half-life upon vesicular neutralization, along with the finding of a normal Golgi-mediated maturation of the CD4 protein, implies that Nef does not mediate endoplasmic reticulum degradation, but appears to route CD4 protein to the lysosome.

XII. Concluding Comments

The procedures discussed permit a characterization of the interactions of the HIV-1 regulatory protein Nef with cellular functions. These effects include surface modulation and alteration of activation pathways. Past studies of retroviral activities suggest these changes are not unique to Nef. For example, CD4 expression on the cell surface is altered by two other HIV proteins, Vpu and gp160. Furthermore, it is common for proviral *env* expression to down-modulate the viral receptor, as gp160 does CD4. The ability to alter cellular biochemical pathways is also not unique to Nef. Retroviral oncogenes are a prime example of a viral protein affecting normal cell function.

What makes Nef unique is the dependence of HIV on Nef expression for the development of immunodeficiency.[27] This implies that Nef possesses an activity that directly or indirectly permits an escape from, or causes the dysfunction and eventual death of, cells that are prominent in the normal *in vivo* functions of the immunological system. Finding that Nef can alter apparently unrelated processes in different cell types and in different biochemical pathways also implies that it has an activity that is integrative into numerous biochemical routes.

References

1. Prentice, H., Sartorelli, V., and Kedes, L., in *Gene and Chromosome Analysis,* Part A, Adolph, K. W., Ed., Academic Press, San Diego, 1993, 303.
2. Kaufman, R. J., *Methods Enzymol.,* 185, 487 (1990).
3. Garcia, J. V. and Miller, A. D., *Nature,* 350, 508 (1991).
4. Sambrook, J., Fritsch, E. F., and Maniatis, T., in *Molecular Cloning: A Laboratory Manual,* Cold Spring Harbor Laboratory Press, Cold Spring Harbor, New York, 1989, 7.3.
5. De, S. K., McMaster, M. T., Dey, S. K., and Andrews, G. K., *Development,* 176, 611 (1989).
6. Lehrach, H., Diamond, D., Wozney, J. M., and Boedtker, H., *Biochemistry,* 16, 4743 (1977).
7. Thomas, P. S., *Proc. Natl. Acad. Sci. U.S.A.,* 77, 5201 (1980).
8. Church, G. M. and Gilbert, W., *Proc. Natl. Acad. Sci. U.S.A.,* 81, 1991 (1994).
9. Denhart, D. T., *Biochem. Biophys. Res. Commun.,* 23, 641 (1966).
10. De, S. K. and Marsh, J. W., *J. Biol. Chem.,* 269, 6656 (1994).
11. Mercep, M., Bluestone, J. A., Noguchi, P. D., and Ashwell, J. D., *J. Immunol.,* 140, 324 (1988).
12. Marrack, P. and Kappler, J., *Nature,* 332, 840 (1988).
13. Gunter, C., Malek, T. R., and Shevach, E. M., *J. Exp. Med.,* 159, 716 (1984).
14. Marsh, J. W., *Genet. Anal. Tech. Appl.,* 11, 20 (1994).
15. Rhee, S. S. and Marsh, J. W., *J. Immunol.,* 152, 5128 (1994).
16. Cobbold, P. H. and Rink, T. J., *Biochem. J.,* 248, 313 (1987).
17. Tsien, R. Y., *Biochemistry,* 19, 2396 (1980).
18. Grynkiewicz, G., Poenie, M., and Tsien, R. Y., *J. Biol. Chem.,* 260, 3440 (1985).
19. Fanatis, A. and Russell, J. T., *Glia,* 5, 95 (1992).
20. Weiss, R., in *RNA Tumor Viruses: Molecular Biology of Tumor Viruses,* Weiss, R., Teich, N., Varmus, H., and Coffin, J., Eds., Cold Spring Harbor Laboratory Press, Cold Spring Harbor, New York, 1984, 209.

21. Crise, B. L., Buonocore, L., and Rose, J. K., *J. Virol.,* 64, 5585 (1990).
22. Jabbar, M. A. and Nayak, D. P., *J. Virol.,* 64, 6297 (1990).
23. Willey, R. L., Maldarelli, F., Martin, M. A., and Strebel, K., *J. Virol.,* 66, 7193 (1992).
24. Guy, B., Kieny, M. P., Riviere, Y., LePeuch, C., Dott, K., Girard, M., Montagnier, L., and Lecocq, J.-P., *Nature,* 330, 266 (1987).
25. Tarentino, A. L., Trimble, R. B., and Maley, F., *Methods Enzymol.,* 50, 574 (1978).
26. Rhee, S. S. and Marsh, J. W., *J. Virol.,* 68, 5156 (1994).
27. Kestler, H. W., Ringler, D. J., Mori, K., Panicali, D. L., Sehgal, P. K., Daniel, M. D., and Desrosiers, R. C., *Cell,* 65, 651 (1991).

Chapter 7

Detection, Quantitation, and Characterization of Human Retroviruses

Suraiya Rasheed

Contents

0-8493-4412-3/96/$0.00+$.50

I. Introduction

Two major groups of pathogenic human retroviruses have been discovered to date. The human T cell leukemia/lymphoma viruses type 1 and type 2 (HTLV-1 and HTLV-2) belong to the oncovirinae or tumor virus group and the human immunodeficiency viruses type 1 and type 2 (HIV-1, HIV-2) are members of the lentivirinae or "slow" viruses. The etiological relationship of retroviruses with human cancers and other diseases has only been recognized since 1980, when the first human retrovirus, HTLV-1, was isolated from a Japanese patient with adult T cell leukemia (ATL).[1,2] A related virus, designated HTLV-2 was originally isolated from the blood of a patient with hairy cell leukemia,[3,4] but has subsequently been reported from drug abusers in the U.S., from individuals with neurological symptoms, and from apparently healthy Guaymi Indians living in Panama.[5,6]

In 1983, the highly infectious HIV was isolated from patients with the acquired immunodeficiency syndrome (AIDS),[7] a disease that has no boundaries globally and has reached epidemic proportions in many countries, including the U.S. Since 1981, when the AIDS epidemic was first recognized in the U.S.,[8] an estimated 5 million people in the U.S. and 20 million people globally have been infected by HIV-1 or related virus HIV type 2 (HIV-2).[9] However, a vast majority of HIV-infected individuals do not show any symptoms for a number of years and many have remained asymptomatic for 12 to 15 years postinfection.[10-12] An active HIV infection can lead to immune dysfunction and destruction of cells that provide immune protection (i.e., CD4+ cells), hence the immune deficiency.

Both groups of human retroviruses (i.e., HTLV and HIV) are exogenously acquired and the primary mode of transmission is through sexual contacts with virus-infected persons. However, a large number of individuals have been infected by transfusion of tainted blood or blood products, exchange of blood by virus-contaminated needles or instruments, and perinatally, transplacentally, and postnatally from infected women to their infants through breast milk or contaminated blood. Hemophiliacs have also been infected by HIV from contaminated factor VIII and other blood products used by these individuals.

Infection of susceptible cells by the HTLV or HIV groups of viruses involves a series of steps common to other retroviral infections. The presence of these viruses can be documented at various molecular levels and depends primarily on the stage of virus replication in specific cell types. Some important sequential events in retrovirus replication that are critical for identification, quantitation, and development of tests for diagnosis, prognosis, or antiviral drug testing are briefly outlined below:

1. Binding of viral particles to specific cellular receptors and entry. CD4 is the major receptor on T cells for both HTLV and HIV, but other receptors have also been identified.
2. Transcription of viral RNA into DNA (reverse transcription) via the viral enzyme reverse transcriptase (RT).
3. Integration of the proviral DNA into host chromosomal DNA, facilitated by integrase (IN).

4. The integrated provirus is flanked by the long terminal repeat (LTR) elements that are generated during reverse transcription.
5. Transcription of proviral DNA into RNA.
6. Processing of RNA (spliced or unspliced) and transport from the nucleus to the cytoplasm.
7. Translation of viral gene products.
8. Packaging and assembly of viral RNA and proteins at the cell surface membrane (i.e., virus particle formation).
9. Virus budding, exit, and maturation of infectious virions.

In the past decade, numerous tests have been established to detect and quantitate human retroviruses in body fluids, blood cells, and other tissues.[13-17] Highly specialized, state-of-the art diagnostic tests are now available for the detection of HIV antibodies, lymphocyte subset counts, and retroviral DNA, RNA, or proteins of both HTLV and HIV groups of viruses. These assays have been used for the diagnosis, prediction of disease progression, and for evaluating the effectiveness of the treatment. In this chapter molecular methods for the detection, quantitation, and characterization of human retroviruses are described, with brief discussion on the applications of these techniques in research and clinical settings. Tests that are commercially available and/or are approved by the Food and Drug Administration (FDA) for clinical diagnosis are described only briefly, and tests that are used as "research" procedures are discussed in detail.

II. Detection of Viral Antibodies

The serological identification of antibodies to HIV- and HTLV-related viruses by enzyme immunoassays are the most widely used methods for the diagnosis of these viral infections.[13,15,16] In addition to the sera, external secretions and body fluids, such as seminal fluid, milk, and mucosal secretions from individuals infected by retroviruses, contain various levels of immunoglobins. The secretory immunoglobulins are extremely heterogeneous; for example, IgA produced by the mucosal cells in the respiratory, intestinal, and vaginal secretions may be significantly high in local sites and may induce humoral protective immune responses.[18] Secretory IgA antibodies are also induced from the mucosal surfaces upon stimulation of the mucus and its immune cells. Detection of antibodies in the oral fluid (saliva) and urine have also been useful, particularly in developing countries, where blood collection facilities are restricted.[19,20]

A. Enzyme-Linked Immunosorbent Assay (ELISA)

The enzyme immunoassays (EIA) and enzyme-linked immunoassays (ELISA) are well-established tests for the detection of antibodies in the serum of infected individuals. Several FDA-approved tests are available from various manufacturers (Abbott, Genetic Systems, Coulter, Organon Teknika, etc.) for both HIV and HTLV antibody screening. These tests are based on the identification of the antigen or antibody by colorimetric detection methods and are used routinely in blood banks and clinics for diagnosis. Samples are tested in duplicate according to the instructions provided by the manufacturer, and all reactive samples are confirmed by another test such as the immunoblot/Western blot or radioimmunoprecipitation assay.

For research purposes, the ELISA can be set up as follows:

1. Coat wells of a flat-bottomed 96-well plate with purified HIV or HTLV-1 proteins or a specific antigen if monospecific antibody detection is desired (i.e., individual viral structural or regulatory proteins).
2. Add each test serum to two wells (i.e., test in duplicate). If antigen-related antibodies are present in the test serum, an antigen-antibody complex will be formed in the well.
3. Detect the antigen-antibody complex formed in the well by using an anti-antibody conjugated to an enzyme, such as horseradish peroxidase, or biotin-avidin. If the first antibody in the antigen-antibody complex is human-derived (i.e., human serum), a goat antibody directed against human antibody proteins can be used as the "probe" or the tracking antibody.
4. Add a substrate such as 3-diaminobenzidine/H_2O_2 solution that can bind to the enzyme-conjugated antibody and can be visualized by color development.
5. The intensity of the color is quantitated by recording absorbance of optical density (O.D.) by spectroscopy.

Several peptide-based ELISAs have been developed to allow greater sensitivity of detection by using synthetic peptides from immunodominant regions of *env, gag,* and/or *pol* amino acid sequences.[21] A new test called time-resolved-fluoroimmunoassay (TR-FIA) uses amino acid sequences from all the structural genes and a 20-amino acid peptide (GIWGCSGKLICTTAVPWNAS) from the gp41 region of HIV.[22] Using this assay the authors claim 100% sensitivity for HIV detection.[22] The TR-FIA has also been used for the detection of amplified sequences of HTLV types 1 and 2, with excellent specificity for HTLV-1.[23] However, these assays have yet to be used for screening large numbers of samples in a clinical setting.

B. Immunofluorescence Assay (IFA)

The indirect immunofluorescence assay (IFA) is not recommended as a diagnostic test for HIV or HTLV-1, because the interpretation of results is subjective, and both false-negative and false-positive results are possible. However, this test can be used to identify HIV or HTLV-1 infection in experimental animals or in cell cultures in the laboratory. This technique is straightforward and is described briefly below:

1. Prepare a stock of cell line stably infected with HIV or HTLV-1.
2. Using cytocentrifuge and 0.1% gelatin or histologic glue-coated slides, adhere 10^6 cells in the middle of each slide or use two "sprays" or circles of cells on the same slide.
3. Fix cells in freshly prepared 4% paraformaldehyde in phosphate-buffered saline (PBS) or methanol-acetone for 10 min. Fixed slides can be stored at –20°C indefinitely.

4. Incubate cells with test antibodies at 4°C for 1 h or overnight for convenience. Use at least two to four circles of cells per test.
5. Use polyclonal patient sera or monospecific antisera generated against HIV-1 or HTLV-1 gag antigen as positive control antibodies and pooled normal human serum as negative control.
6. Wash with multiple changes of PBS (pH 7.6).
7. Incubate with fluorescein isothiocyanate-conjugated anti-human antibody (e.g., IgG generated in rabbit or goat (1:200 dilution), 10% Triton X-100 in PBS at room temperature for 60 to 90 min.
8. Wash with multiple changes of PBS over 10 to 15 min.
9. Examine for fluorescence on the cell surface and/or cytoplasm, depending on the antibodies used.
10. Compare positive cells in controls with experimental or test samples.

C. Immunoblot Assay or Western Blot Test

The immunoblot assay (IBA), commonly known as the Western blot test, is based on the detection of antigen-antibody complexes as discrete colored bands on nitrocellulose membranes.[24] Specific viral antigens can be used to test antibody responses to those viruses. The following protocol has been established in our laboratory for the detection of HIV- or HTLV-related antibodies or antigens by Western blots.[17,25]

1. Grow virus in peripheral blood mononuclear cells' (PBMCs) cultures or susceptible T cell lines such as H9 in a biosafety level-3 containment facility, as described previously.[17]
2. Purify virions from the filtered culture fluid using sucrose density gradient (17% to 25% w/v) centrifugation. Large quantities of virus can be concentrated using an ultrafiltration device (Millipore, Bedford, MA) and then pelleting the concentrated virus by ultracentrifugation.
3. Lyse virus pellet by adding a buffer containing 0.5% Titron X-100, 100 m*M* NaCl, 10 m*M* Tris-HCl, pH 7.5, 1 m*M* phenylmethylsulfonyl fluoride (PMSF).
4. Clarify the virus lysate by centrifugation at the highest speed in a microfuge for 20 sec at 4°C. Retain the supernatant and discard any pelleted material.
5. Separate viral proteins from the purified viral lysates by electrophoresis and size-fractionate in 11% polyacrylamide gel containing 0.01% sodium dodecyl sulfate (SDS). Viral proteins produced in the bacteria by recombinant DNA techniques can also be used instead of whole viral lysates.
6. Transfer the banded, i.e., size-fractionated, proteins on to nitrocellulose membranes (0.2 μm) by electroblotting[24] for 3 h at 100 V (with a 15°C recirculating coolant), in buffer containing 20% methanol, 0.192 *M* glycine, 24 m*M* Tris, pH 8.3.
7. Wash blots four times in buffered PBS containing 0.3% Tween 20 (PTB).
8. Store blots between two plastic sheets at 4°C until used. Cut blots (nitrocellulose membranes) containing proteins into 5-mm strips just before testing. Do not touch without gloves.

9. Place each strip in an elongated grooved tray specially designed to hold blots. React each blot separately with the test serum antibodies (20 to 30 μl) in 2 ml PTB containing 4% normal goat serum (PTGB), for 1 to 2 h at 37°C or overnight at 4°C with constant slow shaking.
10. Wash blots three times with PTB for 5 min each and once with PTGB.
11. React with 5 μl biotinylated anti-human goat IgG (Bethesda Research Laboratories, Bethesda, MD) in 2 ml PTGB for 1 h.
12. Wash four times with PTB and incubate with 5 μl streptavidin-conjugated horseradish peroxidase in 2 ml PTB.
13. Wash four times with PTB and react for 30 min with chromogenic substrate 4-chlor-1-naphthol.
14. Rinse with distilled H_2O twice.
15. Visualize reactive antigens as colored "bands" and compare the banding pattern of the test serum with positive and negative control serum reactions.

D. Radioimmunoprecipitation Assays (RIPA)

The RIPA is used primarily as a confirmatory test for the presence or absence of antibodies in the serum.[25] To evaluate HIV and HTLV infections, viral proteins are labeled metabolically by growing virus-infected cells in the presence of radiolabeled amino acids (^{35}S-methionine-cysteine). Alternatively, purified viral proteins can be iodinated separately *in vitro* using ^{125}I.[17] The sensitivity and specificity of RIPA is similar to Western blot, although the HIV gp120 and gp160 are detected more frequently in RIPA than in the Western blot. A protocol that has been used in our laboratory is as follows:

1. Metabolically label virus-producing cells (chronically or acutely infected) with ≥600 μCi/ml of ^{35}S-methionine-cysteine for 3 h in medium lacking cold methionine and cysteine.
2. Harvest labeled cells, and wash in ice cold PBS; discard medium and PBS in biohazard/radioactive container.
3. Add lysis buffer to cell pellet at a ratio of 0.5 ml per 5×10^6 cells. The lysis buffer contains 10 m*M* Tris-HCl, pH 7.5, 50 m*M* KCl, 2 m*M* $MgCl_2$, 1 m*M* dithiothreitol (DTT; add fresh for each use), 1% Triton X-100 and 2 m*M* PMSF. Add PMSF again just before use. Cell extracts can be frozen in small aliquots at –70°C.
4. Dilute one aliquot of extract in buffer A containing 20 m*M* Tris-HCl, pH 7.5, 1 m*M* EDTA, 1% Triton X-100, 0.4 m*M* NaCl, 1 m*M* DTT, 20% glycerol, aprotinin and PMSF 100 U each per milliliter. The dilution of cell extract depends on the amount of antigen expected to be present in cells. It is recommended that 1:2 dilution should be initially tested.
5. Incubate separately with test serum, known positive serum (pooled patients' serum) and negative (seronegative human serum) control sera at ≥1:200 dilution at 4°C for at least 1 h.
6. Add protein A-Sepharose beads to the mixture and incubate at 4°C for at least 1 h.

7. Spin down precipitate and wash three times in buffer A and three times in buffer B containing 100 m*M* KCl, 0.1 m*M* EDTA, 20 m*M* Tris-HCl, pH 7.5, 20% glycerol, and 100 U/ml aprotinin.
8. Dilute washed immunoprecipitate with an equal volume of buffer containing 2% (w/v) SDS, 10% glycerol, 70 m*M* Tris-HCl, pH 6.8, and 0.5 *M* 2-mercaptoethanol.
9. Size fractionate antigen-antibody complexes using SDS-polycrylamide (14%) gel electrophoresis.
10. Dry gels and autoradiograph.
11. Compare reactive antibody pattern of the test serum with the known positive and negative sera reacting with viral proteins. Specific antibodies are identified by positions of the molecular weight protein markers.

E. Detection of Antibodies by Flow Cytometry

Antibody responses to different antigens can be simultaneously detected by flow cytometry (see also Section III.C). Using immunoreactive beads coated with the major structural antigens of HIV, i.e., products of *env* (gp120 and gp41), *pol* (p31), and core protein (p24), and monospecific fluorochrome-labeled antibodies, different sizes of beads can be separated by light scatter. When the beads bind to specific fluorochrome-labeled antibodies, they fluoresce and different bead populations can be measured against the normal serum and a known HIV antibody-positive serum as the negative and positive controls, respectively. This technique has an advantage that both IgG and IgM responses can be evaluated. However, the instrument used for flow cytometric analysis is quite expensive and most clinical and research laboratories cannot afford to maintain this sophisticated equipment. Therefore, for the detection of antibodies, the ELISA screening is still the most cost-effective and rapid test.

F. Serological Distinction of Dual Infections

Several peptide-based assays are now available to distinguish HIV-1 antibodies from HIV-2 or HTLV-1 antibodies from HTLV-2.[21,25] We have used an EIA (SELECT HIV™; Coulter, Hialeah, FL) to distinguish HIV-1, HIV-2, or dual infection in African sera.[26] All repeatedly reactive sera are then confirmed by immunoblot or Western blots made specifically for testing these two viruses. The HTLV-1 EIA or ELISA-reactive samples are also confirmed by Western blots using HTLV-1 antigens. However, since purified HTLV-2 whole viral antigen-based assays are not able to distinguish HTLV-2 antibodies from those of HTLV-1, peptide-based assays are used to distinguish HTLV-1 antibodies from HTLV-2 by using SELECT-HTLV test (Coulter, Hialeah, FL). These reactions are then confirmed by polymerase chain reaction (PCR; see Section V.C). Using these assays, a high prevalence of antibodies to HIV-1, HIV-2, HTLV-1 and HTLV-2 was detected in various combinations, in sera of apparently healthy women in Ibadan, Nigeria.[26] These results indicate a co-infection with two or more different types of retroviruses. However, an alternative explanation would be the possibility of generating novel recombinant viruses in those individuals co-infected with two or more retroviruses.

III. Detection of Viral Antigens

Several proteins are encoded by the viral structural genes *gag, pol,* and *env.* However, the major core protein encoded by HIV *gag* gene is a highly immunogenic antigen of 24,000 Da molecular weight (p24). Although antigenetically distinct, the molecular weight of the HTLV-1 core protein is also 24,000 to 25,000 Da, and it is referred to as p25. Currently available assays for the detection of these viral antigens are described below.

A. "Regular" HIV-1 p24 Antigen

The antigen capture assays allow detection of HIV or HTLV antigens. Several commercial tests are available from Abbott, Dupont, Organon Teknika, Coulter and other manufacturers. The "regular" HIV antigen tests are specific, quality controlled, and quantitative. However, the sensitivity of the test varies according to the disease course, the virus load, and the amount of antigen that may be complexed with HIV antibodies.

Measurement of HIV-1 p24 antigen in the serum or plasma of infected individuals indirectly indicates an increased virus load, with or without symptomatic illness. Studies of the levels of serum p24 as a surrogate marker in asymptomatic HIV-infected patients with baseline CD4+ cell counts between 200 and 500 indicate that 20% of patients have measureable p24 antigen at baseline (i.e., before antiviral treatment) and changes in p24 levels (i.e., decrease from baseline or elevation after treatment) are not significantly associated with survival or the decreased risk of disease progression.[27]

Various software programs are available to calculate and quantitate the amount of p24 in the serum, plasma, and HIV culture fluids from the optical densities directly obtained from the ELISA reader, thus avoiding any manual handling and chances of errors. These programs are available from various manufacturers and allow quantitation of values from standard curves generated by testing six serial dilutions of purified p24 antigen with each run.

B. Antigen Detection by Immune-Complex Dissociation (ICD)

The cell-free HIV-1 p24 antigen is present in the plasma of most symptomatic and some asymptomatic HIV-infected individuals. This antigen can be complexed with antibodies present in the same patient's plasma. These complexes cannot be detected by the "regular" p24 detection assays, because free antigen is not available for binding the antibody-coated beads (Abbott, Chicago, IL) or plates (Coulter, Hialeah, FL) used for these tests. Thus, the patient is falsely considered to be p24 antigen-negative. Currently there are several HIV-p24 antigen tests designed to provide increased sensitivity, to detect p24 antigen after immune complex dissociation (ICD) based on pretreatment of sera or plasma by acid (Abbott), acid (HCl)-glycine (Coulter), or high pH reagents (Organon Teknika, Durham, NC).

Data on the ICD-p24 detection in sera of HIV-seropositive individuals indicate that acid treatment or acid-glycine treatment, followed by neutralization with NaOH, increases the sensitivity of this test from 22–30% to 50–60%.[28a] Thus, sera from HIV-infected asymptomatic individuals who are p24-negative by the regular p24 test, yield positive values in the ICD p24 test. Patients' samples that are initially positive for p24 antigen show from two- to fivefold increase in the picogram values per milliliter of the p24 antigen.[28-30] This test is also useful for early diagnosis of HIV-1 infection in infants.[28]

C. Detection of HIV Infection by Flow Cytometry

Over the past two decades, flow cytometry, also known as fluorescence-activated cell sorter (FACS) technology, has been improved significantly and is currently considered an extremely efficient tool for the routine phenotypic analysis and identification of specific cell types in a mixed population of peripheral blood cells. By the use of monoclonal antibodies and multicolor immunofluorescence staining, many cytoplasmic and surface antigens can be stained to identify cells involved in carrying out important immunological and cell differentiation functions. For example, using flow cytometric methods, cell surface receptors have been identified for both HTLV and HIV.[31-38] Since several million cells can be analyzed in a few minutes, these techniques have been particularly useful in monitoring CD4+ lymphocyte cell numbers, as the depletion of these cells is considered a marker for disease progression in HIV-infected individuals. Thus, evaluation of CD4+ cell numbers in HIV-infected patients helps in the clinical diagnosis and prognosis of the disease. Antibodies to cell surface antigens identify the cell type and another color of the fluorescent dye-labeled virus can show the binding of virus particles to specific antigens. Alternatively, virus can also be conjugated to biotin and the biotinylated virus can be bound to cells by treatment with fluorescein-isothiocyanate (FITC)-labeled avidin.

Flow cytometric methods have also been used to demonstrate inhibition of HIV binding by the treatment of HIV-infected cells with antiviral compounds that specifically interact with viral gp120 glycoproteins such as dextran sulfate and other polyanionic compounds.[32-34]

The amount of cell-associated and cell-free HIV present in blood or plasma can also be analyzed by flow cytometry. Since approximately 3×10^6 cells per minute can be analyzed rapidly, this technique can screen a large number of cells in a very short time. However, because this test uses monoclonal antibodies to p24, only p24 antigen-positive cells are detected. The sensitivity of this technique, however, is lower (1 in 10^4 cells) than other molecular methods, such as enzyme immunoassays (EIA) for p24 antigen or the detection of DNA or RNA by the polymerase chain reaction (PCR), which can detect one virus-infected cell in 10^6 cells. Further, the specificity of detecting antigen-producing cells in HIV-infected patients by flow cytometry has yet to be standardized for clinical diagnosis or prognosis of the disease.

Recently, the flow cytometry instrumentation has also been used for the detection of HIV-DNA, particularly to measure virus-induced programmed cell death or apoptosis. The light scatter techniques of flow cytometry have been used in two- or three-color fluorescence to quantitate the number of HIV-infected cells undergoing apoptosis.[35-38] Separated lymphocytes or whole blood can be used to label the 3′-OH of the DNA by digoxigenin-dUTP and deoxynucleotidyl transferase. FITC-conjugated antibodies to digoxigenin are then used to quantitate the labeled cells. However, these methods have not been standardized yet and are still in experimental stages.

D. Host Cellular Proteins Associated With Virions

The host-cell proteins play major roles in the formation of infectious virions. Prior to budding, the HIV proteins assemble near the plasma membrane of an infected cell, together with a cellular protein called cyclophilin-A, which attaches to the Gag protein and is carried out in the virus particles.[39,40] Recently, it was shown that, although the virus particles with or without cyclophilin appear similar in morphology, those without cytophilin are less able to infect and replicate in the CD4+ T lymphocytes.[39,41] However, cyclophilin is not the only protein that is incorporated in virions. For example, actin, ubiquitin, and the human lymphocyte antigen-type DR/(HLA-DR) and β2-microglobulin have been shown to be associated with the newly formed HIV virions in cultured cells.[42] It is speculated that HIV may use these cellular proteins to "repair" or mark as signals for recognizing cell surface proteins as the virus enters or exits the cells.

IV. Detection of Reverse Transcriptase (RT) Activity

The hallmark of all retroviruses is the presence of an RNA-directed DNA polymerase, or reverse transcriptase, encoded by the retroviral *pol* gene. This enzyme catalyzes the transcription of single-stranded viral RNA to yield double-stranded DNA before the viral genome is integrated into the host chromosomal DNA. The RT activity can be detected in tissue culture supernatants, body fluids, or cellular extracts by an extremely sensitive biochemical test. However, it is important to note that the measurement of RT activity in a sample only indicates the presence of this retroviral enzyme, it does not identify the type of virus infection. Thus, the RT assay cannot distinguish between HIV, HTLV, or related simian viruses. This distinction must be made by using specific antigen detection assays or nucleic acid-based tests to identify specific viruses (see Sections III and V). Many laboratories involved in testing clinical specimens prefer to use the p24 test for HIV and HTLV-1 specific protein detection assays, because these have been standardized for use in multicenter clinical trials.

A. Radioisotopic Detection of RT Activity

The most commonly used RT assay utilizes a synthetic homopolymer template poly(rA):oligo(dT) in the presence of divalent cation (Mn^{2+} and Mg^{2+}) and ^3H-thymidine triphosphate. In our laboratory the RT assay is performed as follows:[17]

1. Clarify virus from culture fluid or plasma by centrifugation at 10,000 rpm.
2. Remove supernatant in tubes that can be used for ultracentrifugation using a swinging bucket rotor (SW-28); centrifuge clarified supernatant at 28,000 rpm for 1½ h. Alternatively, concentrate virus by the addition of polyethylene glycol 6,000 (10% final concentration).
3. Discard the supernatant culture fluid and drain completely by inverting the tube on a paper towel, which should be subsequently discarded in liquid bleach.
4. Disrupt the virus pellet on ice in 100 μl lysis buffer containing 50 m*M* Tris-HCl, pH 7.9 (at 37°C or 8.2 at 25°C), 0.025% Triton X-100 or 0.1% NP-40, 5 m*M* KCl, and 0.2 m*M* EDTA.
5. In separate tubes add RT reaction mixture containing, in a final volume of 100 μl, 50 m*M* Tris-HCl (pH 7.9 at 37°C or 8.2 at 25°C), 100 m*M* KCl, 5 m*M* $MgCl_2$, 5 to 8 m*M* dithiothreitol, 20 μCi ^3H-thymidine triphosphate (600 Ci/mmol), 0.05 units of exogenous template primer poly(rA):oligo$(dT)_{12\text{-}18}$, and 25 μl of virus lysate (containing putative RT enzyme). For controls, use nonspecific template using dG-dC, cellular DNA polymerase reactions without template, a known virus-negative sample, a known RT positive sample.

Note: *An important consideration in the performance of the RT assay is the requirement of a divalent cation, particularly when using a synthetic homopolymer template. Both HTLV- and HIV-related viruses can use Mg^{2+}. However, HTLV-RT catalysis is more efficient in the presence of Mn^{2+} and the HIV enzyme prefers Mg^{2+}.*

6. Incubate the mixture at 37°C for 60 min in a shaking water bath and terminate the reaction by the addition of 5% cold trichloroacetic acid (TCA) containing 0.1 *M* sodium pyrophosphate, 10 m*M* EDTA. Quench on ice for an additional 15 to 20 min.
7. Separately collect the precipitate from each sample on glass fiber filters (GF/A or GF/C) presoaked with 100 μl of 10 m*M* ATP.
8. Wash each filter twice with decreasing concentrations of sodium pyrophosphate (0.05 *M*) and 5% TCA with a final wash of 5% TCA alone, followed by a wash in 95% ethanol.
9. Air-dry washed filters or dry under a heat lamp and place in vials containing liquid scintillation fluid (Liquifluor).
10. Count the incorporated radioactivity in a scintillation counter. Average enzyme activity is computed from duplicate samples by deducting the background counts (i.e., controls:

nonspecific templates dG-dC, cellular DNA polymerase and reactions without homopolymer template poly(rA):oligo(dT), counts obtained from control cultures, or samples known to be negative by previous tests). Adjust the dilution factor for the amount of virus lysate used.

B. Nonradioactive RT Assay (The RT-Detect™ Test)

A nonradioactive, 96-well microplate-based assay has been developed by Dupont Co. (Boston, MA) for the detection of HIV-1 RT activity. This method utilizes a specific, heteropolymeric, 89-base RNA template (instead of homopolymeric poly(rA):oligo(dT), a 20-base enzyme-linked oligonucleotide primer, and a horse-radish peroxidase (HRP)-conjugated detector probe. The enzymatic activity of RT contained in a test sample generates a specific cDNA using biotinylated oligomer as the primer. The cDNA product is detected by a sandwich hybridization assay, termed enzyme-linked oligonucleotide sorbent assay or ELOSA.[48] The assay is performed in streptavidin-coated microplate wells, which capture the specific cDNA. By using a substrate for the HRP-labeled 20-base oligonucleotide detector probe, a color is developed.

The intensity of reaction is computed from the optical densities obtained by spectrophotometry. The use of a positive control DNA standard allows computation of the amount of the cDNA produced in each reaction. This value is directly proportional to the enzymatic activity of the RT present. The sensitivity of ELOSA is equivalent to the radioactive RT detection method.

V. Detection of Viral Nucleic Acids (DNA and RNA)

Infection of susceptible cells by either the HTLV or HIV group of viruses involves a series of steps. One of the most important steps is the integration of reverse-transcribed proviral DNA into the host chromosomes. The proviral DNA can be detected by Southern blotting or polymerase chain reaction (PCR)-based techniques. Likewise, the transcription of proviral DNA into RNA can be tested by Northern blot or reverse transcription PCR (RT-PCR).

A. Southern Blot Analysis of DNA

Southern blot analysis of DNA obtained from T cells of patients with AIDS and ATL indicate that HIV and HTLV-1 DNAs, respectively, have integrated in cells of these patients. Although there is no single site for the integration of proviral DNA in the chromosome, monoclonal integration of HTLV-1 in the infected T cells does give these cells a growth advantage which results in expansion of T cells and enhances the chances of virus replication cell transformation.[43] Using the Southern blotting technique, genomic diversity of HIV was initially recognized.[44] However, one of the disadvantages of this technique is that it is useful only when higher copy numbers of retroviral sequences are present in the DNA. Further, with the advances in PCR-based technologies,[49] Southern blotting is rarely used for the detection of human proviral DNAs in cells of infected individuals.

B. Northern Blot Analysis of RNA

The Northern blot analysis allows identification of viral RNA by hybridizing fractionated cellular RNA with specific probes for HIV or HTLV genes. RNA is isolated from virus-infected peripheral blood mononuclear cells or tissues (lymph node and spleen are the best sources of virus), using the guanidinium thiocyanate-phenol-chloroform extraction procedure,[50] as described in Section V.F.1. The whole RNA is quantitated by absorbance at 260 and 280 nm, and is analyzed by Northern blotting as follows:

1. Dissolve purified RNA pellet in buffer containing 20 m*M* morpholinepropanesulfonic acid (MOPS), pH 7.0, 50 m*M* sodium acetate, 10 m*M* EDTA, 6% formaldehyde, and 50% formamide. *Use sterile, double-distilled water in all solutions. Pretreat water for 18–20 h with 0.02% diethylpyrocarbonate (DEPC) and autoclave before use.*
2. Heat mixture at 65°C for 5 min and electrophorese through 1% agarose gel containing 2% formaldehyde. Use 28S and 18S rRNA as markers.
3. Transfer RNA to nitrocellulose membranes, as described previously.[17]
4. Mark locations for 28S and 18S RNA on the nitrocellulose blot and bake it at 80°C for 2 h.
5. Prehybridize for 2 h and hybridize overnight at 50°C with ^{32}P-labeled probe of choice (specific HIV or HTLV gene or full-length proviral cloned DNA; see Section V.J. for details of hybridization solutions).
6. Use β-actin probe as control.
7. Wash blot in 2 × SSC with 50% formamide followed by a wash with 2 × SSC..
8. Autoradiograph, compare with controls, and identify virus-specific bands.

C. Polymerase Chain Reaction (PCR) for Proviral DNA Detection

The PCR is one of the most sensitive tests available today for the detection of any gene whose DNA or protein sequence is known. Because PCR detects specific nucleic acid sequences directly by exponential amplification of oligonucleotide primers annealed to the positive and negative DNA strands, this technique provides a greater sensitivity for the detection and quantitation of viral genes than other conventional tests.[45-74] The exponential expansion of a genetic sequence such as that of a viral gene enhances the sensitivity of its detection regardless of the number of virus-infected cells in a sample or immune responses of the host. Thus, the presence of a virus in a person's body can be identified before seroconversion. Further, it is also possible to detect viral DNA when the virus is defective for replication (i.e., is latent),[50,51] and the antibody and/or culture tests are inconclusive. For example, a PCR test is essential when the HIV antibody test is not helpful, in the following situations:

1. Early stage of infection, i.e., before seroconversion
2. Diagnostic screening of infants born to infected mothers, to rule out the presence of passive maternal antibodies in the infant
3. Resolution of indeterminate Western blot results
4. Presence or absence of viral sequences in individuals exposed to virus by needle stick or other means, but who may remain seronegative after exposure (i.e., not seroconverted)

Partial sequences of HTLV-1 proviral DNA have been detected in several lymphoproliferative disorders using PCR primers from the gag, pol, and pX regions of the HTLV-1 genome.[47] Using PCR, monoclonal integration of HTLV-1 proviral DNA was detected in lymphocytes and other cells present in the saliva.[53]

In the following section, a step-wise PCR protocol is provided for the detection of HIV. An identical protocol has been used in our laboratory for HTLV-1 and -2 with the exception of HTLV-1 or -2 specific primers and probes.

The PCR nucleic acid amplification method involves the following basic steps:

1. Primer selection
2. Sample preparation
3. Distribution of sample aliquots into a reaction mixture
4. Addition of enzyme and target amplification, by repeating cycles of denaturation, annealing, and chain extension
5. Hybridization of the amplified product(s) to specific probe(s) and detection of the product:
   a. Isotopic
   b. Nonisotopic

1. Selection of primers

Oligonucleotides to be used as primers and probes are selected from the gene of interest. Highly conserved regions are preferred from both the 3′ and the 5′ ends of the gene. If mutations are known to cause changes in the amino acid sequences, these primers or probes are used to identify mutant alleles. Computer programs are used to facilitate selection of oligonucleotides containing appropriate sequences to be used as primers or probes. Hybridization with oligonucleotide probes that are centered around codons of interest are utilized to localize these genes. Primer pairs may also be selected from genetic changes known to be associated with drug resistance. Some of the commonly used primers and probes for the detection of human retroviruses are listed in Table 7.1.[49]

TABLE 7.1
Commonly Used Primers and Probes for the Detection of Human Retroviruses

Region	Primer/ Probe	Sequence	Specificity
ltr	SK29 (primer)	ACTAGGGAACCCACTGCT	HIV-1
	SK30 (primer)	GGTCTGAGGGATCTCTA	
	SK31 (probe)	ACCAGAGTCACACAACAGACGGGCACACACTACT	
	SK89 (primer)	AGGAGCTGGTGGGGAACG	HIV-2
	SK90 (primer)	GTGCTGGTGAGAGTCTAGCA	
	SK91 (probe)	TTGAGCCCTGGGAGGTTCTCTCCAGCACTAGCAGGTAC	
gag	SK38 (primer)	ATAATCCACCTATCCCAGTAGGAGAAAT	HIV-1
	SK39 (primer)	TTTGGTCCTTGTCTTATGTCCAGAATGC	
	SK19 (probe)	ATCCTGGGATTAAATAAAATAGTAAGAATGTATAGCCCTAC	
tax	SK43 (primer)	C/TGGATACCCA/CGTCTACGTGT	HTLV-1/2
	SK44 (primer)	GAGCC/TGAT/CAACGCGTCCATCG	
	SK45 (probe)	ACGCCCTACTGGCCACCTGTCCAGAGCATCAGATCACCTG	
pol	SK110 (primer)	CCC/ATACAAT/CCCA/CACCAGCTCAG	HTLV-1/2
	SK111 (primer)	GTGGTGA/GAG/TC/TTGCCATCGGGTTTT	HTLV-1/2
	SK112 (probe)	GTACTTTACTGACAAACCCGACCTAC	HTLV-1
	SK188 (probe)	TCATGAACCCCAGTGGTAA	HTLV-2
	SK115 (probe)	CAT/AAGCCCTA/TTGGACA/TA/CTCAAC/TCAC/GC	HTLV-1/2

2. Sample preparation from blood

Any DNA sample prepared from tissues, PBMCs, or cultured cells can be used. Prepare samples for proviral DNA detection directly from the whole blood collected in EDTA or acid citrate dextrose (ACD or CPT) tubes. Blood may be stored at 4°C to 25°C overnight and processed as follows:

1. Mix whole blood with Specimen Wash Solution containing 0.05% sodium azide (Roche Molecular Diagnostic Systems, Alameda, CA) to lyse red blood cells. The presence of low levels of hematin (0.8 μ*M*) has been shown to inhibit Taq polymerase.[24] Transferrin, a naturally occurring serum protein, suppresses the inhibitory effect.[54]
2. Centrifuge the mixture to pellet cells and discard supernatant.
3. Extract DNA from the cell pellet by using a lysis buffer containing 200 μg/ml proteinase K, 7.5 m*M* $MgCl_2$, 0.45% NP-40, 0.45% Tween 20, 10 m*M* Tris-HCl, pH 8.3. The DNA released in the cell lysate can be used directly without further purification.
4. DNA samples can also be purified by conventional phenol/chloroform extraction procedures. However, these organic solvents must be removed by two or more ethanol precipitation steps, because traces of phenol/chloroform in the DNA may interfere with the primer annealing steps.

3. Sample preparation from archival or fresh tissues

Surgically resected tissue (fresh or frozen at –70°C or liquid nitrogen), frozen whole blood[55] and archival specimens fixed in a variety of fixatives and/or embedded in paraffin can be used for gene amplification and detection of human retroviral infections. Thus, tissue sections have been successfully used for the identification of both HIV-1 and HTLV-1 infections[56,57] by PCR. The following procedure is used:

1. Using a pair of fine scissors, chop surgically resected fresh tissues. Homogenize larger size tissues gently in the presence of liquid nitrogen in a ground-glass Dounce tissue grinder with conical bottom. Freeze-thaw archival tissue or tissue section in liquid nitrogen or in acetone-alcohol bath after grinding the tissue.
2. For archival specimens embedded in paraffin, remove paraffin first by treatment with xylene and wash with alcohol gradients (95%, 70%) before extraction of DNA.
3. Extract DNA by one of the following two methods:
   a. Incubate for 1 h with lysis buffer containing 0.6% SDS in 10 m*M* Tris-HCl, 10 m*M* EDTA (pH 7.5), and 50 μg/ml RNase A. Incubate the mixture at 37°C for 16 to 20 h before extraction of DNA with phenol/chloroform, followed by precipitation with three volumes of ethanol in the presence of 5 *M* sodium acetate.
   b. Suspend tissue in TE buffer (10 m*M* Tris-HCl, 1 m*M* EDTA) containing 4 *M* guanidium isothiocyanate (Sigma Chemical, St. Louis, MO). If tissue is still undissolved, these are further homogenized.
      i. Transfer the mixture to an Eppendorf tube and add an equal volume of phenol-chloroform-isoamyl alcohol. Transfer the aqueous phase to a new tube and repeat extraction procedure followed by chloroform-isoamyl alcohol (24:1) extraction.
      ii. Precipitate DNA by adding 0.2 vol of 5 *M* ammonium acetate and 3 vol of cold ethanol.
   c. Centrifuge the precipitate containing DNA and wash with cold 70% ethanol.
   d. Air-dry pellet and suspend in TE. Quantitate DNA concentration by GeneQuant™ (Pharmacia Biotech, Milwaukee, WI).

4. Amplification of DNA by PCR

Specific forward and reverse primers selected for the unique viral genetic sequence are used for annealing and DNA amplification. Both the negative and positive DNA strand primers are annealed to the denatured template of the test DNA and a DNA copy is made by primer extension in the rightward and leftward directions, respectively. One or both primers can be labeled with radioisotope ^{32}P. When unlabeled primers are used, the amplified product is hybridized and detected by a specific labeled probe. Some of the important primers and probes for HIV- and HTLV-1-related viruses are listed in Table 7.1.

The reaction mixture used for gene amplification in our laboratory consists of 1 μ*M* of each primer, 200 μ*M* of each of the four nucleotide triphosphates (dATP, dCTP, dGTP, and dTTP), 10 m*M* Tris-HCl, pH 8.4, 50 m*M* KCl, 1.5 m*M* $MgCl_2$, and 0.01% (w/v) gelatin, and 2 units of a thermostable DNA polymerase (Taq) per 100 μl reaction mix. For the detection of most conserved viral sequences, the DNA is heated to 94°C for 30 sec, annealed with the primers at 55°C for 30 sec and primers extended at 72°C for 60 sec. However, minor adjustments in the denaturation and annealing temperatures may be necessary for the amplification of certain sequences. After the amplification is carried out for 30 to 40 cycles in an automated thermal cycler (GeneAmp™ 9600, Perkin-Elmer, Foster City, CA), the amplicons or amplified DNA product is processed as described below.

5. Hybridization and detection of the PCR amplified product

When unlabeled primers are used, amplicons are heated to 94°C for 10 sec and hybridized with ^{32}P-labeled probes suitable to detect specific viral genes (listed in Table 7.1), and the hybridized product is separated on 11% polyacrylamide gel and autoradiographed.

Amplicons generated by using labeled primers are cleaved with restriction enzymes that cleave at specific sites in the viral DNA and separated on the gel. The size of the band is compared with the DNA standards and quantitated if desired.

Several *nonradioisotopic,* PCR-based tests for HIV-1 are available from Roche, Dupont, GenProbe, and other manufacturers. A comparison of these assays with standard radioisotopic methods indicates that all currently available tests are 96 to 100% sensitive and specific.[49]

For Roche's Amplicor HIV-1 Detection™ assay, follow manufacturer's instructions and proceed as follows:

1. Denature the PCR amplified DNA by the addition of 1.6% sodium hydroxide in EDTA.
2. Add single-stranded DNA of biotin-labeled amplicons to the oligonucleotide "capture" probe on the plate.
3. After hybridization, wash excess DNA and add avidin-horseradish peroxidase conjugate. The avidin of the conjugate binds to the biotin-labeled amplicons, which is bound to the probe on the plate.

4. Remove unbound conjugate, wash, and react with peroxidase (0.01% H_2O_2) and substrate containing 0.1% TMB (3,3′,5,5′-tetramethylbenzidine) in 40% dimethylformamide.
5. Stop the reaction by adding 4.9% sulfuric acid.
6. Read absorbance or optical density (O.D.) in the ELISA plate reader.
7. Positive or negative values are calculated automatically by a computer program provided by the manufacturer.

D. Distinguishing HTLV-1 from HTLV-2

Using HTLV-2 specific *pol* primer pairs and HTLV-1 *tax* primer pairs, the two viruses can be distinguished (Table 7.1). For detection of dual infections or for distinguishing HTLV-1 from HTLV-2, DNA samples are tested by two separate steps of PCR amplification and hybridization of amplicons. First the generic primer pair SK43 + SK44, which identifies sequences that are common between these two viruses are used from the *tax* region of both HTLV-1 and HTLV-2. Samples showing an HTLV generic reaction are tested in a second PCR reaction using SK110 and SK111 primers that are known to flank the *pol* gene region that distinguishes HTLV-1 from HTLV-2. The amplicons from the second step are separately hybridized to HTLV-1 and HTLV-2 specific probes. The SK45 probe is common for the Tax region of both HTLV-1 and HTLV-2 and SK112 specifically hybridizes to HTLV-1 sequences, while the SK188 probe specifically identifies HTLV-2.

The DNA sample is mixed with a reaction mixture that contains 1.5 m*M* $MgCl_2$, 2.5 m*M* each of the four deoxynucleoside triphosphates, 5 units of Taq DNA polymerase (Boehringer Mannheim, Indianapolis, IN) and the buffer, as described in Section V.C.4. The DNA is heated at 94°C for 30 sec and amplified using HTLV-1 and -2 generic or specific primers (Table 7.1) in a thermal cycler (Perkin-Elmer, Foster City, CA), 52°C for 30 sec and 72°C for 60 sec. After 30 to 35 cycles of amplification, DNA is heated to 94°C for 10 sec and 20 μl of the amplicon is hybridized *separately* with three different oligonucleotide probes SK45, SK112, and SK188 labeled with ^{32}P or nonisotopic labels. The hybridization product is subjected to size fractionation by electrophoresis on 11% polyacrylamide gel and exposed to X-ray film for autoradiography.

Dual infection by both HTLV-1 and HTLV-2 has been detected in Japanese blood donors.[47] Using techniques described above, our laboratory has also detected both HTLV-1 and HTLV-2 in the DNA of several healthy subjects living in Nigeria.[58]

E. Distinguishing HIV-1 from HIV-2

HIV-1 and HIV-2 dual infections have been detected serologically in sera from African patients.[26,59] Simultaneous isolation of HIV-1 and HIV-2 from an AIDS patient has also been demonstrated.[59] Dual infections have been confirmed by PCR using specific primers (Table 7.1) that can distinguish between the two infections.[60,61]

F. Reverse Transcription Polymerase Chain Reaction (RT-PCR)

The reverse transcription (RT) PCR (RT-PCR) technique is similar to DNA PCR, except that a cDNA copy is made from the genomic viral RNA prior to its amplification. Since HIV- or HTLV-specific viral RNA is used as template, it minimizes the nonspecific amplification of other genetic sequences in the RNA prepared from blood cells, or other tissues which can also be used for the detection of HIV or HTLV infections by this technique. Viral RNA is obtained from purified virions present in the plasma, tissue culture supernatants, or RNA is extracted from other body fluids, such as cerebrospinal fluid, cervical secretions, semen, etc.[73,74]

Several commercial tests are now available to quantitate HIV-RNA levels for the correlation of disease progression in HIV-seropositive individuals. The RT-PCR test called, Amplicor HIV Monitor™ by the Roche Molecular Systems Inc., Somerville, NJ, is helpful in monitoring the efficacy of antiviral therapy, particularly as it relates to the development of drug-resistant HIV strains, symptomatic illness or AIDs in HIV-infected patients. This RT-PCR test is performed essentially according to the instructions provided by the manufacturer. The quantitative RT-PCR method described in this section can be used for the amplification and quantitation of any viral or cellular gene transcripts (spliced or unspliced), using radioisotopic or nonisotopic probes.

1. Isolation of RNA

1. Mix thoroughly equal portions of 4 *M* guanidinium isothiocyanate (with DTT) and buffered phenol.
2. Add 200 μl of the above mixture to 100 μl plasma, other cell-free body fluid or cells. Vortex for several seconds.
3. Add 100 μl chloroform to each specimen and vortex for several seconds. Place on ice for 10 min.
4. Centrifuge this mixture for 10 min at 10,000 to 15,000 × *g*.
5. Carefully push disposable pipette tip below the resultant protein interface layer and remove as much of the bottom liquid as possible. Discard this liquid.
6. To aqueous solution recovered above, add an equal volume (should be about 100 to 150 μl per specimen) chloroform. Vortex for several seconds and place on ice for 10 min.
7. Centrifuge at 10,000 to 15,000 × *g* to produce liquid/liquid interface.
8. Carefully recover top (aqueous) phase again, as in step 5 above.
9. Place recovered aqueous phase in a new, sterile polypropylene microcentrifuge tube; add 1 to 2 μl (10 to 20 μg) glycogen. Mix well.
10. Precipitate RNA by adding one-half volume cold 7.5 *M* ammonium acetate. Mix well.
11. Add two and a half volumes cold ethanol. Mix, spin at 14,000 rpm for 5 min. Dry the pellet, dissolve it in autoclaved ddH_2O treated with diethylpolycarbonate (DEPC) before the reverse transcription.

2. Single tube reverse transcription and gene amplification

1. Make a master-mix for reverse transcription and gene amplification in a single tube containing the following components in 50 μl: an aliquot of purified RNA (from cells, tissues, or body fluids), 50 m*M* KCl, 10 m*M* Tris-HCl (pH 8.3), 1.5 m*M* $MgCl_2$, 0.1% (w/v) gelatin, 1 μ*M* of each primer (see Table 7.1), 200 μ*M* of each dNTP, 1 unit Taq DNA polymerase (Perkin-Elmer), 2.5 units AMV reverse transcriptase (Boehringer Manheim, Indianapolis, IN), and 6 units RNasin (Promega, Montgomeryville, PA).
2. Test controls of HIV or HTLV RNA copy numbers (10^1, 10^2, 10^3, 10^4, 10^5, 10^6 RNA copies) using identical protocol.
3. Program the PCR cycler as follows: 42°C for 45 min followed by 94°C for 30 sec, 55°C for 30 sec, 72°C for 30 sec for 35 cycles, then allow it to run 10 min at 72°C.

3. Amplicon detection

1. Amplicons are subjected to liquid hybridization with a 5′ end-labeled [^{32}P] oligonucleotide probe. If labeled primers are used, this step is unnecessary.
2. Hybridization products are separated on 11% polyacrylamide gel by electrophoresis (PAGE) at 250 to 300 V for 30 min.
3. Wrap the gel in Saran Wrap and expose to the X-ray film directly for 2 to 16 h.
4. Develop the film.

4. Analysis of results

The amount of RNA molecules is determined by the absorbance at 260 nm and confirmed by comparing the intensity of ^{32}P-labeled RNA transcripts with that of the known copy numbers of viral RNA or DNA. Proviral DNA can be prepared by titrating DNA of a cell line containing a single copy of the HIV genome per cell. To establish the number of HIV viral RNA molecules in patient samples, serial dilutions of HIV of known titers and HIV RNA copy numbers (10^1, 10^2, 10^3, 10^4, 10^5, 10^6 RNA copies) are processed as standards with samples.[62] Quantitations of HIV-1 specific RNA in plasma are based on the comparison of intensities of the autoradiographic signals from patient samples with those derived from parallel analyses of serially diluted standards. Specific primers are used to amplify one of the three different housekeeping genes (HGPRT, β-globin, and β-actin genes) as controls for RNA quantity.

G. Prevention of Contamination of PCR Product or Amplicon "Carry-Over"

The PCR is routinely used in many research laboratories. One of the primary concerns of using the PCR is to distinguish false positive samples from the true positives. PCR primer pairs have been designed that can determine if a positive result originated from the intended target DNA sequence or from the so-called "carry-over" contamination of the previously amplified DNA. In our laboratory we have physically isolated the PCR facility, which is equipped with a biosafety hood containing a UV-germicidal lamp to destroy DNA left behind. The blood is handled in a safety cabinet solely dedicated for this task and reactions are set up in another room. Separate areas are also dedicated for the synthesis of oligonucleotides and gene amplification. To avoid cross-contamination of samples, all pipetting operations are conducted using Gilson pipettors fitted with "aerosol-resistant tips," or positive displacement pipettes. Further, the use of appropriate positive, negative, and "no DNA" controls is important for identifying any false positive samples.

To avoid "carry-over" contamination of clinical specimens, the enzyme uracil *N*-glycosylase (UNG),[52] which catalyzes the destruction of uracil-containing DNA, but not thymine-containing DNA, can also be added to the reaction mixtures. Since uracil is not present in the natural DNA, but is present in the amplicons due to the incorporation of uracil in place of thymine as one of the dNTPs in the reaction mixture, the UNG catalyzes the cleavage of an oligonucleotide at a uracil base by opening the deoxyribose chain at position I. When the reaction mixture is heated for thermal cycling, the alkaline pH of this solution causes the amplicon DNA to break at positions where uracil was incorporated. Thus, the amplicon cannot contaminate the other samples and be reamplified.

H. Quantiplex RNA Assay or Branched DNA (bDNA)

The bDNA assay has been developed by the Chiron Corporation, Emeryville, CA. Unlike the PCR assay in which the target DNA or RNA sequence is amplified, the bDNA technique amplifies the detection signal. The starting material can be RNA or DNA isolated from cells or virions and the assay is performed according to the protocol provided by the manufacturer.

The test developed for HIV RNA is called Quantiplex and it is based on capturing the target nucleic acid sequence in a microwell by using a set of specific oligonucleotide probes that bind to the *pol* gene of the viral RNA. To ensure that only HIV-specific sequences are hybridized, a second HIV-specific probe is used to hybridize to the captured viral RNA and the bDNA amplifiers. Next, multiple copies of an alkaline phosphatase-labeled probe are hybridized to the immobilized complex and signals are detected by incubating with a chemiluminescent substrate and measuring the light emission in a luminometer. A curve is generated from standards with known luminescent counts, and concentrations of HIV RNA in specimens are computed from the standard curve. This test is useful for quantitating viral RNA load in clinical specimens. Recently, investigators in the AIDS Clinical Trials Group (ACTG) compared bDNA and Roche RNA quantitation method (RT-PCR) by Amplicor HIV Monitor™, and found both to be equivalent.[68]

I. Nucleic Acid Sequence Based Amplification (NASBA)

A new technique, called the nucleic acid sequence based amplification (NASBA), has been developed by Organon Teknika Corporation, Durham, NC, for the detection and quantitation of RNA. The NASBA assay is performed according to the instructions provided by the manufacturers. The principles of this assay are as follows.

The NASBA system relies on the simultaneous action of three enzymes: reverse transcriptase (RT), RNase H, and T7 RNA polymerase.[63,64] Although both RNA and DNA can be used as starting material, only RNA is cyclically amplified. If the starting material is RNA, a ssDNA is produced by using RT and a primer containing the T7 promoter. After treatment with RNase H to degrade the RNA strand, a second primer is used to synthesize a dsDNA. Next, the dsDNA is denatured and RNA is synthesized by the T7 RNA polymerase in cyclic manner, i.e., newly synthesized DNA to RNA. The reaction is continuous, homogeneous, and isothermal, i.e., at a constant temperature of 41°C. Further, the entire reaction takes place in a tube without the need for specialized equipment.

The detection of amplified RNA is performed manually by using ^{32}P-labeled probes or by automated systems using electrochemiluminescence technology. The amplified product can also be detected by using an enzyme-linked gel assay based on specific hybridization with a horseradish peroxidase-labeled probe. The hybridized product is separated by polyacrylamide gel electrophoresis. Computer controlled software provided by the manufacturer yields quantitative results.

J. *In Situ* Hybridization

Although the *in situ* hybridization technique was described almost two decades ago using antigen/antibody reaction,[65] only recently have the protocols been standardized for the use of molecular probes.[66,67] The *in situ* hybridization technique has the advantage of detecting both latently and productively virus-infected cells, as well as identifying the phenotype of these cells (T cells [CD4, CD8], B cells, monocytes, etc.). For a number of years, our laboratory has standardized *in situ* hybridization using RNA probes and we have developed a sensitive *in situ* RNA detection technique to quantitate the number of HIV-infected cells in the blood of patients (Figure 7.1).[17,69] However, this technology is not as sensitive as other standardized virologic procedures currently available and many problems continue to interfere with the sensitivity of detection, necessary for quantitation of virus load and for the identification of cell types simultaneously.

Specific cell types infected by the virus can be identified by cellular morphology and by using monoclonal antibodies (mAb) to CD4, macrophage/monocytes, granulocytes, Langerhans cells, or other cell types (Dako, Carpinteria, CA). To identify cell types, cells are first incubated overnight with monoclonal antibody for 18 h, washed, and treated with biotinylated rabbit antimouse antibody (Zymed Lab, San

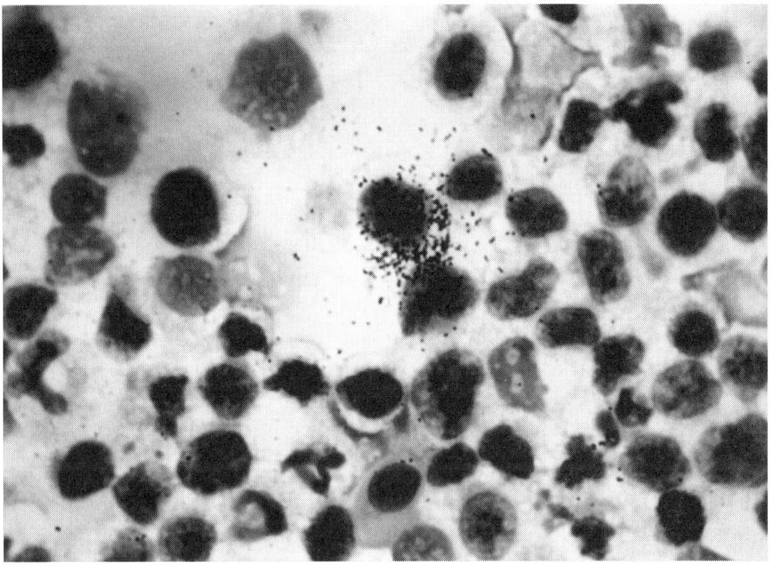

FIGURE 7.1
In situ hybridization of peripheral blood from a patient with AIDS. A 9 kb HIV proviral DNA fragment was inserted in 3′ to 5′ orientation in transcription vector pSP64 and transcribed *in vitro* using ^{35}S-UTP and T7 polymerase. Cells were hybridized on the slide and stained with hematoxyline.

Francisco, CA). The primary and secondary antibody complexes are detected by reacting with streptavidin-horseradish peroxidase with a 3′-3′-diaminobenzidine/H_2O_2 solution as the substrate. To intensify staining, cells are treated for 10 sec with 1% osmium solution and counter-stained with 1% hematoxylin. Following is a stepwise procedure for *in situ* hybridization.

1. Assay procedure

1. Tissue sections or single cell suspensions can be used for *in situ* hybridization.
2. Single cells are spread evenly on the slides or coverslips by cytospin and fixed in freshly prepared RNase-free 2% paraformaldehyde at 4°C for 10 min and then washed with PBS and stored at –70°C until used. For better adherence, glass slides or coverslips are treated with 0.2 *M* HCl for 30 min at 70°C, washed in SSC (1X SSC = 0.15 mol/l sodium chloride and 0.015 mol/l sodium citrate, pH 7.2) for 15 min at 37°C and ethanol/ether for 1 min. Slides are then coated with 0.1% gelatin or histologic glue (ORTHO, Raritan, NJ).
3. For *in situ* hybridization, rehydrate cells on slides in PBS and treat sequentially with 10 m*M* dithiothreitol (DTT) in PBS at 45°C for 30 min, 10 m*M* iodoacetimide at ambient temperature for 10 min.
4. After washing twice with PBS, place slides in 0.5% acetic anhydride in 0.1 *M* triethanolamine buffer, pH 8.0.
5. After 10 min, wash slides again in saline sodium citrate (SSC).
6. Prehybridize cells in solution containing 4% deionized formamide, 2X SSC, Denhardt's solution (0.02% [w/v] Ficoll, 0.02% polyvinylpyrrolidone, 1 mg/ml bovine serum

albumin) in 10 mM Tris-HCl (pH 7.5), 1 mM EDTA, and 600 mM NaCl, 10% dextran sulfate, 50 mM phosphate buffer, 50 mM DTT, 500 μl/ml polyadenylic acid, 100 μg/ml of sonicated salmon sperm or calf thymus DNA and Randomer (Dupont, Boston, MA), 0.05 μM/ml.

7. After 1 h, remove excess solution and add virus-specific RNA, cloned DNA, or oligonucleotide probe. Synthesize RNA probes as follows:
   a. Use at least 1 μg of cloned plasmid DNA template.
   b. Incubate template for 2 h at 37°C in the reaction mixture containing transcription buffer (0.4 M HEPES-KOH, pH 7.5, 60 mM $MgCl_2$, 10 mM spermidine, and 20 mM DTT), 25 mM ATP, CTP, and GTP. Add only 0.5 mM uridine triphosphate. Add alpha [^{35}S]uridine triphosphate 600 to 60 μCi, RNA polymerase SP6 30 units (SP6 can be replaced by T3 or T7 polymerase). Our laboratory has used RNA of an HIV clone (pSP64 plasmid) using [^{35}S]uridine triphosphate and SP6 polymerase. This *in vitro* transcribed probe yields excellent sensitivity with approximately 4% positive in uncultured, freshly obtained PBMCs from HIV-seropositive individuals.[69] We have also used digoxigenin-labeled dUTP probes with similar results.[14]
   c. Add 2.5 units of RNase-free DNase-I.
   d. Incubate for an additional 10 min.
   e. Stop reaction by the addition of EDTA (0.1 M).
   f. Remove enzyme by phenol-chloroform (1:1 v/v) extraction. (Vortex to mix and centrifuge to separate organic and aqueous phases.)
   g. Transfer aqueous phase to new tube and add 10 μl of 5 M ammonium acetate and 300 μl cold ethanol.
   h. Precipitate at –20°C for 1 h or leave in freezer overnight.
   i. Pellet RNA by centrifugation (highest speed in the microfuge).
   j. Wash pellet with cold ethanol to remove salt.
   k. Remove ethanol and resuspend dried pellet in DEPC-treated autoclaved water with 0.1 mM EDTA.

Note: *RNA should not be stored in aqueous solution. To save excess RNA, reprecipitate using ammonium acetate and cold ethanol. Store at –20 or –70°C.*

8. Hybridize with the RNA probe at 42° to 50°C for 3 h or overnight.
9. Rinse cells in 50% formamide and 2X SSC at 52°C, followed by digestion of the probe with RNase at 37°C for 30 min.
10. Rinse slides again, with 2X SSC, drain and immerse in Kodak NTB-2 emulsion. Drain again in a dark container overnight and expose to film for 24 h or longer.
11. Dip slides in Dektol (Kodak) for developing, dry, and stain with Giemsa stain.

K. PCR *In Situ* Hybridization

The conventional *in situ* hybridization methods are not sufficiently sensitive to quantitate HIV- or HTLV-infected cells in blood or tissue sections. Although PCR is an extremely sensitive method, it does not provide information at a single cell level and cytologic and histologic features cannot be correlated directly with PCR data. Thus, an ideal detection system would be to combine PCR with *in situ* hybridization.[70-72] This technique has been developed in several laboratories, but the sensitivity of HIV-positive cells in freshly obtained blood tissue sections from patients is lower than the regular PCR or virus culture methods.[72a]

The *in situ* PCR technique is based on the amplification and detection of target sequences in cells adhered to glass slides. Both paraffin-embedded archival histologic material and single cell suspensions can be used. Briefly, cells are fixed on slides or coverslips as described for the *in situ* hybridization (Section V.J). The target DNA is amplified using Taq DNA polymerase and virus-specific primers in the reaction mixture as described for PCR above (Section V.C and Table 7.1), except that the slides are placed in an aluminum "boat" which can be placed directly on the heat block for gene amplification.

Digoxigenin (dUTP) has been used for labeling amplified DNA according to manufacturers' recommendations using alkaline phosphatase-conjugated anti-digoxigenin antibody.[66] Viral DNA or RNA can be detected using specific probes for HIV or HTLV. Immunohistochemical identification of cells after PCR manipulation *in situ* is difficult and the cellular integrity is lost in most cases.[72a] It is therefore recommended that the cells are stained first for cell surface antigens followed by PCR amplification of viral DNA. Because of the high virus burden in lymph nodes, the PCR *in situ* hybridization technique appears to be quite sensitive for the identification of HIV-infected cells in these tissues, but its use in the quantitation of virus load directly in the blood and other cell types has many limitations. Compared with quantitative cell culture and PT-PCR, the *in situ* PCR is less sensitive.[72a] Several commercial companies are currently developing instruments to automate many labor-intensive procedures of the *in situ* PCR hybridization technique. However, until the sensitivity and specificity of this test is increased, its use in the quantitative evaluation of HIV- or HTLV-1-infected cells in clinical specimens appears limited.

VI. Human Retroviruses in Clinical Specimens Other Than Blood

In addition to the blood, HIV can be isolated from a large number of tissues, including brain, lymph nodes, skin, heart, lung, kidney, gastrointestinal epithelium, seminal fluid, cerebrospinal fluid, cervical secretions, and many other tissues[17,73,74] (Table 7.2). In tissues such as the brain, the virus may enter a cell but not replicate because it does not divide or due to other host cellular and humoral factors. Latently infected cells are present in most organs and tissues *in vivo* and the presence or absence of retroviruses in these tissues can be detected using DNA- or RT-PCR (Table 7.2). Since the majority of T cells and monocytes/macrophages are nondividing *in vivo,* the persistent infection may be lifelong. If the nonproductive and persistently infected cells are activated *in vivo* or *in vitro,* virus production ensues and it continues its life cycle. However, genetic and phenotypic variations in viruses present in blood and tissues of HIV-infected people and the relationships between pathogenic potential of these strains locally in various organs and that present in the blood are not well understood.[74]

TABLE 7.2
Detection and Isolation of HIV from Clinical Specimens

Specimen Type	Methodology	Ref.
Tissues		
Bone allograft	PCR	126,127
Bone marrow	Culture, PCR	127–129
Bowel epithelium	PCR	130,131
Brain	Culture	132
Cutaneous blister fluid	Culture	133
Conjunctival tissue	Electron microscopy	134
Lymph node	PCR	90,135,136
Retina	Immunohistochemistry	137
Testes/prostate	PCR, immunohistochemistry	138,139
Body Fluids		
Whole blood (cells and plasma)	Culture, PCR	Numerous, 14,17,46,140
Breast milk	Culture	141–143
Bronchoalveolar lavage	Culture, PCR	17,144,145
Cerebrospinal fluid	Culture, PCR	17,146
Cervicovaginal secretion	Culture, PCR	74,147,148
Cutaneous blister fluid	Culture	149
Dried blood spot	PCR	150
Ear secretions	PCR	151
Saliva	Culture, PCR	18,152
Seminal fluid/cell	Culture, PCR, *in situ*	17,73,138,153,154
Serum/plasma	Culture, PCR	Numerous, 73,155
Tears	Culture	156
Urine	PCR	157
Others		
"Wastewater"	PCR	158
"Syringes"	PCR	159

To study the biological or biochemical properties of the virus, it must be grown in susceptible cells *in vitro*. Human T cells and monocytic cells are the most susceptible cells for both HTLV- and HIV-related viruses. Since most of the HTLV-1 remains cell-associated or latent, its detection in most clinical specimens has been by PCR only.[53,75-80]

VII. Morphological Characterization of Human Retroviruses

Although morphological characterization of the known human retroviruses is not essential for diagnosis, all new viruses must be characterized morphologically when initially isolated. Electron microscopy is useful for morphological characterization of virus-infected cells and for the identification of some antigenic determinants by immunogold staining. The progress in molecular biology of human retroviruses has been instrumental in the development of antiviral drugs and potential vaccines. However, electron microscopy is still the only tool to visualize the morphology of the virus particles (Figure 7.2). This technique is also useful for quantitative analysis or estimation of the number of physical virus particles present in a cell or fluid by negative staining of viral pellets.

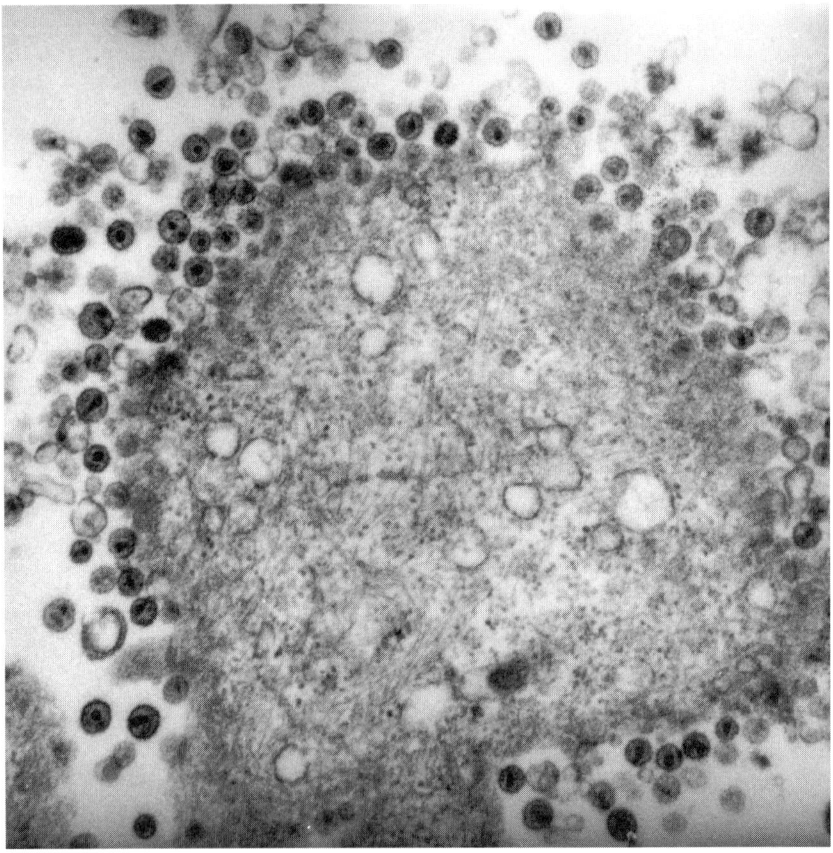

FIGURE 7.2
Electron micrograph of a new HIV-1 strain of subtype A isolated from Nigeria.

VIII. Biological Characterization of Human Retroviruses

Both the HTLV and HIV groups of human retroviruses grow preferentially in stimulated human T cells. Discovery of T cell growth factor or interleukin-2 (IL-2) produced by the activated T cells[81] has been instrumental in the isolation of the first human retrovirus, the HTLV-1, from patients with adult T cell leukemia.[2] Similarly, almost all strains of HIV-1 or HIV-2 have been successfully grown in activated human T cells, monocytes, neuronal cells, and Langerhans cells.

High HIV burden or viral load in the blood of HIV- or HTLV-1-infected individuals has been related directly to the replication kinetics of the virus and disease progression. A low virus load is correlated with nonprogressive HIV infection. However, the replication kinetics of both HTLV and HIV depend on a variety of parameters, including the phenotypic and genotypic properties of the virus, immune responses of the host, and cellular factors produced before and after virus infection. The dynamics of HTLV replication *in vivo* are not well understood and infection of new cells by HTLV-related viruses is much more efficient by cell–cell contact than by cell-free virus infection. In contrast, both cell-associated and cell-free HIV isolates can infect new cells efficiently *in vitro* or *in vivo*. Depending on the host cell that a virus infects, several phases of infection can be identified in both HTLV and HIV groups of viruses using a variety of *in vitro* techniques:

1. Steady-state: The virus enters the cell, integrates, replicates, and "lives" in harmony or in "symbiosis" with the infected cell. Thus, both cell and virus replicate and survive at a steady state. The state of virus infection can be assessed by culturing the virus-infected cells and isolating virus *in vitro*.
2. Latent infection: Virus enters the cell, integrates in the host chromosomal DNA (or may even be present as an episomal DNA). This virus may not replicate or may replicate transiently because of changes in the internal and external cellular milieu (i.e., due to activation, stimulation of DNA synthesis by cytokines, growth factors, etc.). The latent virus infection can be identified by PCR as described in Section V.C.
3. Cell fusion and syncytia induction: The HIV-1 envelope proteins gp120/41 have epitopes similar to those present on activated CD4+ cells. Abundance of activated or CD4+ cells induces fusion of cell membranes. This process can be recognized with or without virus replication in the CD4+ cells (see Section VIII.E. for syncytium-inducing assay).
4. Cytopathic effects: Single cell death and apoptosis are commonly observed in many HIV-infected cells. Cytopathic effects are visualized microscopically and the apoptotic process can be analyzed by examining DNA.[35-38]

A. Isolation of Human Retroviruses *In Vitro* (Qualitative Cultures)

Although a virus can be cultured or isolated from a large number of tissues or body fluids, the most convenient source of HIV or HTLV is the peripheral blood mononuclear cells (PBMCs) from virus-infected individuals. The latent, defective, and transiently or fully replicating viruses are detectable by PCR; virus culture is the only test that can detect the *in vitro* infectivity of the virus, qualitatively and quantitatively. Consequently, this test will always remain a "gold standard" for the detection of infectious virus, although its sensitivity varies according to the stage of the disease and treatment regimens of the HIV-infected individuals.

For virus culture or virus isolation, freshly harvested PBMCs from HIV-seronegative donors are used. Leukocyte-enriched blood collected by leukophoresis at a blood bank is useful when large numbers of patient samples are processed in a laboratory. For culture of HTLV-related viruses, umbilical cord blood is most suitable. Cells used for co-cultivation must be negative for HIV and HTLV-1 genes by PCR. The following protocol has been standardized in this laboratory and is used routinely for virus isolation from the blood or other cells.

1. Preparation of phytohemagglutinin-stimulated donor cells

1. Collect donor blood (peripheral blood, umbilical cord, or leukophoresis) in 10- to 15-ml vacutainers containing acid-citrate-dextrose (ACD) as anticoagulant. Other anticoagulants such as heparin can also be used to collect blood, but heparinized plasma is not suitable for use in PCR unless treated with heparinase or the virus particles are first pelleted by ultracentrifugation.
2. Separate cells and plasma from the whole blood by centrifugation for 10 min at 400 × *g*.
3. Transfer plasma to a new tube and centrifuge again to remove platelets. Aliquot plasma in small tubes and freeze at –70°C for future use.
4. Dilute blood cells (buffy coat) with saline or PBS in a volume equal to the plasma removed.
5. Layer the diluted blood over the Ficoll-Hypaque solution or the lymphocyte separation medium (Organon Teknika, Durham, NC, or Pharmacia, Milwaukee, WI) and centrifuge at 400 × *g* for 20 min. If CPT tubes (citrated plasma separation tubes from Becton and Dickinson Labware, Bedford, MA) are used for the collection of blood, follow manufacturer's instructions for separation of PBMCs.
6. Collect PBMCs from the interface and wash with PBS or Hanks balanced salt solution without Ca^{2+} or Mg^{2+}. Cells are enumerated manually in a hemocytometer or in an automated cell counter (Coulter, Hialeah, FL) and used for culture real time. Alternately, fresh PBMCs can be cryopreserved in medium containing 80% fetal bovine serum, 10% RPMI 1640, and 10% DMSO for future use (freeze slowly, drop temperature approximately 1° per minute).
7. Use freshly separated PBMCs (9 to 10 × 10^6) from step 6 above, to stimulate in a culture flask for 24 to 48 h in 10 ml of complete medium containing RPMI 1640, 15 to 20% fetal bovine serum, 2 m*M* glutamine, 5% delectinated human IL-2 (natural, not recombinant), 50 μg/ml gentamicin, and 5 μg/ml phytohemagglutinin-P (PHA-P) (Difco, Detroit, MI or Sigma, St. Louis, MO).

2. Co-cultivation of stimulated donor cells with patient's cells

1. Collect patient's blood (10 to 15 ml), isolate PBMCs as described above and mix ≥5 × 10^6 patient's PBMCs in a 25-cm^2 flask (T-25) with approximately the same number of PHA-prestimulated PBMCs from a seronegative donor and co-cultivate (i.e., culture together) in complete medium as described in step 7 above. The number of patient cells used for co-culture is not fixed and as many as 10 × 10^6 can be used in a T75 flask. However, in newborn infants it is difficult to get sufficient blood and therefore the number of co-cultivated cells can be adjusted accordingly.

Note: *On the first day of co-cultivation, 2 μg polybrene is also added to the complete medium to facilitate virus adsorption to stimulated cells.*[17] *Complete medium consists of RPMI 1640, 15 to 20% fetal bovine serum, 2 mM glutamine, 5% human IL-2, and 50 μg/ml gentamicin. The use of antibiotics in our laboratory is discouraged, but when culturing patient samples, it is preferred to use gentamicin (50 μg/ml) than penicillin and streptomycin, which are slightly inhibitory to the growth of the lymphocytes.*

2. Incubate cultures at 37°C in an atmosphere of 5% CO_2 in air.
3. After 2 or 3 d (or twice a week), carefully remove half of the culture medium from the top without disturbing the cells at the bottom of the flask and add, as described in step 7 above, fresh medium. In addition, once a week add freshly prestimulated HIV-seronegative donor PBMCs (5 × 10^6 cells) with complete medium to maintain a continuous viral growth in these cells.
4. After 7 d (5 × 10^6) test culture supernatants for reverse transcriptase (RT) activity, indicative of the presence of a retrovirus (HTLV or HIV). If the source of the material is expected to be from a HIV-seropositive individual, test culture supernatant for HIV p24 antigen.
5. If culture supernatants are negative for HIV p24 and RT activity, test culture supernatant again after 14 to 28 d, while maintaining the media change twice a week and addition of prestimulated PBMCs once a week as in step 3. If cultures are HIV p24 positive at 7 d, repeat the test again after 14 d. The values of p24 at day 14 should double (or quadruple) the p24 values on day 7. This will be an indication that virus is replicating and virus has been "isolated" *in vitro*.

Although a negative culture may be truly negative for the infectious virus at day 14 or 28, in rare cases virus has been detected after 30 to 35 d of culture.[161] If only the RT assay is positive (i.e., p24 negative and RT positive), it is highly likely that the virus is HTLV-1 or another related human retrovirus. Since some latent viruses may not be expressed in cultured patient cells, these should be tested by another molecular test, such as PCR, to confirm the presence or absence of viral genes.

B. Quantitation of Cell-Associated Virus Load (Quantitative Cell Culture)

An important consideration in the pathogenesis of AIDS is the viral load in the blood or other tissues of HIV-infected individuals. Whereas the frequency of HIV-infected cells isolated from lymph node tissues of patients in early stages of HIV infection are similar to those found in the blood of symptomatic individuals, the virus titers, also called virus load or virus burden, in the PBMCs is reversed. Thus, the virus load in symptomatic individuals is higher than in asymptomatic people.

The method by which the cell-associated virus load is assessed in the peripheral blood is by co-cultivating serial fivefold dilutions of PBMCs with a constant number of PHA-stimulated PBMCs from seronegative donors. The protocol used for evaluation of virus load is as follows.

1. Isolate, wash, and enumerate patient PBMCs as described above and resuspend in complete medium at a concentration of 2×10^6 cells per milliliter. Approximately 4 to 5×10^6 cells are required for quantitation. Although stimulated PBMCs are sufficient for the growth of most human retroviruses, mononuclear cells from the umbilical cord are superior, particularly for HTLV-1.
2. Starting with 1×10^6 cells per milliliter in the first dilution, make six fivefold serial dilutions of patient cells in complete culture medium as described above. Add duplicate samples of each dilution (i.e., two wells, each), in the 24-well culture plates. If fewer cells are available, the first dilution of 1×10^6 can be tested as singleton.
3. Add an equal number (1×10^6) of PHA-prestimulated PBMCs from seronegative donors in an equal volume (approximately 1 ml) of complete medium in each well. *Note* the total volume of medium in the co-culture (i.e., patient cells plus donor cells) should not exceed 2.4 ml per well in a 24-well plate.

Note: *If a larger number of cells are to be tested, larger plates or T-25 flasks should be used. If only a small number of cells are available, such as from some newborn pediatric patients, 48-well plates are recommended. The volume of the medium should then be adjusted according to the capacity of the well (approximately 1 ml). As long as the number of patient cells cultured in each dilution is known, the virus-infected cells can be quantitated or titrated against a constant number of prestimulated donor PBMCs.*

4. Test supernatant from each of the duplicate wells for reverse transcriptase activity for HTLV, or the HIV p24 antigen if HIV seropositivity is known or suspected.
5. HIV-1 p24 is quantitated using any of the several commercial kits available from Abbott, Coulter, Dupont, Organon Teknika, etc. Each manufacturer has a computer program that allows quantitation of the p24 values in the culture supernatant and a value of ≥30 pg/ml in culture fluid is indicative of HIV replication. Although the positive or negative p24 measurements depend on the cut-off values defined by each manufacturer, the values of p24 antigen obtained by different kits can be standardized using various algorithms and computer programs available to laboratories participating in the AIDS Clinical Trials (ACTG).[27]

The sensitivity of the HIV culture test in symptomatic individuals who are not treated with antiviral drugs is almost 100%.[17] In most cases virus can be detected *in vitro* within 3 to 5 d, and >95% of cultures are positive by 10 to 14 d, unless the subject is being treated by antiviral drugs.[17,72a]

C. Quantitation of Cell-Free Virus (Plasma Viremia)

The level of plasma viremia is a useful marker of HIV disease progression, because it increases over time in symptomatic or mildly symptomatic patients rather than in AIDS. This method of virus isolation and quantitation is not suitable for evaluation of HTLV-1 infections because most of the infectious virus is cell associated. Since plasma virus comes from various organ systems into the blood circulation, its increase in the blood indirectly implies an augmentation of the number of infected cells throughout the body. Thus, although the absence of virus in the plasma or blood cells in response to treatment represents a change in the virus load, it may or may not correlate with the clinical improvement. Further, assessment of HIV burden by plasma culture is not as sensitive as the evaluation of viral RNA load by RT-PCR, bDNA, NASBA, or Roche's Amplicor HIV Monitor tests.[73,74]

Several cell culture methods have been evaluated for the isolation and quantitation of HIV from plasma. Comparison of the flask cultures or macroculture to microculture systems in 96-, 48-, and 24-well plates indicated that both 48- and 24-well plates are useful and cost-effective, but cells can be maintained for long periods (up to 60 d) in 24-well plates.

The following protocol is used in our laboratory for the quantitation of plasma HIV by culture:

1. Draw whole blood (10 to 30 ml) in tubes containing acid citrate, dextrose (ACD) or citrate plasma separation tubes (CPT) containing lymphocyte separation medium (Becton and Dickinson, Bedford, MA).
2. Separate plasma immediately or within 4 h by centrifugation at 400 to 800 × *g* for 20 min at ambient temperature (20 to 24°C).
3. Transfer plasma to a new centrifuge tube and centrifuge again at 400 × *g* to remove platelets and cellular debris.
4. Culture plasma fresh or dispense in small aliquots of 500 μl to 1 ml and freeze at –70°C for future testing. To determine an accurate plasma virus titer by tissue culture, it is preferable to test freshly collected plasma.
5. Any volume of plasma can be tested for the presence of infectious virus particles in cultures as long as it is not toxic to the prestimulated lymphocytes. We have tested 1 μl of plasma obtained from newborns to 1 ml from adults per 24-well plate. Since greater volumes of some human plasmas can be toxic to human lymphocyte cultures and small volumes such as 1 μl may not be sufficient to culture HIV from most plasma samples, we typically use 400 to 500 μl to culture virus.
6. Prepare prestimulated PBMCs in medium containing PHA, as described earlier for virus isolation (see Section VIII.A.1). Culture medium consists of RPMI-1640 with

20% heat-inactivated fetal bovine serum, 2 m*M* glutamine, 5% natural (delectinated) IL-2, and 50 μg/ml gentamicin.

7. Add approximately 2×10^6 cells in each of the 24-well plates in 1.5 ml complete medium containing 2 μg/ml polybrene and add 500 μl of clarified plasma in each of the two wells as the first dilution. Make serial fivefold dilutions of the same plasma aliquot (1:5 to 1:3125 or 1:6250) and add 500 μl of each of the plasma dilutions in two wells each (i.e., test in duplicate).
8. Incubate plates in a plastic plate holder "Modular Incubator Chamber" (Billups-Rothenberg Inc., Del Mar, CA) and place it in a 37°C incubator with 5% CO_2 atmosphere.
9. After 24 h, remove half of the medium and replace with fresh plasma culture medium as described in step 6 above. In case there is any evidence of cytotoxicity or cell death in wells containing highest volumes of plasma, then add 1×10^6 freshly stimulated PBMCs to these wells and all other wells to maintain identical growth conditions. This step is necessary to allow virus released from "dying" cells to be taken up by newly added, prestimulated cells in the medium.[160]
10. Continue incubation as in step 8. Change one half of the medium twice a week and add freshly stimulated 5×10^5 PBMCs once a week.
11. On the 14th day, harvest culture supernatant from each well and test for the presence of the p24 antigen.
12. Culture supernatants from plasma dilutions testing positive against standardized p24 antigen are scored positive.
13. The infectious titer is expressed as the cell culture infectious unit (CCIU), tissue culture infectious dose (TCID), or simply as infectious units per milliliter of undiluted plasma tested. An estimated titer is expressed as the reciprocal of the highest dilution containing at least one infectious unit and showing positive p24 antigen levels.

D. Host Range Specificity or Cell Tropism of Human Retroviruses

The natural host of HIV is the global human population, but the HTLV-related viruses have a somewhat restricted demographic distribution globally.[9]

An important phenomenon that affects the pathogenicity of human retroviruses relates to the rate of replication and their preferential choices of specific cell types in which to replicate (i.e., cell- or tissue-tropism or "host range"). Although both the HTLV and HIV groups of viruses preferentially infect T cells bearing the cell surface cluster of differentiation antigen No. 4 (i.e., CD4+ cells), both viruses have also been shown to replicate in cells that lack CD4+ marker.[82-87]

The changes in the tropism or the host range of a virus is a complex biological phenomenon, which is affected by a variety of changes in the intracellular and intercellular environment *in vivo*. Macrophages and monocytes are the most readily infected cells *in vivo*. HIV replication is upregulated or suppressed in these cells by cytokines produced by other HIV-infected or uninfected cells *in vivo*. In addition to viral regulatory controls, genetic variations, conformational changes in the proteins due to specific mutations, or other factors are important for virus replication and pathogenesis. These changes are exhibited by HIV biological phenotypes, in its ability to grow, induce cytopathic effect, or be neutralized by serum antibodies.

Epidermal Langerhans cells have been shown to be permissive to HIV infection.[88,89] Langerhans cells originate from bone marrow and then migrate into the peripheral epithelium of the skin and mucous membranes. These cells are present in the genital and rectal mucosa and they may be the first cell types to be infected by HIV *in vivo*. The role of lymphoid organs in HIV pathogenesis is also well established.[90] The involvement of lymph nodes is particularly important in the early phases of HIV infection or asymptomatic individuals when the number of peripheral blood cells appears to have a fewer number of infected cells compared to the symptomatic individuals. During this clinically latent period, HIV is sequestered in the lymph nodes, where it replicates actively until the dendritic lymph node tissue is destroyed and free HIV particles are released into the periphery, i.e., plasma.

The genome of HIV is complex and the antigenic determinants, particularly in the third hypervariable region (V3) of the external viral envelope glycoprotein (gp120), have been shown to have important epitopes for virus neutralization. After going through several replication cycles in a particular host cell type, the V3 epitopes of HIV change. These mutated viruses are not recognized by the antibodies present in the host and thus they escape neutralization. The "escaped" virus usually changes its host tropism and is capable of infecting other cell types, such as macrophages and astrocytes. Thus, viruses that are able to infect only T cells *in vitro* or *in vivo* can change the host range after mutations are induced in the envelope gene and spread the infection to different organs and sites. The opposite is also true; a virus that changes tropism will also escape neutralizing antibodies made against the unmutated V3 loop. HIV strains with broad host ranges (i.e., lymphotropic as well as monocyte/macrophage tropic) have been associated with high virulence (i.e., pathogenicity) *in vivo*. To evaluate the growth kinetics and macrophage tropism of HIV-1, we have established the following protocol.

1. Macrophage/monocyte tropism

An important consideration in the development of the adherent macrophage culture protocol in our laboratory, has been the fact that if PBMCs are plated in the PHA medium (see step 3 below), the growth factors, including macrophage colony-stimulating factor (M-CSF), produced from the nonadherent lymphocytes help in the differentiation of macrophages and the culture yields a good monolayer of mature macrophage/monocyte cells within a short period of 48 h. Since growth kinetics of viruses are studied in the 48-h PHA-stimulated PBMCs, our group prefers to use the same batch of PBMCs from which macrophages are isolated, as a study control for virus growth in adherent and nonadherent cells of the same individual. When using 48-h PHA-stimulated PBMCs for macrophage/monocyte isolation, the need for special supplemented medium is also eliminated for routine screening of cell tropism. However, studies addressing the role of macrophage differentiation factors involved in the regulation of molecules produced by HIV-infected or uninfected cells may require different techniques. Thus, for growth factor analysis our laboratory prefers to grow macrophage/monocyte cells in synthetic medium (AMES Medium,

Gibco BRL, Grand Island, NY) with no serum supplement, so that the effect of each factor can be evaluated independently without the collective influence of many growth factors, hormones, and peptides present in normal human or animal sera.[161]

Note: *Some investigators prefer to infect adherent macrophage cultures 7 to 14 d after plating PBMCs in the dish. This procedure allows time for the maturation of monocytic cells grown in medium containing M-CSF (1000 units per milliliter, Cellular Products, Buffalo, NY). In addition, pooled human serum from both blood groups type A and B is also added (10 to 15%) to supplement the growth of adherent cells. It is also the experience of some investigators that the yield of adherent macrophage/monocyte cells is relatively higher if no serum is used for the first 1 to 2 h in culture. Some laboratories use only 2 to 5% fetal bovine serum in the macrophage adherent medium. The following protocol is used in our laboratory:*

1. Collect human peripheral blood in tubes containing ACD or in citrated plasma/lymphocyte separation tubes (CPT; from Becton Dickinson Labware, Bedford, MA) or use leukophoresed blood from a blood bank.
2. Separate peripheral blood mononuclear cells (PBMCs) by Ficoll-Hypaque gradient centrifugation at 400 × *g* or by simply centrifuging blood collected in CPT tubes (400 × *g* for 20 min) according to the manufacturer's instructions provided with these tubes.
3. Wash PBMCs twice with Ca^{2+}- Mg^{2+}-free Hanks balanced salt solution and resuspend in complete growth medium (RPMI 1640, 15% fetal bovine serum, 5 μg/ml IL-2, 2 m*M* glutamine, 5 μg/ml PHA, and 50 μg/ml gentamicin).
4. Plate 4×10^6 PBMCs per 2 ml medium (as in step 3 above) in 24-well plates.
5. Incubate overnight at 37°C in 5% CO_2 atmosphere.
6. Aspirate medium together with all nonadherent cells from wells.
7. Wash each well of adherent cells with at least 2 ml of complete medium to remove any nonadherent cells.
8. Make six serial fivefold dilutions of stock virus and add each dilution (in duplicate) to wells. Use nonadherent PBMCs as controls. Starting dose of the virus should be at least 100 cell culture (or tissue culture) infectious unit or dose (CCIU or TCID). Culture adherent cells with virus dilutions in the presence of 2 μg/ml polybrene in the growth medium with FBS but *no IL-2,* as described in Section VIII.B. Addition of 10 m*M* HEPES buffer may also be used to stabilize pH in the medium, but it is not necessary when a CO_2 incubator is used.

Note: *Polybrene is added in the growth medium only on the day of virus infection.*

9. Incubate at 37°C in an atmosphere of 5% CO_2 in air.
10. Maintain cell growth by changing half of the medium twice a week. Do not add polybrene in culture medium.

Note: *Nonadherent cells (i.e., PBMCs) are grown in complete medium containing IL-2 and the adherent cells are grown without IL-2, with 20% FBS with or without M-CSF.*

11. Culture supernatants are harvested and tested for HIV-1 p24 antigen at day 7 and day 14 post-infection.
12. The kinetics of viral infection is calculated from the amount of HIV produced in successive virus dilution-infected cultures. Several dilutions of the known p24 standards are used to generate a standard curve using computer programs supplied by the manufacturers of various p24 kits. The amount of p24 antigen produced is calculated from the standard curve and the test values are corrected from the optical densities. The p24 antigen assay sensitivity of positive cultures by some tests is as low as 7.8 pg/ml p24 antigen (Coulter, Hialeah, FL), but 25 to 30 pg/ml of p24 has been suggested for evaluation of clinical specimens.

E. Phenotypic Characterization of HIV Strains

Phenotypic analysis of HIV-1 can be tested as follows:

1. Ability of HIV to induce cell fusion or syncytia formation in an HTLV-1 infected MT-2 cell line.[93]
2. Ability of the virus to induce cytopathic changes and/or apoptosis in individual cells.
3. Susceptibility or resistance to HIV infection (change in the host range and cell tropism, i.e., T cell, macrophage, monocyte, neuronal cell, dendritic cells, etc.).
4. Susceptibility or resistance to neutralization by naturally occurring antibodies or monoclonal or polyclonal antibodies made to distinct viral epitopes.

1. Syncytium-inducing assay

In addition to changes in host cell tropism, the generation of mutations in the *env* gene may also lead to phenotypic changes, such that the HIV-infected cells fuse with neighboring CD4+ infected and uninfected cells and form giant balloon-shaped syncytia. The syncytium induction (SI) property of HIV has been associated with higher replication rates and pathogenicity, and the non–syncytia-inducing (NSI) HIV strains have been related to low replication rates, macrophage tropism, and asymptomatic period of disease. However, these properties are not uniformly identified among many pathogenic HIV strains and both SI and NSI viruses have been isolated from patients with full-blown AIDS.[91,92]

A clonal cell line, which is nonproductively infected by HTLV-1 (designated MT-2),[93] has been shown to be susceptible to induction of syncytial plaques when exposed to certain strains of HIV or HIV-infected cells.

Although the significance of SI/NSI variants of HIV is not completely understood, production of HIV strains that are highly resistant to antiviral nucleoside drugs such as 2′-3′-deoxy 3′-azidothymidine (AZT or Zidovudine) have been associated with SI phenotypes.[92]

The following protocol has been standardized for the MT-2 syncytium assay in our laboratory:

1. Grow MT-2 cells in RPMI 1640 medium containing 2 m*M* glutamine and 15% FBS. These cells do not need IL-2 for their growth. The original MT-2 cells are T cell lymphoma tumor cells and therefore have a tendency to "clump". It is important to maintain these cells in the log phase of growth by subculturing 1:5 (or more) per week and using only those cells that form a monolayer when plated at a density of approximately 50,000 cells per well in flat-bottomed 96-well culture plates. Because frozen cells clump together, it is advised that when frozen stocks are used they should be allowed to go through several replicative cycles before using in MT-2 assay.
2. To maintain sufficient humidity and a balanced pH in all cultures, the first row of wells, both horizontal and vertical (i.e., all around), and one center row of wells should be filled with sterile PBS.
3. Two wells in each inner corner (i.e., in the second row) of the plate should be used for the known SI-producing virus as a positive control and the negative control well will have no virus. Thus, altogether four positive and four negative control cultures are used (i.e., one set of four wells in each inner corner of the plate).
4. Add (50 to 100 μl) HIV test samples (the amount of virus used depends on virus titer). The total volume of medium per well of the 96-well plate should not exceed 200 μl.
5. Incubate at 37°C in an atmosphere of 5% CO_2 in air in a module.
6. Less than a third to one half of the medium is removed on day 5 and is replaced with an equal amount of fresh, complete medium. This step is repeated on day 10 of the culture.
7. On days 2, 4, 7, 10, and 14, examine cells and note if syncytia formation has occurred. Some high syncytia-inducing strains can form multinucleated giant balloon-shaped cells within 24 h of exposure to the virus.
8. Terminate cultures on day 14. High syncytia-scoring cell cultures can be terminated earlier.

Note: *The MT-2 assay can also be performed in a 48-well plate. The number of plated cells should be increased to 100,000 per well and the total amount of culture medium plus virus inoculum should not exceed 800 μl per well. The schedule for media changes should be the same as described above and syncytia are scored as described in step 7 above. Syncytia and giant cell formation can also be seen in some HIV-infected T cell lines other than MT-2 cells (Figure 7.3).*

F. Virus Neutralization Assay or Neutralizing Antibody Test

An important parameter of all virus infections is that soon after virus enters the body the host immune system starts to "react" and produce different classes of antibodies (IgG, IgM, and IgA) against all viral proteins. These antibodies attach to the virus and can inactivate or neutralize it. The *env* gene products gp120 and gp41 are major

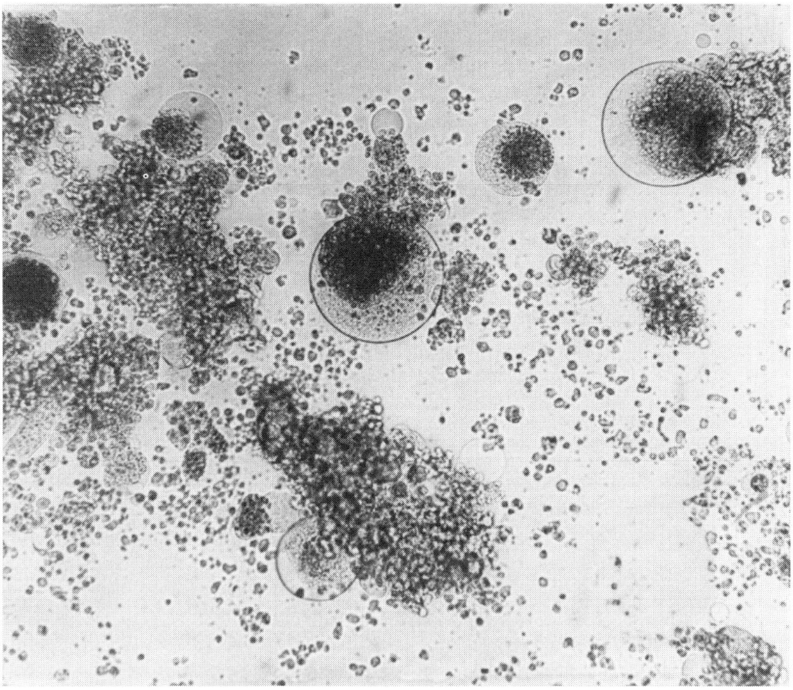

FIGURE 7.3
Syncytia induction and cytopathic effects in a human T cell line exposed to HIV.

targets that are involved primarily in virus neutralization. Individuals exposed to HIV produce antibodies initially but as the infection progresses mutational changes in the neutralizing epitope allow the virus to escape the neutralizing activity. Thus, testing for virus neutralizing antibodies in HIV-infected people will not provide clinically useful data.[94] However, this test becomes critical in patients who are given a vaccine or treated with immune-modulating agents. The neutralizing antibody test is also essential to identify monoclonal and polyclonal antibodies that may be able to neutralize different HIV isolates.[95-97] This test will also elucidate the relationship between neutralization of autologous virus and the emergence of viral variants that cannot be neutralized by the same sera.

The following protocol has been used in our laboratory:

1. Prepare quantities of virus stocks that can be used for all comparative experiments. The virus stocks must be titrated using the plasma viremia protocol described in Section VIII.C. This will determine how much virus to use for the neutralization tests. Freeze all virus in liquid nitrogen or at –70°C.
2. Prepare freshly PHA-stimulated PBMCs as described in Section VIII.A.
3. Inactivate complement by heating the test serum to 56°C. Many investigators believe that heating to 56°C is not necessary, but we prefer to heat because complement sometimes interferes with neutralizing activity and therefore it is helpful.

4. Using fivefold dilutions of the antibody to be tested, add 200 μl of each antibody dilution to an equal volume of virus stock containing 100 cell culture infectious units ($CCIU_{50}$ or $TCID_{50}$) in 5 ml culture tubes with appropriate labels to identify the dilution and the antibody used. Mix well and incubate for 1 h at 37°C.
5. Use normal (HIV-seronegative) human serum as a negative control and a known neutralizing antibody as a positive control serum. After 1 h incubation of virus-serum mixture, add 2×10^6 prestimulated PBMCs in 2 ml culture medium (RPMI 1640, 2 m*M* glutamine, 20% fetal bovine serum, 5% natural, delectinated IL-2 (Boehringer Mannheim, Indianapolis, IN), 2 μg/ml polybrene, and 50 μg/ml gentamicin sulfate).
6. After overnight incubation at 37°C in a 5% CO_2 atmosphere, centrifuge all tubes, remove medium, and wash cells twice with complete medium for Hanks balanced salt solution.
7. Resuspend cells in 2 ml fresh culture medium (same as in step 4, but *without* polybrene) and transfer to 24-well plate (1 tube to 1 well). Mark all wells to identify appropriate antibody dilution and virus used.
8. On day 3 remove half of the medium from each well and replace with equal volume of fresh medium containing 5×10^5 prestimulated PBMCs.
9. On day 7 harvest (1 ml) culture supernatant and test for the presence of the HIV-1 p24 protein using regular HIV p24-antigen assay.
10. On day 14, remove 1 ml of medium from each well and save at –70°C for p24 testing. *Test day 7 harvest first.*
11. Test the day 7 culture supernatant for p24 antigen and calculate the log inhibitory dose (ID_{50}). The antibody dilution showing ≥90% neutralization of virus infectivity, as evident by the absence or reduction of p24 antigen in the culture supernatant, is considered as a neutralizing antibody titer.

Note: *It is expected that an effective neutralizing antibody will show inhibition or reduction of virus infection by day 7 in culture. The day 14 results may show overgrowth of the virus, indicating a lack of virus neutralizing antibodies. In contrast, an ideal HIV antiserum would be that which completely abolishes virus infection in culture. Unfortunately, this serum is not available yet for HIV neutralization.*

IX. Molecular Characterization of Human Retroviruses

Molecular characterization of the known viruses is based on the structural and regulatory genes encoded by the virus. The human retroviruses share many properties with animal retroviruses, but the genetic structures of human and other primate retroviruses are much more complex than those of other vertebrates. Consistent with the genomic organization of animal retroviruses, three major structural genes have been identified in all human retroviruses. These are, from 5′ to 3′ end, the *gag* gene, which encodes group specific antigens, including the major structural proteins of the virus; the *pol* gene, which is responsible for the viral enzyme (i.e., reverse transcriptase) necessary for virus replication, and the *env* gene, which encodes the envelope glycoprotein or the coat protein of the virion. In addition to the structural genes, the human retroviruses and their primate counterparts also encode proteins that are critical for the regulation of transcription, translation, and other important steps in their life cycle. The viral regulatory or accessory genes encode at least seven or eight additional proteins that regulate HIV gene expression and interact with cellular factors such that the viral replication, infectivity, host cell tropism, and pathogenesis are affected.[98-100]

A. Identification of Drug Resistant Viruses

The resistance of pathogenic microorganisms to the antimicrobial drugs is a well-known phenomenon and it has been studied extensively in bacteria. Thus, development of drug resistant viruses is not uncommon and HIV is no exception. Most patients treated with antiviral agents, such as AZT or Zidovudine, which is a nucleoside analogue, can develop resistance to the drug because it inhibits reverse transcription of the HIV RNA and the virus mutates. The genotypic changes or mutations caused in the reverse transcriptase or RT gene also affect the phenotypic properties of the virus. Other nucleoside analogues didanosine (2′,3′-dideoxyinosine or ddI) and dideoxycytosine (ddC) also show antiviral effects and these drugs have been used in patients with advanced HIV disease in AZT-intolerant patients (i.e., patients who have developed resistance to AZT). Thus, both ddI and ddC are being used to treat HIV-infected individuals in combination with AZT, protease inhibitors or with other drugs. Although many mutations have been identified to be associated with AZT drug resistant HIV strains, mutations at codons 41, 67, 70, 74, 181, 215, and 219 are common, with 215 being the most stable mutation. Currently several protease inhibitors are also being used as antiviral agents, but specific mutations associated with resistance to these drugs have not been defined yet.

Several assays are used to identify HIV strains that are drug susceptible or drug resistant. In one assay, virus is isolated by co-culture of the patient's PBMCs and tested for drug susceptibility or resistance *in vitro*. A second approach is to identify point mutations by "selective" PCR directly in the DNA of the patient's PBMCs with or without culture. Both methods are described briefly in this and the next section. The method for identification of drug-resistant HIV by *in vitro* culture is as follows:

1. Isolate HIV *in vitro* by culturing up to 10^9 cells from patients with equal numbers of PHA-stimulated PBMCs from HIV-seronegative donors (qualitative cultures), as described in Section VIII.A. Since the amount of blood (and hence the number of cells) obtained from babies is limited, it is recommended to culture whatever number of cells are available, with an equal number of prestimulated donor PBMCs.
2. Determine the *in vitro* infectivity titer ($CCIU_{50}$ or $TCID_{50}$), of *each* clinical isolate (i.e., primary cultures of the patient's blood) prior to its use in the drug susceptibility assay.
3. Stimulate 5×10^6 donor PBMCs with PHA, as described in Section VIII.A.1. After 48 h, pellet cells, wash with complete medium, and incubate cells with HIV at a concentration

of 1000 $TCID_{50}$ per 10^6 cells for 1 h at 37°C in a CO_2 incubator. Mix well and shake tube occasionally for even distribution of virus while incubating.

4. Remove unabsorbed virus by centrifugation at 400 × *g* and resuspend cells in 2.5 ml fresh medium, i.e., a concentration of 2×10^6 cells per milliliter of medium.
5. Prepare stock solutions of Zidovudine (ZDV), ddI, and other antiretroviral drugs and store in borosilicate glass vials at –20°C. To prevent photosensitive degradation of the drug it is recommended to wrap foil around the vial.
6. Make 10-fold drug dilutions in culture medium such that the dilution of the drug (i.e., before addition of the cell) is 2X concentrated. Thus, the final dilution of the drug is achieved *after* the addition of cells. Drug dilutions are expressed in molar concentrations or μg/ml.
7. It is important that all drugs are tested for their cytotoxicity to the PBMCs *prior* to testing for the virus susceptibility. At least four wells per drug dilution are tested. Cell viability is compared with the viability of control untreated cells, which should not be less than 85% viable.
8. Add 100 μl of each drug dilution to four wells each in a 96-well plate. To maintain sufficient humidity, the first row of wells all around the plate (i.e., both vertical and horizontal rows) should be filled with sterile PBS and should not be used for experimental cultures.
9. Add virus-infected cells (from step 4) at a concentration of 200,000 cells per 100 μl to each well except control wells with no virus, which will receive equivalent numbers of PHA-stimulated, *uninfected* cells.

Note: *The following controls of four wells each must also be added: (A) PBMC + drug dilutions, (B) PBMC + virus, (C) PBMC + virus + drug, (D) PBMC alone in medium (i.e., no drug, no virus).*

10. After 7 days, test for HIV-1 p24 antigen in the cell-free culture supernatant of each well, except those wells that did not receive virus.
11. Observe cells for cytotoxic effects and count cell numbers in A and D.
12. Determine susceptibility to AZT (or other antiviral drugs) by calculating 50% inhibitory dose (ID_{50}) of virus used to infect 1×10^6 cells compared to the control, ie., without drug but with virus. The drug concentration required to inhibit 90% production of HIV-1 p24 antigen compared to drug-free culture is the inhibitory dose 90 (ID_{90}).

B. Identification of Point Mutations

Drug resistant HIV-1 strains can be identified in cellular DNA or RNA of HIV-infected individuals regardless of the treatment they have received. Several protocols have been established for identifying HIV-1 strains with specific mutations that confer resistance to AZT (Zidovudine) or other nucleoside analogues.[101] Using the following protocol, genetic mutations involved in drug susceptibility or resistance have been identified.[102]

1. Isolate DNA from patient PBMCs before or after culturing, according to methods described in Section V.C.3. To avoid selection of viral strains *in vitro,* it is preferred to use uncultured cells. Viral RNA extracted from the plasma can also be analyzed for mutant viruses after the synthesis of cDNA by using reverse transcription PCR, as described in Section V.F.1.
2. Amplify the RT region of the *pol* gene, which is expected to contain mutations that confer drug resistance to HIV. For identifying mutations, the genetic sequence is amplified in two separate PCR reactions (i.e., nested PCR).
3. *First PCR:* Select primers that flank a larger region than is expected to contain mutations. For example, to identify mutation at codon 215 in the RT gene of HIV, the primers outside of this region are used in the first PCR. The two commonly used primers for the RT genes are (A′ (35) 5′-TTG (GTTGCACTTTAAATTTTCCCATTAGTCCTATT-3′ and NEI(35)-5′-CCTACTAACTTCTGTATGTCATTGACAGCCAGCT-3′).[102]
4. Anneal DNA for the first PCR using 250 μg of each of the A′ and NEI primers in 100 μl of the reaction mixture, as described in Section V.4, except use 2.5 m*M* $MgCl_2$ and 2.5 U of Taq polymerase.
5. After 35 cycles, remove mineral oil, extract PCR mixture with chloroform and fractionate on 1% SeaKem agarose gel in Tris-borate-EDTA buffer containing 1 μg of ethidium bromide.
6. Confirm the size of the 805 bp PCR product by running on 1% agarose gel. Dilute the PCR product in ddH_2O (1:100) and use 10 μl of the diluted product in the *second PCR*.
7. *Second PCR:* Select a primer pair that flanks the conserved region but inside the 805 bp sequence that was amplified by the first PCR. This procedure is based on the differential priming and DNA synthesis using wild-type and mutant oligonucleotide primers. These primers are used in separate reaction tubes with an appropriate common primer and a specific wild-type or the mutant primer.[101,102] Several primer pairs have been described for the identification of specific codon changes that confer AZT resistance to HIV.[101,102] For the identification of codon 215, Larder et al. recommend the use of B, 3 W or 3 M primers.[101,102] The basic common primer is B = 5′-GGATGGAAAGGATCACC-3′, 3W is a wild-type 5′-ATGTTTTTGTCTGGTGTCGT-3′ and 3M primer 5′-ATGTTTTTGTCTGGTGTGAA-3′ contains specific mutation for codon 215. Thus, specific PCR product is generated only when the 3′ end of the selective (mutant) primer exactly matches the target sequence.
8. For each mutation to be identified or analyzed, use the diluted PCR product from the first PCR in two separate tubes. One tube will contain the wild-type primer (3W) and the common primer (B) and the other tube will contain the mutant primer (3M) with the same common primer (B). Controls should include known wild-type and mutant DNA templates, negative amplification controls, reagent controls without DNA, and diluted products of the first PCR without selective primers.
9. Amplify DNA for 30 cycles in the presence of 1.5 m*M* $MgCl_2$.

Note: *If 1.5 mM $MgCl_2$ is used in the buffer, the annealing temperature should be 45°C. If 2.25 mM $MgCl_2$ is used in the buffer, the annealing temperature should be 48°C.*

10. Identify fragment size by electrophoresis in 3% agarose gel containing 0.5 μg/ml ethidium bromide in 1X TBE (0.09 m Tris-borate, 0.002 m EDTA). Use 10 to 20 μl of the product to test.
11. Identify size-specific bands. Samples that yield a 210 bp product with the mutant primer (3M) are considered to have a mutation at codon 215. Samples yielding specific size products (i.e., 210 bp for 215 codon and 178 bp for codon 74) with wild-type primers are considered to have a wild-type sequence. Samples with mutant primers are considered mutants. Using these techniques several mutations in codons 41, 67, 70, 74, 219 and others have also been identified.[101-104] If both wild-type and mutant primers yield products of the same intensity, the experiments should be repeated. Mutants are confirmed by sequencing for identifying drug resistance.

Recently several new assays for identification of *pol* gene resistance and point mutations have been developed using microtiter formats, cloned recombinant viruses, and other modifications.[105-109] These assays are considered rapid and more specific than the PCR-based assay described above (for review see Chapter 4 by Prasad et al.).

1. Heteroduplex mobility assay (HMA)

A quasispecies is defined as a heterogeneous population of HIV which may be dominated by a consensus or master genetic sequence.[110] Sequence heterogeneity in HTLV-1 is not as common as in HIV, but genetic variations have been reported in HTLV-1 strains from brain and other organs of HTLV-1 infected individuals.[111] Quasispecies have been identified by DNA heteroduplex formation between a fragment of known DNA sequence in a region expected to show mutations, such as the HIV *env* gene and the reverse transcriptase (RT) gene. The genetic changes in the *env* gene sequences of HIV-1 are more prevalent than observed in the *gag* or *pol* genes of most isolates. Of the five variable domains identified in the HIV gp120 (designated V1, V2, V3, V4, V5), V3 contains the principal neutralizing domain, which is also considered as the sequence associated with the macrophage cell tropism.

Based on the *env* gene variations, all known HIV strains have been grouped in several families or clades designated A, B, C, D, E, F, G, H, and O. Thus, not only do different patient's HIV isolates show variations, the virus from the same individual at different stages of HIV disease progression may vary significantly. First developed by Delwart and coworkers in 1993,[112,113] the heteroduplex mobility assay (HMA) has proven to be an extremely useful technique for genetic typing of HIV. A recent survey by the World Health Organization (WHO) network on HIV isolation and characterization indicated that 100% (54/54) of HIV strains could be accurately genotyped by HMA when compared to the nucleotide sequence analysis.[114] The HMA is also useful for the identification of intra- and inter-quasispecies. Thus, HMA is the recommended method of choice for genotypic screening of large numbers of HIV strains. The technique is based on annealing the PCR-amplified, denatured, partially complementary DNA strands and analyzing mobility shifts of the heteroduplex due to sequence variability by polyacrylamide gel electrophoresis (PAGE). The mismatched (or unpaired) nucleotide or mutation causes a gap and influences the structure of the heteroduplex, which results in reduced mobility.

It is anticipated that identification of HIV-1 genetic variants will greatly impact vaccine development. The subtyping may also lead to a better definition of antigenic subgroups of HIV-1 than is currently possible. The procedure for subtyping HIV strains by HMA is according to those described earlier.[113]

1. Purify DNA as described in Section V.C.2. or use detergent-lysed cellular DNA from various tissues or red cell-depleted whole blood.[115]
2. Using primers from well-conserved *env* or *gag* genes and nested PCR (some env and other primers are available from the AIDS Repository, NIH), amplify 1.2 to 1.6 kb fragments (see Section V.C. for PCR amplification, reaction mixtures, etc.). All test samples and reference DNA samples must be amplified in separate tubes, but under *identical* conditions.
3. Quantitate DNA as described in Section V.C.3.d. Mix 5 μl of PCR-amplified DNA (100 to 250 ng) from the unknown test sample separately with 5 μl of each of the known sequences. Add 1.1 μl of different HIV-1 clades.
4. Add 1.1 μl of 10X buffer (10X = 1 *M* NaCl, 100 m*M* Tris pH 7.8, 20 m*M* EDTA), heat to 94°C for 2 min then add $^1/_5$ volume Ficoll-Orange G loading dye (5X = 25% Ficoll, 1% Orange G) on wet ice.
5. Electrophorese the amplified product on 5% neutral polyacrylamide gel at 100 V for 1 h in buffer containing 89 m*M* Tris-borate, 89 m*M* boric acid, and 8 m*M* EDTA. Reference and unknown samples should be run on the *same* gel.
6. Stain gel with ethidium bromide (0.5 μg/ml in water) for 30 to 60 min. Heteroduplexes formed between the test sample and most closely related reference sequence are expected to exhibit fastest mobility and these are visualized by ultraviolet transillumination.
7. Photograph the gel for pattern recognition and future analysis.
8. First analyze the heteroduplex pattern of *each* PCR amplified sample. Heteroduplexes formed between the known reference fragment and those deliberately formed (controls) are then distinguished from the unknown. It is therefore recommended that the patterns from the known or test sample be compared with *each reference strain* by using primers available from the AIDS Repository, National Institute of Allergy and Infectious Diseases, National Institutes of Health.

Although several subtypes or clades of HIV have been identified within a given geographic region, one or more predominant subtypes can exist in a particular region. For example, subtype B is predominant in North American and some western European countries, but in Thailand both B and E subtypes are present.[116] Our laboratory has recently isolated a new strain of clade-A subtype which is present in western Africa.[117] Thus, the use of local reference strains must initially be used for HMA analysis with the unknowns. This allows an intra-subtype comparison and assignment of strain with confidence. Phenotypic analysis and relative mobility patterns of heteroduplex are estimated from the standard curves generated from reference standards. Several programs (Fitch-Margoliash, FITCH and Felsensteins PHYLIP) are available for calibration and comparison of patterns and analysis.[147]

Note: *Since related DNA products can coamplify from divergent templates of the quasispecies present in an individual, genetic relationships between various mutant viruses can be evaluated by HMA. Heteroduplex tracking assay (HTA) is a similar technique, except one PCR product is used as a ^{32}P-labeled probe and is mixed with a 100-fold excess of unlabeled PCR for heteroduplex formation.*

2. Trouble shooting HMA

1. Ambiguous results are obtained when larger fragments are used or when a genetic "outlier" within a known subtype is analyzed. Major deletions in the env glycoprotein V1, V2, or V4 and V5 regions may cause some difficulty in heteroduplex pattern identification, and these can be resolved by determining the DNA sequence of that region.
2. It is also recommended to include a cloned proviral DNA template as a control because it will only form homoduplex. This is a good control, since a heteroduplex from such a template would result only when polymerase errors are made during transcription.
3. Heteroduplexes can migrate as sharp bands or can form a smear in the polyacrylamide gel. The DNA of samples yielding a smear can be serially diluted prior to nested PCR and heteroduplex formation. This allows a better distribution of DNA in the reaction mixture. However, we recommend testing several aliquots of this DNA, as the quasispecies will be divided into these aliquots.
4. Bands that migrate with a mobility of 40% that of homoduplexes could be due to ssDNA fragments that failed to reanneal with complementary strands.[118]
5. Some very slowly migrating bands could represent nonviral DNA amplification products. These should be confirmed by reamplification of original DNA and heteroduplexing.
6. To compare data across experiments, it is critical to standardize electrophoresis, voltage/current, buffer, DNA amounts, gel conditions, etc.

C. DNA Sequence Analysis

Although DNA nucleotide sequencing is not essential for the detection or identification of viral genomes in human DNA, it is a critical tool for characterizing the genetic structure and for confirming specific mutations. Generally, the strategy of Sanger's dideoxy DNA sequencing technique is used. This technique is based on a sequence extension using synthetic oligonucleotides and universal primers. Manual sequencing reactions are performed using Sequenase enzyme, except in rare cases when ambiguous areas arising as a result of GC-rich secondary structures cannot be resolved by sequencing both strands, in which case we use Taq DNA polymerase instead.

Recent advances in the automated analysis of DNA sequencing have also encouraged investigators to look into the possibility of utilizing this technique for the rapid diagnosis of specific mutations in the proviral DNA's of retroviruses. For automated sequencing using Applied Biosystems Inc. (ABI), Foster City, CA, sequencer, Taq polymerase is more useful than Sequenase. The direct sequencing procedure is also capable of analyzing both strands of DNA. An advantage of this procedure is that "erroneous" PCR products generated by nucleotide misincorporation will not interfere with the sequence determination. The following protocol is used in our laboratory:[117]

1. Use purified, RNA-free plasmid DNA, which yields best results. Although PCR amplified cellular DNA has been sequenced directly without cloning it is recommended that both plasmid and genomic DNAs be purified by gel filtration chromatography using a ChromaSpin-400 column (Clontech Laboratories, Inc., Palo Alto, CA) and centrifugation. After purification, analyze DNA prior to sequencing by agarose gel electrophoresis to estimate concentration of the DNA.
2. Denature 5 to 10 μg of purified DNA by adding 0.1 vol of 2 *M* NaOH/2 m*M* EDTA. Incubate at 37°C for 30 min.
3. Neutralize by the addition of 0.1 vol 3 *M* sodium acetate, pH 5.2, and immediately transfer samples to ice.
4. Precipitate DNA by adding 3 vol of cold 100% ethanol. Incubate 15 to 30 min at –80°C.
5. Pellet DNA by centrifugation in a microfuge, 20 min at 4°C. Carefully wash the pellet with 70% ethanol. Air-dry pellet.
6. Resuspend pellet in 7 μl water by vigorously pipetting 50 to 60 times. Add 2 μl "5 X Sequenase Reaction Buffer" (provided in sequencing kit from United States Biochemical/Amersham Life Sciences, Cleveland, OH), and 1 μl primer, approximately 100 to 150 pmol. Concentration of primer in pmol/μl = $OD_{260}/(0.01 \times N)$, N = number of nucleotides in primer.
7. Heat samples to 65°C for 2 min. Immediately place samples at 37°C for 30 min to anneal primer to the template.
8. Place samples on ice until needed.
9. Follow the sequencing protocol according to the instructions provided by the manufacturer (United States Biochemical, Cleveland, OH). The following DNA sequencing protocol is used in our laboratory:[117]
   a. Dilute "Labeling Mix" (provided in sequencing kit) fivefold with deionized, distilled water.
   b. Dilute "Sequenase, version 2.0" enzyme (provided in sequencing kit) 1:8 in ice-cold "Enzyme Dilution Buffer" (provided in sequencing kit). Store on ice.
   c. To the annealed template-primer add 1.0 μl 0.1 *M* DTT (provided in sequencing kit), 1.0 μl diluted labeling mix, 5.0 μl [α-thio^{35}S]dATP, 0.5 μl pyrophosphatase (provided in sequencing kit), and 2.0 μl diluted Sequenase enzyme.
   d. Mix and incubate at room temperature for 2 to 5 min.
   e. Label four microfuge tubes G, A, T, and C for each template/primer combination.
   f. Aliquot 2.5 μl of the appropriate dideoxy chain termination mixture (i.e., ddGTP to the tube labeled G, etc.) to each labeled tube.
   g. Prewarm tubes at 37°C for ≤1 min.
   h. Remove 3.5 μl after completion of labeling reaction, and transfer it to one of the labeled tubes, mix, and place tube at 37°C. In similar manner, transfer labeling reaction mixture to each of the three remaining tubes.
   i. Incubate at 37°C for 3 to 5 min.

j. Add 4 μl Stop Solution (provided in sequencing kit) to each of the termination reactions, mix thoroughly, and store on ice.
k. Heat termination reactions at 75 to 80°C for 2 min immediately prior to loading (2 to 3 μl) on sequencing gel.
l. Electrophorese on a 5 to 7% polyacrylamide gel.
m. Run the gel at 1500 V for 3 to 6 h depending on the sequence of interest.

For automated sequencing, dissolved 2 to 4 μg of DNA in H_2O and perform sequencing according to the Applied Biosystems' chain terminator protocol, as follows:

1. Add 0.8 pmol/μl primers to 7.5 μl of reaction mixture containing 400 m*M* Tris-HCl, 10 m*M* $MgCl_2$, 100 m*M* ammonium sulfate, pH 9.0, 750 μ*M* dITP, 150 μ*M* dATP, 150 μ*M* dTTP, 150 μ*M* dCTP, and fluorescene-tagged dideoxy chain terminators (ddGTP, ddTTP, ddATP, and ddCTP) (Applied Biosystems) and 4 U Taq DNA polymerase.
2. Remove excess dideoxy chain terminators by centrifugation through G50 Sephadex spin column (Pharmacia, Milwaukee, WI).
3. Precipitate the DNA with an equal volume of 5 *M* ammonium acetate and 300 μl ethanol.
4. Dissolve the precipitate in a mixture of 1 μl of 50 m*M* EDTA and 5 μl deionized formamide.
5. Heat the mixture, denature, and load on a 6.5% acrylamide, 8 *M* urea sequencing gel.

Direct sequencing of amplified mRNA product is useful for identification of variants with single mutations, insertions, or deletions. This method is not limited to previously sequenced transcripts and requires only two adjacent or partially overlapping specific primers from only one side of the region to be amplified. Poly(A)+ regions of mRNA are targeted with oligo(dT) and the amplified product can be sequenced directly. Comparison between structural motifs of a conserved normal or natural sequence and the mutant alleles of viruses can readily be made.

Attachment of a phage promoter on to one or both primers increases the transcription by PCR and provides abundant single-stranded template for reverse transcriptase-mediated dideoxy sequencing. By the use of an end-labeled gamma ^{32}P reverse transcriptase primer complementary to the desired sequence, the additional specificity required for unambiguous sequence data is generated. Further, co-amplification and co-transcription of multiple regions further increase the rate of sequence acquisition of a fragment.

X. Computer Analysis of Nucleotide Sequences and Proteins

A variety of sequence analysis software packages are available that allow investigators to translate DNA into protein, identify restriction sites, draw protein hydropathy and hydrophilicity plots, search sequence similarities, select primers, and align multiple sequences for structural and functional analysis of genes. These searches include distance matrices corrected for multiple mutations, nucleotide substitutions such as transitions and transversions, unweighted pair-group method, neighbor-joining method, and the Fitch-Margoliash methods.[147]

"Sequencer" is a package of advanced software for DNA sequences for Gencodes Corp., Annenberg, MI. The FAST alignment algorithm detects overlapping ends in sequencing fragments and assembles them into a consensus sequence. The GENBANK, EMBL, PIR, and NBRF database programs are available. Sequence comparison of both sense and antisense strands in single-strand format is also available. Open reading frame (ORF), 3-frame translations in single and triple letter codes, stop codons, amino acid compositions, hydrophilicity plots, etc. are important considerations for comparison of viral or cellular genes.

XI. Retroviral Inhibition Mediated by Gene Transfer (Gene Therapy)

One of the important molecular techniques that is being seriously considered for the inhibition of HIV-1 infection is gene therapy.[119-121] The virus is inactivated or inhibited by transferring a gene in the virus-infected cell, which would destroy viral RNA or inhibit virus production in the cells. Several genetic antiviral approaches to therapeutic interventions have been identified.[120,122-125] These include ribozymes, antisense RNA, RNA-decoy and dominant negative mutants of HIV-1 Gag, Tat, and Rev proteins. These strategies are based on the regulated expression of these proteins in the infected cells. By using constructs with Rev-responsive elements the expression can be suppressed or eliminated. Thus, it is feasible to use ribozyme technology or similar approaches to confer intracellular inhibition of virus production.[125]

Acknowledgments

I would like to thank my entire research group for standardizing most of the tests related to the identification and characterization of human retroviruses. Particular thanks are due to Dr. Thomas Howard for critical reading, Zhiliang Li for his help with the bibliography and diagrams, and Laura Garibay for the preparation of the manuscript. Grant supports T32-AI07383, AI27673-09-C1, and AI27673-09-C3 from the National Institute of Allergy and Infectious Disease are also acknowledged.

References

1. Poiesz, B. J., Ruscetti, F. W., Reitz, M. S., Kalyanaraman, V. S., and Gallo, R. C., *Nature*, 294, 268 (1981).
2. Poiesz, B. J., Ruscetti, F. W., Gazdar, A. F., Bunn, P. A., Minna, J. D., and Gallo, R. C., *Proc. Natl. Acad. Sci. U.S.A.*, 77, 7415 (1980).

3. Kalyanaraman, V. S., Sarngadharan, M. G., Guroff, M. R., Miyoshi, I., Golde, D., and Gallo, R. C., *Science,* 218, 571 (1982).
4. Rosenblatt, J. D., Golde, D. W., Wachsman, W., Giorgi, J. V., Jacobs, A., Schmidt, G. M., Quan, S., Gasson, J. C., and Chen, I. S., *N. Engl. J. Med.,* 315, 372 (1986).
5. Lee, H., Swanson, P., Shorty, V. S., Zack, J. A., Rosenblatt, J. D., and Chen, I. S., *Science,* 244, 471 (1989).
6. Lairmore, M. D., Jacobson, S., Gracia, F., De, B. K., Castillo, L., Larreategui, M., Roberts, B. D., Levine, P. H., Blattner, W. A., and Kaplan, J. E. *Proc. Natl. Acad. Sci. U.S.A.,* 87, 8840 (1990).
7. Barre Sinoussi, F., Chermann, J. C., Rey, F., Nugeyre, M. T., Chamaret, S., Gruest, J., Dauguet, C., Blin, C. A., Brun, F. V., Rouzioux, C. et al., *Science,* 220, 868 (1983).
8. Gottlieb, M. S., Schroff, R., Schanker, H. M., Weisman, J. D., Fan, P. T., Wolf, R. A., and Saxon, A., *N. Engl. J. Med.,* 305, 1425 (1981).
9. World Health Organization, *Global AIDS News,* 1 (1995).
10. Greenough, T. C., Somasundaran, M., Brettler, D. B., Hesselton, R. M., Alimenti, A., Kirchhoff, F., Panicali, D., and Sullivan, J. L., *AIDS Res. Hum. Retroviruses,* 10, 395 (1994).
11. Gao, Y., Qin, L., Zhang, L., Safrit, J., and Ho, D. D., *N. Engl. J. Med.,* 332, 201 (1995).
12. Levy, J. A., *Am. J. Med.,* 95, 86 (1993).
13. Rudolph, D. L., Khabbaz, R. F., Folks, T. M., and Lal, R. B., *Diagnostic Microbiol. Infect. Dis.,* 17, 35 (1993).
14. Figueroa, M. E. and Rasheed, S., *Am. J. Clin. Pathol.,* 95, S8 (1991).
15. Ishikawa, E., Hashida, S., Kohno, T., Hirota, K., Hashinaka, K., and Ishikawa, S., *J. Clin. Lab. Anal.,* 7, 376 (1993).
16. Bylund, D. J., Ziegner, U. H., and Hooper, D. G., *Clin. Lab. Med.,* 12, 305 (1992).
17. Rasheed, S., Norman, G. L., and Su, S., in *Viral Hepatitis and AIDS,* Villarejos, V. M., Ed., Hermanos Trejos, San Jose, Costa Rica, 1987, 31.
18. Archibald, D. W., Johnson, J. P., Nair, P., Alger, L. S., Hebert, C. A., Davis, E., and Hines, S. E., *AIDS,* 4, 417 (1990).
19. Tamashiro, H. and Constantine, N. T., *Bull. WHO.,* 72, 135 (1994).
20. Sun, D., Archibald, D. W., and Furth, P. A., *AIDS Res. Hum. Retroviruses,* 6, 933 (1990).
21. Lal, R. B., Heneine, W., Rudolph, D. L., Present, W. B., Hofheinz, D., Hartley, T. M., Khabbaz, R. F., and Kapplan, J. E., *J. Clin. Microbiol.,* 29, 2253 (1991).
22. Lombardi, V. R., Libonatti, O., Alimandi, M., Ansotegui, I., Scarlatti, G., Moschese, V., Wigzell, H., and Rossi, P., *Eur. J. Epidemiol.,* 8, 298 (1992).
23. Iitia, A., Dahlen, P., Nunn, M., Mukkala, V. M., and Siitari, H., *Anal. Biochem.,* 202, 76 (1992).
24. Tsang, V. C., Peralta, J. M., and Simons, A. R., *Methods Enzymol.,* 92, 377 (1983).
25. Hartley, T. M., Khabbaz, R. F., Cannon, R. O., Kaplan, J. E., and Lairmore, M. D., *J. Clin. Microbiol.,* 28, 646 (1990).
26. Olaleye, D. O., Ekweozor, C. C., Sheng, Z., and Rasheed, S., *Int. J. Epidemiol.,* 24, 198 (1995).
27. DeGruttola, V., Beckett, L. A., Coombs, R. W., Arduino, J. M., Balfour, H. H., Jr., Rasheed, S., Hollinger, F. B., Fischl, M. A., and Volberding, P., *J. Infect. Dis.,* 169, 713 (1994).

28. Palomba, E., Gay, V., de Martino, M., Fundaro, C., Perugini, L., and Tovo, P. A., *J. Infect. Dis.,* 165, 394 (1992).
28a. Rasheed, S., unpublished data.
29. Kageyama, S., Yamada, O., Mohammad, S. S., Hama, S., Hattori, N., Asanaka, M., Nakayama, E., Matsumoto, T., Higuchi, F., Kawatani, T. et al., *J. Virol. Methods,* 22, 125 (1988).
30. Ascher, D. P., Waecher, N. J., Ottolini, M. A., Raszka, W. V., Moriarty, R., and Robb, M. L., *Pediatric AIDS and HIV Infection,* 5, 319 (1994).
31. Burke, D. S., *Vaccine,* 11, 883 (1993).
32. Schols, D., Pauwels, R., Desmyter, J., and De Clercq, E., *Cytometry,* 11, 736 (1990).
33. Schols, D., Pauwels, R., Desmyter, J., and De Clercq, E., *Virology,* 175, 556 (1990).
34. Schols, D., Baba, M., Pauwels, R., and De Clercq, E., *J. AIDS,* 2, 10 (1989).
35. McSharry, J. J., *Clin. Microbiol. Rev.,* 7, 576 (1994).
36. Hotz, M. A., Gong, J., Traganos, F., and Darzynkiewicz, Z., *Cytometry,* 15, 237 (1994).
37. Cameron, P. U., Pope, M., Gezelter, S., and Steinman, R. M., *AIDS Res. Hum. Retroviruses,* 10, 61 (1994).
38. Carbonari, M., Cibati, M., Cherchi, M., Sbarigia, D., Pesce, A. M., DellAnna, L., Modica, A., and Fiorilli, M., *Blood,* 83, 1268 (1994).
39. Franke, E. K., Yuan, H. E., and Luban, J., *Nature,* 372, 359 (1994).
40. Luban, J., Bossolt, K. L., Franke, E. K., Kalpana, G. V., and Goff, S. P., *Cell,* 73, 1067 (1993).
41. Thali, M., Bukovsky, A., Kondo, E., Rosenwirth, B., Walsh, C. T., Sodroski, J., and Gottlinger, H. G., *Nature,* 372, 363 (1994).
42. Arthur, L. O., Bess, J. W., Jr., Sowder, R. C., Benveniste, R. E., Mann, D. L., Chermann, J. C., and Henderson, L. E., *Science,* 258, 1935 (1992).
43. Wong Staal, F. and Gallo, R. C., *Nature,* 317, 395 (1985).
44. Wong Staal, F., Shaw, G. M., Hahn, B. H., Salahuddin, S. Z., Popovic, M., Markham, P., Redfield, R., and Gallo, R. C., *Science,* 229, 759 (1985).
45. Boni, J. and Schupbach, J., *Mol. Cell. Probes,* 7, 361 (1993).
46. Aono, Y., Imai, J., Tominaga, K., Orita, S., Sato, A., and Igarashi, H., *Virus Genes,* 6, 159 (1992).
47. Hamakado, T., Matsumoto, T., Koyanagi, Y., Hayashi, K., Fukada, K., Kouchiyama, T., Yoshida, T., and Yamamoto, N., *Virus Genes,* 6, 119 (1992).
48. Ranki, M., Palva, A., Virtanen, M., Laaksonen, M., and Soderlund, H., *Gene,* 21, 77 (1983).
49. Kwok, S. and Snisnky, J. J., in *Diagnostic Molecular Microbiology: Principles and Applications,* Persing, D. H., Smith, T. F., Tenover, F. C., Eds., ASM Press, Washington, D.C., 309, 1993.
50. Chomczynski, P. and Sacchi, N., *Ann. Biochem.,* 162, 156 (1987).
51. Zack, J. A., Arrigo, S. J., Weitsma, S. R., Go, A. S., Haislip, A., and Chen, I. S., *Cell,* 61, 213 (1990).
52. Longo, M. C., Berninger, M. S., and Hartley, J. L., *Gene,* 93, 125 (1990).
53. Taniguchi, S., Maekawa, N., Yashiro, N., and Hamada, T., *Br. J. Dermatol.,* 129, 637 (1993).
54. Aisen, P. and Listowsky, I. X., *Annu. Rev. Biochem.,* 49, 357 (1980).

55. Lee, T. H., Diaz, R., and Busch, M. P., *2nd Natl. Conf. Hum. Retroviruses Relat. Infect.*, 29, 16 (1995).
56. Honma, M., Ohara, Y., Murayama, H., Sako, K., and Iwasaki, Y., *J. Clin. Microbiol.*, 31, 1799 (1993).
57. Dube, D. K., Dube, S., Erensoy, S., Jones, B., Bryz Gornia, V., Spicer, T., Love, J., Saksena, N., Lechat, M. F., Shrager, D. I. et al., *Virology,* 202, 379 (1994).
58. Olaleye, O. D., Ekwozor, C. C., Li, Z., Opala, E., Sheng, Z., Onyemenem, T. N., and Rasheed, S., *Arch. Virol.,* 141, 345 (1996).
59. Rey, F., Salaun, D., Lesbordes, J. L., Gadelle, S., Henry, F. O., Barre Sinoussi, F., Chermann, J. C., and Georges, A. J., *Lancet,* 2, 1391 (1986).
60. Calabro, M. L., Luparello, M., Grottola, A., Del Mistro, A., Fiore, J. R., Angarano, G., and Chieco Bianchi, L., *J. Infect. Dis.,* 168, 1273 (1993).
61. Rayfield, M., De Cock, K., Heyward, W., Goldstein, L., Krebs, J., Kwok, S., Lee, S., McCormick, J., Moreau, J. M., Odehouri, K. et al., *J. Infect. Dis.,* 158, 1170 (1988).
62. Lin, H. J., Myers, L. E., Yen Lieberman, B., Hollinger, F. B., Henrard, D., Hooper, C. J., Kokka, R., Kwok, S., Rasheed, S., Vahey, M. et al., *J. Infect. Dis.,* 170, 553 (1994).
63. van Gemen, B., Kievits, T., Schukkink, R., van Strijp, D., Malek, L. T., Sooknanan, R., Huisman, H. G., and Lens, P., *J. Virol. Methods,* 43, 177 (1993).
64. Kievits, T., van Gemen, B., van Strijp, D., Schukkink, R., Dircks, M., Adriaanse, H., Malek, L., Sooknanan, R., and Lens, P., *J. Virol. Methods,* 35, 273-286 (1991).
65. Gall, J. G. and Pardue, M. L., *Methods Enzymol.,* 38, 1649 (1971).
66. Gentilomi, G., Zerbini, M., Musiani, M., Gallinella, G., Gibellini, D., Venturoli, S., Re, M. C., Pileri, S., Finelli, C., and La Placa, M., *Mol. Cell. Probes,* 7, 19 (1993).
67. Zeller, R., Rogers, M., and Watkins, S., in *Current Protocols in Molecular Biology,* Ausubel, F. M., Brent, R., and Kingston, R. E., Eds., John Wiley & Sons, New York, 1991, p. 14.0.1.
68. Yen, L. B., Brambilla, D., Jackson, B., Palumbo, P., Herman, S., Todd, J., Lin, H. J., Rasheed, S., Bremer, J., and Coombs, R., et al., *2nd Nat'l Conf. Hum Retroviruses Relat. Infect.,* 116, 92-102 (1995).
69. Rasheed, S., Gowda, S., Gill, P. S., Meyer, P. R., and Levine, A. M., *Int. J. Immunother.,* 3, 81 (1987).
70. Bagasra, O., Seshamma, T., and Pomerantz, R. J., *J. Immunol. Methods,* 158, 131 (1993).
71. Nuovo, G. J., Margiotta, M., MacConnell, P., and Becker, J., *Diagnostic Mol. Pathol.,* 1, 98 (1992).
72. Komminoth, P., Long, A. A., Ray, R., and Wolfe, H. J., *Diagnostic Mol. Pathol.,* 1, 85-87 (1992).
72a. Rasheed, S., unpublished data.
73. Rasheed, S. Li, Z., and Xu, D., *J. Reprod. Med.,* 40:747 (1995).
74. Rasheed, S., Li, Z., Xu, D., and Kovacs, A., *Am. J. Obstet. Gynecol.,* 175, *in press* (1996).
75. Loughran, T. P., Jr., Coyle, T., Sherman, M. P., Starkebaum, G., Ehrlich, G. D., Ruscetti, F. W., and Poiesz, B. J., *Blood,* 80, 1116 (1992).
76. Nagafuji, K., Harada, M., Teshima, T., Eto, T., Takamatsu, Y., Okamura, T., Murakawa, M., Akashi, K., and Niho, Y., *Blood,* 82, 2828 (1993).
77. Sherman, M. P., Dube, D. K., Saksena, N. K., and Poiesz, B. J., *Cancer Treatment Res.,* 64, 79 (1993).

78. Hasunuma, T., Nakajima, T., Aono, H., Sato, K., Matsubara, T., Yamamoto, K., and Nishioka, K., *Clin. Immunol. Immunopathol.,* 72, 90 (1994).
79. Lessin, S. R., Vowels, B. R., and Rook, A. H., *Dermatol. Clin.,* 12, 243 (1994).
80. Chan, W. C., Hooper, C., Wickert, R., Benson, J. M., Vardiman, J., Hinrichs, S., and Weisenburger, D., *Diagnostic Mol. Pathol.,* 2, 192 (1993).
81. Morgan, D. A., Ruscetti, F. W., and Gallo, R. C., *Science,* 193, 1007 (1976).
82. Klatzmann, D., Champagne, E., Chamaret, S., Gruest, J., Guetard, D., Hercend, T., Gluckman, J. C., and Montagnier, L., *Nature,* 312, 767 (1984).
83. Dewhurst, S., Sakai, K., Bresser, J., Stevenson, M., Evinger Hodges, M. J., and Volsky, D. J., *J. Virol.,* 61, 3774 (1987).
84. Cheng Mayer, C., Rutka, J. T., Rosenblum, M. L., McHugh, T., Stites, D. P., and Levy, J. A., *Proc. Natl. Acad. Sci. U.S.A.,* 84, 3526 (1987).
85. Chiodi, F., Fuerstenberg, S., Gidlund, M., Asjo, B., and Fenyo, E. M., *J. Virol.,* 61, 1244 (1987).
86. Filice, G., Cereda, P. M., and Varnier, O. E., *Nature,* 335, 366 (1988).
87. Levy, J. A., Cheng Mayer, C., Dina, D., and Luciw, P. A., *Science* 232, 998 (1986).
88. Giannetti, A., Zambruno, G., Cimarelli, A., Marconi, A., Negroni, M., Girolomoni, G., and Bertazzoni, U., *J. AIDS,* 6, 329 (1993).
89. Dezutter, D. C. and Schmitt, D., *Immunol. Lett.,* 39, 33 (1993).
90. Pantaleo, G., Graziosi, C., and Fauci, A. S., *AIDS,* 7, S19 (1993).
91. Koot, M., Keet, I. P., Vos, A. H., de Goede, R. E., Roos, M. T., Coutinho, R. A., Miedema, F., Schellekens, P. T., and Tersmette, M., *Ann. Intern. Med.,* 118, 681 (1993).
92. St. Clair, M. H., Hartigan, P. M., Andrews, J. C., Vavro, C. L., Simberkoff, M. S., and Hamilton, J. D., *J. AIDS,* 6, 891 (1993).
93. Harada, S., Koyanagi, Y., and Yamamoto, N., *Science,* 229, 563 (1985).
94. Rasheed, S., Norman, G. L., Gill, P. S., Meyer, P. R., Cheng, L., and Levine, A. M., *Virology,* 150, 1 (1986).
95. Okuda, K., Kaneko, T., Yamakawa, T., Tanaka, S., Shigematsu, T., Yamamoto, A., Hamajima, K., Nakajima, K., Kawamoto, S., and Phanuphak, P., *J. Mol. Recognition,* 6, 101 (1993).
96. Niedrig, M., Harthus, H. P., Hinkula, J., Broker, M., Bickhard, H., Pauli, G., Gelderblom, H. R., and Wahren, B., *J. Gen. Virol.,* 73, 2451 (1992).
97. Mascola, J. R., Louwagie, J., McCutchan, F. E., Fischer, C. L., Hegerich, P. A., Wagner, K. F., Fowler, A. K., McNeil, J. G., and Burke, D. S., *J. Infect. Dis.,* 169, 48 (1994).
98. Hammarskjold, M. L., Heimer, J., Hammarskjold, B., Sangwan, I., Albert, L., and Rekosh, D., *J. Virol.,* 63, 1959 (1989).
99. Felber, B. K., Cladaras, M. H., Cladaras, C., Copeland, T., and Pavlakis, G. N., *Proc. Natl. Acad. Sci. U.S.A.,* 86, 1495 (1989).
100. Paskalis, H., Felber, B. K., and Pavlakis, G. N., *Proc. Natl. Acad. Sci. U.S.A.,* 83, 6558 (1986).
101. Larder, B. A., Kohli, A., Kellam, P., Kemp, S. D., Kronick, M., and Henfrey, R. D., *Nature,* 365, 671 (1993).
102. Larder, B. A. and Boucher, C. A. B., in *Diagnostic Molecular Microbiology: Principles and Applications,* Persing, D. H., Smith, T. F., and Tenover, F. C., Eds., American Society for Microbiology, Washington, D.C., 1993, 527.

103. Kozal, M. J., Shafer, R. W., Winters, M. A., Katzenstein, D. A., and Merigan, T. C., *J. Infect. Dis.,* 167, 526 (1993).
104. Mayers, D. L., McCutchan, F. E., Sanders Buell, E. E., Merritt, L. I., Dilworth, S., Fowler, A. K., Marks, C. A., Ruiz, N. M., Richman, D. D., and Roberts, C. R., *J. AIDS,* 5, 749 (1992).
105. Kaye, S., Loveday, C., and Tedder, R. S., *J. Med. Virol.,* 37, 241 (1992).
106. Kellam, P., Boucher, C. A., and Larder, B. A., *Proc. Natl. Acad. Sci. U.S.A.,* 89, 1934 (1992).
107. Kellam, P. and Larder, B. A., *Antimicrob. Agents Chemother.,* 38, 23 (1994).
108. St. Clair, M. H., Martin, J. L., Tudor Williams, G., Bach, M. C., Vavro, C. L., King, D. M., Kellam, P., Kemp, S. D., and Larder, B. A., *Science,* 253, 1557 (1991).
109. Prasad, V. R. and Goff, S. P., *J. Biol. Chem.,* 264, 16689 (1989).
110. Connor, R. I., Notermans, D. W., Mohri, H., Cao, Y., and Ho, D. D., *AIDS Res. Hum. Retroviruses,* 9, 541 (1993).
111. Kira, J., Koyanagi, Y., Yamada, T., Itoyama, Y., Tateishi, J., Akizuki, S., Kishikawa, M., Baba, E., Nakamura, M., Suzuki, J. et al., *Ann. Neurol.,* 36, 149 (1994).
112. Delwart, E. L., Sheppard, H. W., Walker, B. D., Goudsmit, J., and Mullins, J. I., *J. Virol.,* 68, 6672 (1994).
113. Delwart, E. L., Shpaer, E. G., Louwagie, J., McCutchan, F. E., Grez, M., Rubsamen Waigmann, H., and Mullins, J. I., *Science,* 262, 1257 (1993).
114. Delwart, E. L., Herring, B., Rodrigo, A. G., and Mullins, J. I., in *PCR Methods and Applications,* Cold Spring Harbor Laboratory Press, New York, 1995, 4, S202.
115. Mercier, B., Gaucher, C., Feugaes, O., and Mazurier, C., *Nucleic Acids Res.,* 18, 5908 (1990).
116. *Human Retroviruses and AIDS,* Myers, G., Korber, B., Wain-Hobson, S., Smith, R. F., and Paviakis, G. N., eds., *A Compilation and Analysis of Nucleic Acid and Amino Acid Sequences,* Los Alamos National Laboratory, Los Alamos, NM, 1993.
117. Howard, T. M., Olaylele, D. O., and Rasheed, S., *AIDS Res. Hum. Retroviruses,* 10, 1755 (1994).
118. Jensen, M. A. and Straus, N., *PCR Methods Appl.,* 3, 186 (1993).
119. Rossi, J. J., Elkins, D., Zaia, J. A., and Sullivan, S., *AIDS Res. Hum. Retroviruses,* 8, 183 (1992).
120. Bevec, D., Dobrovnik, M., Hauber, J., and Bohnlein, E., *Proc. Natl. Acad. Sci. U.S.A.,* 89, 9870 (1992).
121. Carroll, R., Peterlin, B. M., and Derse, D., *J. Virol.,* 66, 2000 (1992).
122. Torrence, P. F., Maitra, R. K., Lesiak, K., Khamnei, S., Zhou, A., and Silverman, R. H., *Proc. Natl. Acad. Sci. U.S.A.,* 90, 1300 (1993).
123. Prochaska, H. J., Yeh, Y., Baron, P., and Polsky, B., *Proc. Natl. Acad. Sci. U.S.A.,* 90, 3953 (1993).
124. Smythe, J. A., Sun, D., Thomson, M., Markham, P. D., Reitz, M. S., Jr., Gallo, R. C., and Lisziewicz, J., *Proc. Natl. Acad. Sci. U.S.A.,* 91, 3657 (1994).
125. Ojwang, J. O., Hampel, A., Looney, D. J., Wong-Staal, F., and Rappaport, J., *Proc. Natl. Acad. Sci. U.S.A.,* 89, 10802 (1992).
126. Salzman, N. P., Psallidopoulos, M., Prewett, A. B., and O'Leary, R., *Clin. Orthop. Relat. Res.,* 292, 384, 390 (1993).

127. Fideler, B. M., Vangsness, C. T., Jr., Moore, T., Li, Z., and Rasheed, S., *J. Bone Joint Surg. Am. Vol.,* 76, 1032 (1994).
128. Kaczmarski, R. S., Davison, F., Blair, E., Sutherland, S., Moxham, J., McManus, T., and Mufti, G. J., *Br. J. Haematol.,* 82, 764 (1992).
129. Marche, C., Tabbara, W., and Matthiessen, L., *Ann. Med. Intern.,* 143, 191 (1992).
130. Nelson, J. A., Wiley, C. A., Reynolds Kohler, C., Reese, W., Margaretten, W., and Levy, J. A., *Lancet,* 1, 259 (1988).
131. Rodgers, V. D. and Kagnoff, M. F., *West. J. Med.,* 146, 57 (1987).
132. Cheng Mayer, C., Weiss, C., Seto, D., and Levy, J. A., *Proc. Natl. Acad. Sci. U.S.A.,* 86, 8575 (1989).
133. Supapannachart, N., Breneman, D. L., and Linnemann, C. C. J., *Arch. Dermatol.,* 127, 1198 (1991).
134. Dugel, P. U., Gill, P. S., Frangieh, G. T., Rasheed, S., and Rao, N. A., *Am. J. Ophthalmol.,* 110, 86 (1990).
135. Shibata, D., Brynes, R. K., Nathwani, B., Kwok, S., Sninsky, J., and Arnheim, N., *Am. J. Pathol.,* 135, 697 (1989).
136. Schuurman, H. J., Krone, W. J., Broekhuizen, R., and Goudsmit, J., *Am. J. Pathol.,* 133, 516 (1988).
137. Freeman, W. R., Chen, A., Henderly, D. E., Levine, A. M., Luttrull, J. K., Urrea, P. T., Arthur, J., Rasheed, S., Cohen, J. L., Neuberg, D. et al., *Am. J. Ophthalmol.,* 107, 229 (1989).
138. Nuovo, G. J., Becker, J., Simsir, A., Margiotta, M., Khalife, G., and Shevchuk, M., *Am. J. Pathol.,* 144, 1142 (1994).
139. da Silva, M., Shevchuk, M. M., Cronin, W. J., Armenakas, N. A., Tannenbaum, M., Fracchia, J. A., and Ioachim, H. L., *Am. J. Clin. Pathol.,* 93, 196 (1990).
140. Falk, L. A., Jr., Paul, D., Landay, A., and Kessler, H., *N. Engl. J. Med.,* 316, 1547 (1987).
141. Nduati, R. W., John, G. C., and Kreiss, J., *Lancet,* 344, 1432 (1994).
142. Mok, J., *Lancet,* 341, 930 (1993).
143. Thiry, L., Sprecher Goldberger, S., Jonckheer, T., Levy, J., Van de Perre, P., Henrivaux, P., Cogniaux LeClerc, J., and Clumeck, N., *Lancet,* 2, 891 (1985).
144. Landay, A. L., Schade, S. Z., Takefman, D. M., Kuhns, M. C., McNamara, A. L., Rosen, R. L., Kessler, H. A., and Spear, G. T., *J. AIDS,* 6, 171 (1993).
145. Ziza, J. M., Brun Vezinet, F., Venet, A., Rouzioux, C. H., Traversat, J., Israel Biet, B., Barre Sinoussi, F., Chermann, J. C., and Godeau, P., *N. Engl. J. Med.,* 313, 183 (1985).
146. Ho, D. D., Rota, T. R., Schooley, R. T., Kaplan, J. C., Allan, J. D., Groopman, J. E., Resnick, L., Felsenstein, D., Andrews, C. A., and Hirsch, M. S., *N. Engl. J. Med.,* 313, 1493 (1985).
147. Fitch, W. M. and Margoliash, E., *Science,* 155, 279 (1967).
148. Vogt, M. W., Witt, D. J., Craven, D. E., Byington, R., Crawford, D. F., Schooley, R. T., and Hirsch, M. S., *Lancet,* 1, 525 (1986).
149. Supapannachart, N., Breneman, D. L., and Linnemann, C. C., Jr., *Arch. Dermatol.,* 127, 1198 (1991).
150. Yourno, J. and Conroy, J., *J. Clin. Microbiol.,* 30, 2887 (1992).
151. Levy, J. A., *HIV and the Pathogenesis of AIDS,* ASM Press, Washington, D.C., 1995.

152. Yeung, S. C., Kazazi, F., Randle, C. G., Howard, R. C., Rizvi, N., Downie, J. C., Donovan, B. J., Cooper, D. A., Sekine, H., and Dwyer, D. E., *J. Infect. Dis.,* 167, 803 (1993).
153. Mermin, J. H., Holodniy, M., Katzenstein, D. A., and Merigan, T. C., *J. Infect. Dis.,* 164, 769 (1991).
154. Borzy, M. S., Connell, R. S., and Kiessling, A. A., *J. AIDS,* 1, 419 (1988).
155. Ariyoshi, K., Bloor, S., Bieniasz, P. D., Bourrelly, M., Foxall, R., and Weber, J. N., *J. Med. Virol.,* 43, 28 (1994).
156. Fujikawa, L. S., Salahuddin, S. Z., Palestine, A. G., Masur, H., Nussenblatt, R. B., and Gallo, R. C., *Lancet,* 2, 529 (1985).
157. Li, J. J., Friedman Kien, A. E., Huang, Y. Q., Mirabile, M., and Cao, Y. Z., *Lancet,* 335, 1590 (1990).
158. Ansari, S. A., Farrah, S. R., and Chaudhry, G. R., *Appl. Environ. Microbiol.,* 58, 3984 (1992).
159. Myers, S. S., Heimer, R., Liu, D., and Henrard, D., *AIDS,* 7, 925 (1993).
160. Rasheed, S., unpublished results.
161. Rasheed, S., unpublished observations.

Chapter 8

Quantitative Methods to Monitor Viral Load in Simian Immunodeficiency Virus Infections

Silvija I. Staprans, Brian C. Corliss, Jessica L. Guthrie, and Mark B. Feinberg

Contents

I. Introduction

Determination of the viral and host factors affecting progression to AIDS is essential for the elucidation of the natural history and pathogenic mechanisms of human immunodeficiency virus (HIV) infection. Given the inherent limitations associated with studies of HIV infection of humans, it is critical to maximally utilize the most relevant animal models for the study of AIDS. Simian immunodeficiency virus (SIV)-infected non-human primates provide important model systems in which to investigate the virus-host interaction. SIV infection of susceptible species, such as rhesus macaques, causes AIDS. This pathogenic model provides an opportunity to study the virological and immunologic aspects of lentivirus disease and the effects of therapeutics. The existence of SIV-infectable, but disease-resistant, primates, such as African green monkeys and sooty mangabeys (the latter being the likely reservoir for HIV-2 infection of humans), provides a unique opportunity to study host disease-resistance factors.[1]

Previous studies of SIV have been limited by the lack of quantitative methods that have been so useful in illuminating essential aspects of HIV pathogenesis. Quantitative HIV-1 assays have significantly elucidated virologic aspects of HIV-1 disease, and have provided objective criteria for assessment of antiviral efficacy. Most of these assays rely on either polymerase chain reaction (PCR) or branched DNA probe techniques.[2-5] These assays have provided great insight into the underlying virology of HIV-1 infection, in particular the appreciation that HIV-1 replication is an active and ongoing process throughout the course of HIV disease.[6,7] Furthermore, quantitative plasma HIV-1 RNA measurements appear to be an important clinical marker of disease progression and antiretroviral drug efficacy.[8,9]

In order to fully study the virologic aspects of SIV disease, we have developed and applied quantitative methods to monitor viral load in SIV infection of macaques (SIVmac), sooty mangabeys (SIVsmm), and African green monkeys (SIVagm). This chapter describes the detailed methodology necessary to perform these assays.

II. Methods

A. Overview (Figure 8.1)

SIV plasma viremia is measured by a quantitative-competitive RNA PCR (QC-PCR) assay adapted from that described by Piatek et al. to measure HIV-1 RNA and DNA levels.[2] In this method, a synthetic competitive RNA template, which is derived from viral sequences but is distinguishable from the real viral RNA by virtue of an internal deletion or insertion, is included in the reverse transcription and amplification steps to provide a stringent internal copy number control. This method allows accurate quantitation of viral load in SIV infections.

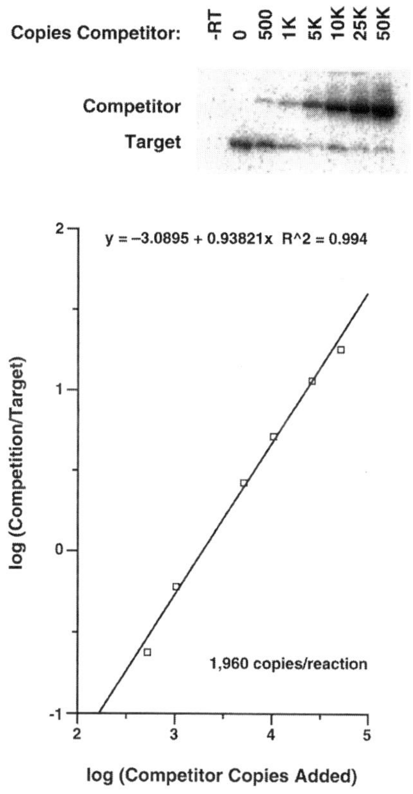

FIGURE 8.1
Application of the QC-PCR assay for quantitation of SIV target sequences. Top: Radiolabeled gel for QC-PCR analysis of input 2,000 copies of wild-type SIVmac gag RNA (pSIVgag, 150 bp). Increasing amounts (0 to 50,000 copies) of competitive template (pSIVgag+40, 190 bp) were added to the reaction series. Bottom: Plot of data obtained by quantitation of radioactivity in competitor and wild-type templates (using procedures described in the text). Determination of the equivalence point (includes correction for relative sizes of products) is in good agreement with the calculated input copy number.

B. Production of Wild-Type Control Plasmids and Competitor Templates (Figure 8.2)

1. SIVmac and SIVsmm

A region in the SIVmac239 *gag* gene, flanked by sequences conserved among almost all HIV-2-related SIV isolates, was selected as the target sequence.[10] A plasmid containing SIVmac239 sequences (pSIVgag), and a plasmid identical to pSIVgag except for an internal 40-bp insertion (pSIVgag+40, constructed by overlap extension PCR[11]) were generated for use as a positive control target sequence and competitor templates, respectively. A 162-bp fragment of the SIV *gag* gene corresponding to positions 129–291 of the SIVmm239 sequence,[12] generated by PCR amplification with primers containing restriction site recognition sequences, was subcloned into pBluescript II SK+ (Stratagene, La Jolla, CA). An analogous fragment containing a 40-bp insertion (between positions 199 and 200) generated by overlap extension PCR was also cloned into the same vector. The SIV sequences in pSIVgag and pSIVgag+40 are under the control of the T7 promoter to allow generation of the corresponding 232-nt and 272-nt *in vitro* transcripts for use as positive control and competitive templates, respectively, in RNA PCR.

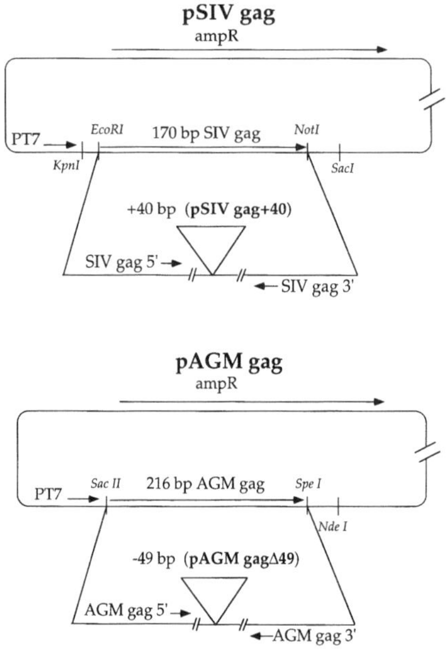

FIGURE 8.2
Plasmids for SIVmac and SIVsmm (pSIVgag), and SIVagm (pAGMgag) QC-PCR assays. The wild-type SIVgag target sequences and corresponding insertion- or deletion-containing competitor templates and PCR primer positions are shown.

The conserved primer sites contained within these control templates are the following:

SIVgag5′ — AGA AAG CCT GTT GGA IAA CAA AGA AGG-3′ (positions 141–167,[12] conserved among all SIVmac and SIVsmm isolates. "I" denotes an inosine residue.)

SIVgag3′ — 5′-AGT GTG TTT CAC TTT CTC TTC TGC GTG-3′ (complement of positions 291–265,[12] conserved among all SIVmac and SIVsmm isolates)

This primer pair amplifies 150-bp and 190-bp fragments for the wild-type and competitor templates, respectively.

2. SIVagm

The different subspecies of African green monkeys harbor diverse SIVagm viruses. Thus, vervets, grivets, tantalus, and sabaeus monkeys each harbor their own subspecies-specific SIVs. To develop a quantitative assay that can detect most SIVagm isolates, a region in the SIVagm *gag* gene, flanked by sequences conserved among the tantalus, vervet, and grivet isolates of SIVagm, was selected as the target sequence. A 216-bp fragment of SIVagm *gag* sequence (corresponding to positions 88–303 of the TYO isolate of SIVagm[13]) was amplified from isolate Kenya Cpa66 (Reagent No. 1012 obtained through the AIDS Research and Reference Reagent Program, Division of AIDS, National Institute of Allergy and Infectious Diseases, National Institutes of Health: Molt 4 Clone 8/SIVagm from Dr. Ronald Desrosiers[14]) by PCR and cloned into the pGEM-T vector (Promega, Madison, WI). An analogous fragment containing a 49-bp deletion (between positions 172 and 219 of the TYO isolate sequence) was generated by overlap extension PCR and cloned into the same vector. The SIV sequences in pAGMgag and pAGMgagΔ49 are under the control of the T7 promoter, allowing the generation of 298-nt and 249-nt *in vitro* transcripts for use as positive control and competitive templates, respectively, in RNA PCR.

The conserved primer sites contained within these control templates are:

SIVagm5′ — 5′-CAA ITI AAA CAT TTA ATA TGG GCA GG-3′ (positions 88–113 TYO isolate sequence,[13] conserved among tantalus, vervet, and grivet SIVagm isolates. "I" denotes an inosine residue.)

SIVagm3′ — 5′-TAC IGC TTC ITC TGT GTC TTT CAC TT-3′ (complement of positions 303–278 TYO isolate sequence,[13] conserved among tantalus, vervet, and grivet SIVagm isolates. "I" denotes an inosine residue.)

This primer pair amplifies 216-bp and 167-bp fragments for the wild-type and competitor templates, respectively.

C. RNA Preparation

1. *In vitro* transcription of control RNAs

RNA is produced from plasmid DNA templates using T7 polymerase according to the Promega RiboMAX™ (Promega, Madison, WI) protocol (Promega Technical Bulletin No. 166 [1993][15,16]). Plasmid templates are linearized prior to *in vitro* transcription by digestion with NotI (pSIVgag) or NdeI (pAGMgag) so that run-off transcripts can be generated. Purified, linear templates are transcribed by incubation with T7 RNA polymerase, rNTPs, RNAsin, and an appropriate buffered solution. Thereafter, the template is removed by digestion with RNase-free DNase. Unincorporated nucleotides are removed from the RNA by size exclusion chromatography through a small RNase-free Sephadex G50 column (5 Prime→3 Prime, Inc., Boulder,

CO) and ethanol precipitation. A portion of the RNA is resuspended in formaldehyde gel denaturing buffer[17] and analyzed by native agarose gel electrophoresis to verify its integrity and size. The RNA concentration is determined by ultraviolet light absorbance at wavelengths of 260 and 280 nanometers. Calculation of the RNA copy number, and QC-PCR reconstruction analysis of copy number controls ("2,000 copy control", see below) are an important validation of the RNA copy number as determined by spectrophotometric absorbance.

2. Isolation of virion RNA from plasma samples

a. Processing of SIV$^+$ plasma (follow appropriate biosafety procedures)

1. Use tubes containing acid citrate dextrose (ACD, solution A) as an anticoagulant (vacutainer, Becton Dickinson, Rutherford, NJ). Blood should be processed as quickly as possible, preferably within 3 h of being drawn.
2. Spin cells out (~1000 *g*, 10 to 15 min). (Alternatively, cells can be Ficoll-purified for analysis of proviral DNA or viral RNA expression in the peripheral blood mononuclear cells [PBMCs]).
3. Centrifuge plasma with a second, harder spin (1600 *g*, 15 min) to remove cellular debris; this is critical to remove any contaminating DNA. Label tubes beforehand to indicate the side of the tube on which the tiny pellet will be expected to form. When removing plasma supernatant, take care not to disturb the "pellet" (i.e., leave some of the last remaining supernatant on top of the pellet so as to avoid any possibility of contaminating the plasma with debris).
4. Plasma should be frozen immediately in 1 ml aliquots in screw cap tubes (Sarstadt, Newton, NC) at –70°C and should not be thawed and refrozen; plasma can be stored in this form for up to several years. Virus can be pelletted and processed for QC-PCR at a later time.

b. Viral RNA extraction

1. Thaw 1 ml aliquots of plasma in a 37°C water bath.
2. Centrifuge 39,500 *g* (22,000 rpm in an HFA22.1 rotor in a Heraeus Centrifuge [Heraeus Instruments, South Plainfield, NJ]), 4°C, 1 h.
3. Aspirate supernatant.
4. Add 0.8 ml of TRIZOL reagent (Gibco/BRL, Grand Island, NY) and 6.5 ng of "carrier" RNA (7.5 kb polyadenylated RNA [Gibco/BRL, Grand Island, NY]) to each sample. (Just prior to lysis, make a fresh master mix of TRIZOL reagent and 7.5 kb RNA.)
5. Vortex. Incubate at room temperature, 5 min.
6. Add 0.2 ml chloroform and hand shake for 15 sec.
7. Incubate at room temperature, 5 min.
8. Centrifuge (39,500 *g*), 4°C, 10 min.
9. Transfer aqueous layer (usually ~450 µl) to polyallomer microfuge tubes (Beckman, Fullerton, CA); add 20 µg glycogen (Boehringer Mannheim, Indianapolis, IN).
10. Add 0.5 ml isopropanol and invert a few times to mix.

11. Centrifuge (39,500 *g*), 4°C, 10 min.
12. Remove supernatant. Wash pellet with 1 ml 70% cold ethanol (optional for high titer RNAs). If pellet becomes dislodged from bottom of tube during wash, recentrifuge (as in step 11) before proceeding.
13. Remove supernatant and air dry 10 to 15 min. Resuspend in 100 µl RNAse-free H_2O.
14. Store RNA at –70°C. In concentrated form, the RNA appears to be stable for at least two years at –70°C.

Note: *Cellular RNA from* $\leq 10^7$ *cells lysed in TRIZOL can be extracted according to same protocol.*

D. Detailed QC-PCR Protocol

Implementation of RNase-free technique and the avoidance of all possible contamination from extraneous PCR products is essential for the successful application of this technique; these precautions cannot be overemphasized. We recommend that all reagents and equipment, including pipetters and aerosol-resistant pipet tips, be kept in a separate "Pre-PCR Room," which should be used solely for performing this assay and should remain free of PCR product contaminants at all times. These reagents and equipment should not be interchanged with other supplies in the lab.

1. Prepare a completed set of QC-PCR Worksheets (see the appendix at the end of the chapter), including the sample numbers to be assayed. This assay is run in a 96-well plate format; therefore, it is necessary to have a thermal cycler that can accommodate 96-well plates (MJ Research, Inc., Watertown, MA). Each specimen to be assayed requires eight sample wells. Use the accompanying QC-PCR Worksheets as a cross-reference when reading this protocol.
2. Thaw reagents and prepare for RT reactions
   a. Thaw at room temperature all reagents for RT MASTER MIXES (10X PCR Buffer [500 m*M* KCl, 100 m*M* Tris-HCl, pH 8.3; Perkin-Elmer, Norwalk, CT], 25 m*M* $MgCl_2$, 20 m*M* each dNTPs, 50 µ*M* Random Primers [Promega, Madison, WI], 100 m*M* DTT) except RNAsin and Reverse Transcriptase (RT) (see Worksheet 1).
   b. Fill one ice bucket, one shallow tray, and one tray for the 96-well plate with ice.
   c. Pour an aliquot of bottled RNase-free H_2O (USB, Cleveland, OH) into a 15-ml tube and place on ice.
   d. Place the 96-well plate (MJ Research, Inc., Watertown, MA) on ice, make sure the tube bottoms are submerged in ice water.
   e. Vortex and spin the thawed reagents; transfer to ice bucket.
3. Make RT Master Mixes and the PCR mix
   a. In microfuge tubes on ice, make up the RT MASTER MIXES *without* the RNAsin and RT, as indicated on Worksheet 1. Alternatively, these master mixes may be made beforehand (without RT and RNAsin), aliquoted in appropriate volumes, and stored at –20°C until use.
   b. Make the PCR MIX (see Worksheet 4) without the isotope and Taq polymerase.

4. Prepare the RNAs
   a. Thaw on ice: wild-type and competitor RNAs and any sample RNAs to be run. (RNAs are stored at –70°C. Wild-type RNA (from pSIVgag or pAGMgag) is prepared at a concentration of 100,000 copies per microliter and stored in 10 μl aliquots. The competitor RNA (from pSIVgag+40 or pAGMgagΔ49) is prepared at a concentration of 500,000 copies per microliter and stored in 10 μl aliquots. Do *not* reuse dilute RNA samples that have undergone repeated rounds of freeze-thawing; a fresh aliquot of RNA should be used for each new assay.)
   b. Place a prechilled metal microfuge tube rack (stored at –20°C) on ice in the shallow tray. Label sterile, RNAse-free 0.5-ml tubes for the indicated wild-type and competitor RNA dilutions and place them in the rack.
   c. Dilute the RNAs. Aliquot the required amounts of water for the dilution series, then dilute as indicated on Worksheet 2, briefly vortexing and spinning the tubes after each dilution. Alternative dilution schemes to the one illustrated may be used when analyzing specimens known to contain lesser amounts of viral RNA.
5. Prepare the reverse transcription reactions
   a. Add the RT (Superscript, Gibco/BRL, Grand Island, NY) and RNAsin (Promega, Madison, WI) to the RT MIXES as indicated (see Worksheet 1); vortex and spin briefly to mix.
   b. Using a repetitive pipetter (Eppendorf Repeater, Brinkmann Instruments, Westbury, NY), add 20 μl of the appropriate RT MIX to the indicated wells (see Worksheet 3), making sure to dispense the mix into the bottom of each well. (The 96-well plates may be RNAse contaminated; adding the mixes to the wells first reduces the possibility of RNA degradation.)
   c. Following the template on Worksheet 3, for each sample to be assayed, aliquot 5 μl of RNase-free H_2O or diluted competitor RNA (from pSIVgag+40 or pAGMgagΔ49) as indicated. This is done with an electronic pipettor (Eppendorf), capable of repeatedly aliquoting small volumes of solution, and a 100 μl aerosol resistant pipet tip. Dispense the 5 μl into the mix.
   d. Add 5 μl of "2,000 copy control" RNA (from pSIVgag or pAGMgag) or each test RNA to the eight wells within the vertical column assigned to that RNA on Worksheet 3. Dispense the RNA directly into the mix with the electronic pipetter as before. Aliquot the RNA sequentially from the top well (no competitor/no RT well) down to the bottom well (highest competitor concentration).
   e. Overlay the reactions with 1 to 2 drops of light mineral oil (Fisher Scientific, Fair Lawn, NJ) using a pasteur pipet, taking care not to splash.
   f. Cover the plate with wide adhesive tape (3M Co., Elk Grove Village, IL) or a 96-well plate sealer. Press lightly to seal (do this while the plate is on the heating block of the thermal cycler). It is best to use a thermal cycler (or heating block) designated for reverse transcription only (no PCR reactions) to avoid the possibility of PCR product contamination of the RT reactions.
6. Reverse transcription
   a. Allow the random hexamer-priming reaction to proceed for 10 min at room temperature.
   b. Start the RT program (see conditions on Worksheet 3); make sure that there is good contact between the wells and the block; use mineral oil in the block wells if necessary.

c. Add the appropriate primers and Taq polymerase (Stratagene, La Jolla, CA) to the PCR MIX as indicated on Worksheet 4; keep on ice until use.

d. While the RT program is underway, prepare to set up the subsequent PCR reaction.

7. Prepare for PCR:

a. Equipment for the PCR set-up should be kept in a designated "clean" area, free of any concentrated DNA or PCR products. Thaw the isotope, prepare radioactive disposal containers, and lay out pipetters, tips, etc.

b. After the RT reaction is complete, place the 96-well tray on ice.

8. PCR reaction

a. Remove the adhesive from the 96-well plate by gently peeling it off and avoiding any splatter; change gloves after the adhesive has been removed to avoid contamination. (Static caused by peeling off the tape may introduce contamination when adding the PCR mix. Use of an antistatic spray [Staticide, ACI, Inc., Elk Grove Village, IL] can prevent this.)

b. Add isotope to the PCR MIX (see Worksheet 4), mix, and return to ice.

c. Using a repetitive pipetter, dispense 30 μl of PCR MIX into the reaction mix in each well.

d. Cover the plate with adhesive; press to seal as before. On ice, bring the plate to the thermocycler.

e. Perform thermocycling as indicated on Worksheet 4: initiate the thermocycling program; when the block temperature reaches 90°C, transfer the 96-well plate from ice to the hot block ("hot start method").

9. Gels: The PCR products can be analyzed by acrylamide gel electrophoresis of radiolabeled products, or agarose gel electrophoresis of ethidium bromide-stained material. The method of choice will depend on the availability of quantitative scanning equipment necessary for the precise quantitation of the wild-type and competitor PCR products. We describe acrylamide gel electrophoresis of radiolabeled products here. The ideal gel apparatus is wide (40 cm width, 20 cm length), capable of accommodating multiple samples in sets of eight (e.g., 40 or 48 wells).

a. Gel: 6% acrylamide (19:1 acrylamide: bisacrylamide) in 1X TBE (10.8 g Tris base, 5.5 g boric acid, 4 ml 0.5 *M* ethylene diamine tetraacetic acid [EDTA, pH 8.0], bring to 1 l with dH_2O).

b. Clean glass plates and spacers. Coat the eared glass plate with Sigmacoat (Sigma, St. Louis, MO) or equivalent to prevent sticking of the gel. Dry and assemble.

c. Combine 50 ml 6% acrylamide, 500 μl 10% ammonium persulfate, and 50 μl TEMED (Sigma, St. Louis, MO). Mix well by swirling. Pour the gel mixture into the glass plate sandwich at a steady rate.

d. Allow the gel to polymerize for 15 to 20 min.

e. Using a repetitive pipetter, add 20 μl loading dye (0.25% bromophenol blue [Fisher Scientific, Fair Lawn, NJ], 0.25% xylene cyanol FF [USB, Cleveland, OH], 15% Ficoll [Type 400, Pharmacia, Uppsala, Sweden] in water) to each well of the 96-well plate and mix well with a multichannel pipettor.

f. Load approximately 8 μl of sample per well.

g. Run the gel at 200 to 300 V for about 1.5 h (until the xylene cyanol dye is almost to the bottom).

h. Remove the gel onto filter paper (Whatman 3MM, Fisher Scientific, Fair Lawn, NJ) and cover with plastic wrap.
i. Dry the gel at 80°C, under vacuum, for 30 min. Place the gel into a cassette and place a Fuji BAS-IIIs Imaging Plate (Fuji Photo Film Co., Japan, or equivalent) on top; close.
j. Expose the plate for 30 to 40 min.
k. Remove the imaging plate in the dark room. Scan the imaging plate using the FUJIX Bio-Imaging Analyzer System (BAS 1000, or equivalent) (Fuji Photo Film Co., Japan) and quantitate the radioactivity in each band.
l. Transfer the information to a computer (Macintosh) for data analysis (see below).

E. Sample Analysis and Quantitation

After the gel is exposed to the Fuji (BAS-IIIs) Imaging Plate (IP) (Fuji Photo Film Co., Japan), the IP is scanned and analyzed by using the FUJIX Bio-Imaging Analyzer System (BAS 1000) with matched software (MacBas v1.0) (Fuji Photo Film Co., Japan). The work station includes a plate scanner, Macintosh Quadra with a magnetic optical disc drive, IP eraser, and printer for hard copies. Once the IP is scanned, the data are transferred to the computer for analysis and quantitation. The density of each DNA band is proportional to the DNA present. Bands of interest are captured on the monitor for analysis, and the density of radioactivity associated with each band is determined by MacBas. We transfer the density quantitation data to Microsoft Excel v5.0 (Microsoft Corp., Redmond, WA) to determine the $\log_{10}$ ratios and then to plot the data and determine the linear regression equation. The $\log_{10}$ number of copies of competitor added to each reaction and the $\log_{10}$ of the corrected ratio of the competitor/wild-type species are plotted on the horizontal and vertical axes, respectively (Figure 8.1).

The wild-type SIVmac and SIVsmm gag PCR product is 150 bp, and the corresponding product for the competing species of SIVmac and SIVsmm gag is 190 bp. Because QC-PCR is based on molar concentrations of product, it is necessary to correct for the size difference between the 190-bp species and the 150-bp species by multiplying the 190-bp species by a correction factor of (150/190). This allows for direct comparison of the corrected density of the 190-bp band with the density of the 150-bp band. The equivalence point is determined by plotting the $\log_{10}$ of the ratio of the corrected density of the 190-bp band over the density of the 150-bp band [$\log_{10}$(corrected density 190/density 150)], as a function of the $\log_{10}$ of the number of copies of competitor added to each reaction. The resulting plot is linear. At the equivalence point, the corrected density of the 190-bp band should equal the measured density of the 150-bp band, and their ratio is equal to 1. When the linear regression equation is solved for $Y = 0$ ($\log_{10} 1 = 0$), it yields the number of copies corresponding to the equivalence point and, therefore, the number of copies in the unknown sample. The unknown sample usually corresponds to 1/20th of the original sample (5 μl from a total of 100 μl of RNA); the final results are reported as "n SIV RNA copies per ml of plasma." Analysis of SIVagm RNA copy number is the same, except that the wild-type and competitor PCR product sizes are 216 and 167 bp, respectively.

F. Simplification of the Assay

Simplification of the QC-PCR assay is in progress and involves the following changes:[18] direct lysis of the plasma omits the virion pelleting step. Direct precipitation of the virion RNA-containing lysis solution omits sample loss associated with organic extraction and sample transfer. Use of rTth DNA polymerase (Perkin-Elmer, Norwalk, CT) for both reverse transcription and PCR with a single buffer solution obviates the need for preparation of a second set of reagents for the PCR reaction. Finally, alternative methods for PCR product quantitation, such as fluorescent probe-based detection, would allow for direct quantitation of the PCR products in a 96-well plate or gel-free analysis (Perkin Elmer TaqMan, ABI, Foster City, CA). These simplifications have not yet been validated in the context of the described SIV QC-PCR assays.

III. Application of the Assay

A. Assay Sensitivity

This assay can reproducibly detect as few as 50 to 100 copies of SIV RNA per reaction. Although we have been able to detect as few as 10 to 20 copies of SIV RNA, we have not optimized the RT-PCR conditions to do so reproducibly.

B. Assay Reproducibility

Assay reproducibility was determined by duplicate testing of a small panel of split plasma specimens. All steps of the duplicate assays were performed on separate days, including RNA extraction from plasma, reverse transcription, PCR amplification, and gel analysis. The reproducibility of the assay was high, as indicated by a small variance between duplicate tests. The variance was 0.24 on a natural log scale, indicating that 95% of repeated measurements will be less than 2.6-fold different from the mean.

C. Assay Validation

We and others have previously validated the HIV-1 QC-PCR method by comparing it to the branched (b)DNA method for quantitation of HIV-1 RNA.[9,9a] As the SIV QC-PCR method described here uses the same methodology, it is likely that the numbers obtained by this assay represent a fairly accurate absolute quantitation of SIV RNA copy number. To verify this, we have performed limiting dilution PCR assays of SIV RNA-containing specimens. Based on our known SIV PCR assay sensitivity (10 to 50 copies SIV RNA per reaction), the estimated SIV copy numbers that we obtain in limiting dilution studies agree with those obtained by the QC-PCR method. The SIV QC-PCR assay has been further validated through demonstration of high correlation of viral load as determined by QC-PCR, quantitative PBMC culture, and p27 antigen levels. Comparison of a limited panel of plasma samples run in parallel using SIVmac QC-PCR and an SIVmac bDNA assay (P. Dailey, Chiron Corp.) indicated high degrees of correlation ($R^2 = 0.96$). Thus, preliminary validation of the SIVmac QC-PCR method by comparison to the SIVmac bDNA assay suggests good agreement between the two methods; further validation is ongoing. SIVsmm QC-PCR did not correlate well with the SIVmac bDNA assay as expected, given the bDNA probes were designed to anneal to SIVmac sequences that were not conserved in SIVsmm.

D. Measurement of Plasma Viremia in SIV Infection

This assay can be applied to cross-sectional or longitudinal studies. Given that interassay variation can be as high as 2.6-fold, it is important to make comparisons of interest within one assay.

Figure 8.3 shows the analysis of plasma viremia in a rhesus macaque undergoing primary infection following intravenous inoculation with SIVmac251;[19] the levels of plasma viremia appear to be one to two orders of magnitude higher than those that have been observed for primary HIV-1 infection.[3] The level of SIV replication appears to be influenced by the route of inoculation,[20] and may differ with different SIV isolates.

While the region of the SIVagm gag chosen for amplification in this assay is best conserved among the tantalus, grivet, and vervet isolates of SIVagm, we have also had success in amplifying SIV RNA from sabaeus African green monkeys. However, due to sequence heterogeneity in the SIVsabaeus gag sequences amplified in the SIVagm QC-PCR assay, this assay may underestimate the SIV RNA copy number present in sabaeus monkeys.

IV. Conclusion

The new methods described here provide sensitive and accurate measures of SIVmac, SIVsmm, and SIVagm load in infected primates. Intense virologic analysis of the simian AIDS model, and comparison to nonpathogenic SIV infection models, will elucidate the underlying pathogenic processes of AIDS. A clearer understanding of the biology of SIV infection will lead to increased understanding and improved treatment of HIV disease.

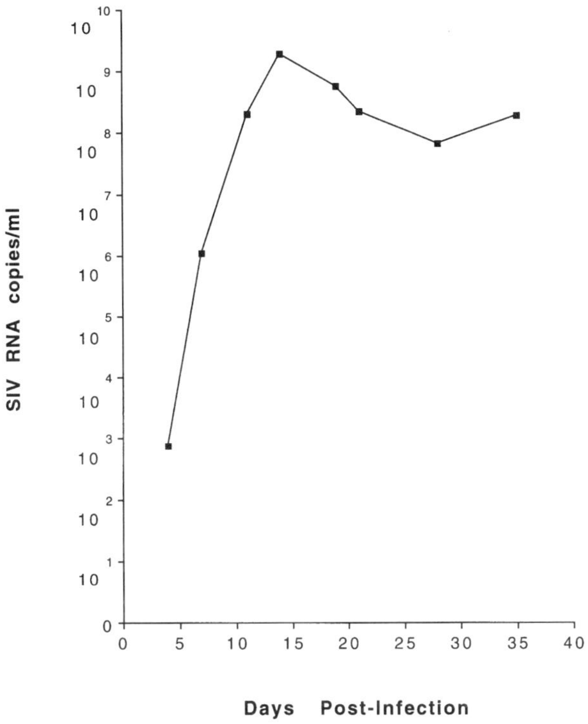

FIGURE 8.3
Measurement of plasma SIV RNA following intravenous inoculation of SIVmac251 into a rhesus macaque, days 4 to 35.

Acknowledgments

Drs. Jeff Lifson and Mike Piatak of Genelabs Technologies, Inc. (Redwood City, CA) generously provided reagents and advice concerning the HIV-1 QC-PCR method, on which the SIV QC-PCR method described here was modeled and developed. Drs. Ann Rosenthal, Nick Lerche, Jim Carlson, and Christopher Miller generously provided the SIV-infected macaque specimens used to validate the QC-PCR assay for SIVmac. Dr. Harold McClure provided SIV-infected sooty mangabey specimens, and Dr. Jonathan Allen provided SIV-infected African green monkey specimens for the validation of the SIVsmm and SIVagm assays, respectively. Dr. Robert M. Grant provided advice on assessing performance characteristics of the assays. This work was supported by the J. David Gladstone Institutes (M.B.F.) and grants from the American Foundation for AIDS Research (AmFar grant No. 02364-17-RG) and the University of California, San Francisco AIDS Clinical Research Center (ACRC grant No. CC93-SF130). S.I.S. was a Pediatric AIDS Foundation Scholar during the time that these assays were developed.

References

1. Gardener, M. B., Endres, M., and Barry, P., in *The Retroviridae,* Vol. 3, Levy, J. A., Ed., Plenum Press, New York, 1994, 133.
2. Piatak, M., Jr., Luk, K. C., Williams, B., and Lifson, J. D., *BioTechniques,* 14, 70 (1993).
3. Piatak, M., Jr. et al., *Science,* 259, 1749 (1993).
4. Pachl, C. et al., *J. AIDS Hum. Retrovirol.,* 8, 446 (1995).
5. Elbeik, T. and Feinberg, M. B., in *The AIDS Knowledge Base,* 2nd ed., Cohen, P. T., Sande, M. A., and Volberding, P. A., Eds., Little, Brown, Boston, 1994 2.4-1.
6. Ho, D. D. et al., *Nature,* 373, 123 (1995).
7. Wei, X. et al., *Nature,* 373, 117 (1995).
8. Mellors, J. W. et al., *Ann. Intern. Med.,* 122, 573 (1995).
9. Cao, Y. et al., *AIDS Res. Hum. Retroviruses,* 11, 353 (1995).

9a. Staprans, S., Guthrie, J., and Feinberg, M., unpublished data.

10. Regier, D. A. and Desrosiers, R. C., *AIDS Res. Hum. Retroviruses,* 6, 1221 (1990).
11. Ho, S. N., Hunt, H. D., Horton, R. M., Pullen, J. K., and Pease, L. R., *Gene,* 77, 51 (1989).
12. Meyers, G. et al., Eds., *Human Retroviruses and AIDS 1994, A Compilation and Analysis of Nucleic Acid and Amino Acid Sequences,* Los Alamos National Laboratory, Los Alamos, New Mexico, 1994.
13. Meyers, G., Wain-Hobson, S., Smith, R. F., Korber, B., and Pavlaki, G. N., Eds., *Human Retroviruses and AIDS 1993 I-II. A Compilation and Analysis of Nucleic Acid and Amino Acid Sequences,* Los Alamos National Laboratory, Los Alamos, New Mexico, 1993.
14. Daniel, M. D. et al., *J. Virol.,* 62, 4123 (1988).
15. Gurevich, V. V. et al., *Anal. Biochem.,* 195, 207 (1991).
16. Cunningham, P. R. and Ofengand, J., *BioTechniques,* 9, 713 (1990).
17. Ausbel, F. M. et al., Eds., *Current Protocols in Molecular Biology,* Vol. 1, John Wiley & Sons, 1993.
18. Mulder, J. et al., *J. Clin. Microbiol.,* 32, 292 (1994).
19. Daniel, M. D., Kirchhoff, F., Czajak, S. C., Sehgal, P. K., and Desrosiers, R. C., *Science,* 258, 1938 (1992).
20. Staprans, S., Miller, C., and Feinberg, M., unpublished data.

Appendix 1: Quantitative Competitive (QC)-PCR Worksheets

Worksheet 1

This assay is performed in a 96-well plate format. Each sample requires eight reaction wells; therefore, the number of samples is multiplied by 8 when calculating the amount of reaction mixes to make. (For convenience, round up to nearest 10.)

Reverse Transcription (RT) Master Mixes

	1X (+RT Master Mix) µl	X____ µl	Check	1X (–RT Master Mix) µl	X15 µl	Check
Bottled, sterile H_2O for irrigation	5.85			6	90	
10X PCR buffer	3			3	45	
$MgCl_2$, 25 m*M*	6			6	90	
dNTPs, 20 m*M* each	1.5			1.5	22.5	
Random primers, 50 µ*M*	1.5			1.5	22.5	
DTT, 100 m*M*	1.5			1.5	22.5	
rRNAsin, 40 U/µl	0.5			0.5	7.5	
Reverse transcriptase, 200 U/µl	0.15			—	—	
Total	20			20	300	

Per 30 µl reaction:(1) 20 µl RT Master Mix
(2) 5 µl competitor RNA or water
(3) 5 µl test RNA or 2,000 copies wild-type control RNA

Worksheet 2

Dilutions of Control RNAs

Wild-type RNA template (from pSIVgag or pAGMgag)

Control transcript: 100,000 copies/μl STOCK, 10 μl aliquots

10 μl STOCK + 90 μl water = 10,000 copies/μl→10 μl + 90 μl water = 1,000 copies/μl

2000 copies/5 μl control

	X1 μl	X20 μl
RNAse-free H_2O	2.7	54
16 μg glycogen + 4 ng 7.5 kb RNA/3 μl (optional)	0.3	6
1000 copies/μl	2	40
Total	5	100

Competitor RNA template (from pSIVgag+40 or pAGMgagΔ49)

Competitor transcript: 500,000 copies/μl STOCK, 10 μl aliquots; dilute with RNase-free H_2O as follows:

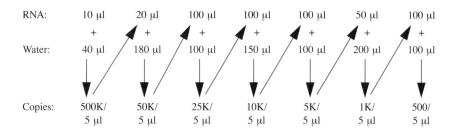

RNA:	10 μl	20 μl	100 μl	100 μl	100 μl	50 μl	100 μl
	+	+	+	+	+	+	+
Water:	40 μl	180 μl	100 μl	150 μl	100 μl	200 μl	100 μl
Copies:	500K/ 5 μl	50K/ 5 μl	25K/ 5 μl	10K/ 5 μl	5K/ 5 μl	1K/ 5 μl	500/ 5 μl

Worksheet 3

SIV RNA QC-PCR Sample Plate

	1	2	3	4	5	6	7	8	9	10	11	12	RT Mix 20 µl	Competitor 5 µl
A													–RT	H_2O
B													+RT	H_2O
C													+RT	500
D													+RT	1K
E													+RT	5K
F													+RT	10K
G													+RT	25K
H													+RT	50K
SAMPLE														

1. Add rRNAsin and RT to Master Mixes, then add 20 µl ± RT mix to each well.
2. Add 5 µl H_2O (rows A–B) or competitor RNA (rows C–H).
3. Add 5 µl 2,000 copy control (lane 1) or test RNA (lanes 2–12).
4. Seal top of plate and incubate 10 min at room temperature.
5. Perform reverse transcription reaction.

RT Reaction Conditions: 42°C, 30 min; 99°C, 5 min; 4°C for minimum of 2 min.

Worksheet 4

PCR Master Mix

	X1 μl	X_____ μl
Bottled sterile H_2O	26	
10X PCR buffer	3	
10 μCi/μl $dCTP^{32}$ (3,000 Ci/mmol)	1 μCi = 0.1 μl/rxn	
SIV gag 5′, 100 pmol/μl Or AGM gag 5′	0.2	
SIV gag 3′, 100 pmol/μl Or AGM gag 3′	0.2	
Taq polymerase, 5 U/μl	0.5	
Total	30	

PCR conditions for SIVmac/smm: 1 cycle: 94°C, 3 min; 45 cycles: 93°C, 30 sec, 59.5°C, 30 sec, 72°C, 55 sec; 1 cycle: 72°C, 1–10 h (prolonged 72°C incubation may reduce heteroduplex formation); 4°C hold.

PCR conditions for SIVagm: same except that the annealing step is carried out at 51°C rather than 59.5°C.

Chapter 9

Using Molecular Genetics to Probe the Immune Response to Vesicular Stomatitis Virus

Carol S. Reiss, Shelli A. Oien, and Shirley M. Bartido

Contents

0-8493-4412-3/96/$0.00+$.50

I. Introduction

The availability of molecular genetic tools and techniques has had a profound impact on the analysis of the immune response to viral infection. This promises to result in the rapid development of vaccines and therapeutics for these important diseases.

In the last decade, the use of cDNA cloning of expressed viral genes has led to the development of libraries of viral genes that have been cloned and sequenced. Subsequently, regulatory regions of genes have been analyzed, as well. These approaches, combined with comparative sequence analysis, have resulted in (1) studies of homologies of sequences, (2) evolutionary analyses of relatedness of pathogens, (3) predictions of active sites or post-translationally modified region(s) of proteins, (4) hypotheses of molecular mimicry in autoimmunity, and (5) even discernment of the capture and reuse of host proteins by a number of viruses for their selective advantage (e.g., the inclusion of interleukin [IL]-10 in Epstein-Barr virus).[1]

In some cases, entire viral genomes have been cloned into one vector (pioneered by Racaniello and Baltimore[2] for polio virus), or in the cases of larger viral size and complexity, into several vectors or plasmids. Coexpression of one or more viral genes has resulted in examination of the function of viral polymerases or replicases, which can permit the development of new classes of antiviral drugs to use as therapeutics.

Examination of functional domain(s) of viral proteins has also been made possible using molecular genetic approaches. For instance, the regions involved in receptor binding or intracellular traffic, such as targeting to different compartments, has been investigated by exchange or deletion of fragments of viral genes and expression of the proteins either transiently or following transfection.[3]

However, while we can learn a lot about the virus and its proteins using this approach, we can use this methodology to probe the immune response(s) to virus as well. The latter can promote development of vaccines, knowing the proteins that are critical for immune recognition, as well as provide basic understandings of the workings of the immune system to real, not hypothetical, antigens. In some cases, idealistic and naive attempts to develop vaccines resulted in more severe infections; specifically, inactivated virus (respiratory syncytial virus and measles virus) vaccines administered in the 1960s resulted in much more severe infections, not protection of the children involved.[4]

The immune response to infection involves innate (nonspecific) responses, such as cytokines and phagocytic cells, as well as antigen-specific humoral and cellular immunity. The cellular immunity can be divided further into two groups: CD4-expressing T cells, which recognize class II major histocompatibility complex (MHC) heterodimers in association with viral peptides, and CD8-expressing T cells, which recognize class I MHC in association with viral peptides. Any structural or nonstructural viral protein can participate in this binding; functional interaction is determined solely by the availability of the relevant peptide bearing a MHC-determined motif of amino acids in specific positions, and its ability to engage the physical pockets of the groove of the MHC heterodimer (reviewed by Engelhard[5]).

Both of these thymus-derived cell populations can secrete cytokines and have cells capable of specifically lysing virally infected cells. This form of host response to viral infection was first described by Rolf Zinkernagel and Peter Doherty in 1975[6] and is exquisitely specific for both host and virus. Both CD4+ and CD8+ T cells have specialized proteins on their surface that are called the T cell receptor (TCR) complex, which includes the variable αβ chains (which bind the MHC thermolecular complex) and the CD3 γδεζη chains (generally referred to as the CD3 complex). A detailed description of how the viral peptides get made and how they are loaded into the MHC, termed "antigen processing and presentation", is beyond the scope of this chapter (interested readers are referred to Reference 7).

This chapter focuses on the experimental procedures that are used to analyze the humoral and cellular immune responses to a virus infection. The one studied in our laboratory is vesicular stomatitis virus, an RNA virus closely related to rabies. This virus has a glycoprotein that is embedded in the lipid envelope, a single strand of RNA, which is coated with a nucleoprotein, phosphoprotein and large protein (which comprise the polymerase complex), and a matrix protein. There are two principal serotypes (Indiana and New Jersey), which are principally distinguished by changes in amino acids in the glycoprotein. In the experimental model used, virus is applied to the nose of the mouse and causes an infection that enters the brain via the olfactory nerve.[8-10] We have shown a critical requirement for T lymphocytes in recovery from the infection. Antibody (neutralizing antibody to the glycoprotein) is protective for secondary challenges, but does not appear until mice have already recovered from infection, nor is it required for the maintenance of a virus-free state in recovered mice.[11] This contrasts with the infection of mice by Sindbis virus.[12]

In general, T cells that recognize internal viral proteins are CD8+ and those that recognize the glycoprotein are CD4+ in this experimental system, and several other viral infections. While it is possible to get a lot of information using bulk populations of lymphocytes from virus-immune mice, analysis of the fine specificity of immune responses is clouded by having multiple clones, which recognize distinct and different viral gene products and parts thereof, in the sample. Therefore, development of monoclonal T cell populations has been invaluable, as have generation of monoclonal antibodies, to map viral epitopes recognized.

There are two principal ways of getting pure clonal populations of T cells: *cloning,* and generating immortalized cells through fusion with an immortal T cell, termed *hybridomas*. We have used both of these approaches, because different questions can be asked with each of the reagents. Clones can be cytolytic or just proliferators/cytokine producers; clones require more fastidious care *in vitro,* but can be used for studies *in vivo* as well. Hybridomas are rarely cytolytic, but grow with ease and will undergo apoptosis, releasing cytokines to the cell supernatant, when appropriately stimulated.

T cell clones and hybridomas can be made from virally immune lymphocytes by *in vitro* cell culture techniques. Mouse T cell clones are derived by repeated stimulation of the lymphocytes with viral antigen, antigen-presenting cells of the same MHC type (histocompatible), and a cytokine-rich supernatant derived from either mouse or rat spleen cells stimulated with conconavalin A (originally described by Braciale et al.[13]). Hybridomas start the same way, but following initial *in vitro* culture are isolated and fused with BW5147 cells using polyethylene glycol and selected using HAT medium.

Clones have many uses *in vitro,* and can also be adoptively transferred to syngeneic hosts for functional *in vivo* studies. In contrast, hybridomas are immortalized, and adoptive transfer to hosts would kill the animals; the hybridomas would present an enormous tumor load in a short time. Clones have been shown by many laboratories, including this one, to home to both peripheral and central nervous system sites. At the appropriate sites, given the specific antigen, distinct clones produce activities ranging from clearance of lethal infections to eliciting pathologic inflammatory reactions. However, like many cultured entities, some clones lose the ability to home.

II. Method: Preparation of Virus-Specific Cytolytic T Cell Clones[14]

1. Groups of five mice (the responder strain is dependent on the system being studied; we use BALB/c or BALB/c-H-2^{dm2} mice) are immunized with virus (route and whether infectious, attenuated, or inactivated depend on the system and the types of effector cells you wish to generate, see below). Generally, at least 10 d later, a single-cell suspension of splenocytes from an individual donor is made, for *in vitro* culture.
   - The route of immunization and the form of the virus or viral antigen can have profound effects on the type of responses engendered. Generally, inactivated or purified protein preparations will be efficient at eliciting CD4+/class II restricted effector cells, but less so with CD8+/class I effectors. It usually requires infectious virus (or attenuated virus capable of one cycle of infection) to elicit CD8+/class I restricted T cells. Both of these populations can be cytolytic on the appropriate target cells, and both will secrete some cytokines, but CD8+ cells are likely to be dependent on a source of exogenous IL-2 for maintenance of growth; CD4+ cells often will be autoendocrine producers of IL-2 or IL-4, though this may not always be sufficient for their growth needs.
2. Antigen-presenting cells (virus sensitized; see below) are prepared, and mixed 1:5 with 5×10^6 primed lymphocytes using 2 ml culture medium (RPMI 1640 medium supplemented with 10% fetal bovine serum, L-glutamine, antibiotics), in 24-well tissue culture plates. After 5 d the cytolytic function of the secondary lymphocytes is tested on histocompatible virus-infected ^{51}Cr-pulsed targets (A20 cells are suitable for BALB/c lymphocytes, for example, because A20 express class I and class II MHC, and LFA-1, ICAM-1, B7.1, etc., and label efficiently; technique below). If there is no activity in the assay, this culture should be discarded and another donor used. If the culture is active, replace half of the medium, rest cells for 5 to 10 d, and then restimulate with antigen-pulsed syngeneic cells.
   - Bear in mind that some viruses are capable of productive infection in antigen-presenting cells (B cells, macrophages, or dendritic cells). Therefore, unless you want to lose your cultures to viral lysis, cytopathic effect (CPE), or persistent infection, use temperature-sensitive mutants, lightly UV-irradiated virus, or either

X-irradiate or fix infected cells in paraformaldehyde. Alternatively, transfect a gene of choice or use inactivated virus (this latter form will likely only induce CD4+/class II responders; the other methods will induce both CD4+ and CD8+ cells).

- Stable transfected cell lines, such as derived in Reference 15, can also be used as antigen-presenting cells. The advantage of a stable transfected cell line is the minimal variability of the reagent from one experiment to another. Multiple sequential changes can be made in the parental cell line, as well.

3. Three days later, add medium supplemented with a source of IL-2 (whether from commercially available cloned material, or from your own preparation of rat ConA supernatant, method follows). Repeat cytolytic T cell (CTL) assay at 5 d of the 3° culture. You only want to proceed with active populations.
4. Limit dilution culture: using 1×10^4 sensitized histocompatible antigen-presenting cells per well, make serial dilutions of the T cell population to yield 0.1 cell per well for a 96-well U bottom plate in a total volume of 200 μl of the IL-2-supplemented culture medium. Examine the plates carefully at 3- to 4-d intervals using an inverted microscope for wells in which T cells are actively growing. Twice a week, aspirate half the medium and replace with fresh IL-2-supplemented culture medium. Isolate growing wells onto another plate with virus-sensitized antigen-presenting cells, and expand wells if, and only if, fewer than one third of the wells on the donor plate are growing (otherwise clonality will be dubious). When you have two or more wells derived from the same source, assay the putative virus-specific CTL clones in a CTL screening assay. Maintain only specific clones.
   - Too much IL-2 will result in loss of specificity, too little will not sustain cytokine-dependent CD8 effector cells. Similarly, periodic stimulation is essential to maintain specificity.
   - Morphology Rorschach: T cells are often described as baseball bats or hand mirrors, by different people. That is, there is generally a large cell body with a psuedopodium of cytoplasm which propels the cell. Cytoskeletal plasticity is essential for locomotion. CTLs "walk" around *in vitro,* seeking targets. This is thought to be analagous to the *in vivo* immune surveillance function of T lymphocytes.
   - Cloning is an art, it is not a cookbook procedure. You need to use tender loving care and flexibility. The power of visual observation is a critical and essential component to successful cloning.
   - Call the clones anything you want; at least initially, they are often identified by their location on the parental plate and the number of the experimental cloning or date of the first culture, etc. Later, a more logical nomenclature may present itself.
5. As the populations grow, stimulate every 10 to 14 d with antigen, as above. As soon as you can, freeze down aliquots of the clones.
   - The generation time of different clones is quite variable, and you may want to maintain only the most rapidly growing clones, which are highly active, unless you are gifted with extraordinary patience.
   - We always avoid the external-most wells on plates, as these are more likely to have varying pH and pCO_2, and are, unfortunately, at greatest risk for inadvertent contamination.
   - It is critical to maintain frozen stocks of clones. The strategy we use is to put down any excess cells, when possible. The best time to freeze cells is about 3 d after the last stimulation.

- Despite one's best efforts, contamination in long-term culture is a fact of life. Some people maintain parallel sets of cell lines and clones in separate incubators. These individuals feed and stimulate the lines on alternate days, using separate bottles of media, etc., to minimize the impact of loss through contamination.
- Many clones proliferate but do not have lytic capacity. These may also secrete cytokines. Previously, the standard measure of such clones was to use ^{3}HTdR incorporation; this technique is not provided here, however, because New York and most other states are now severely constraining the use of long-lived radioisotopes, as there are virtually no acceptable waste disposal sites. Therefore, it is possible to measure cytokine release. Such a method is found below, for the characterization of hybridomas.

6. Now the clones are ready for screening for MHC restriction, viral antigen recognition, and T cell receptor (TCR) expression.

III. Method: Assay for Cytolytic Activity[16]

1. Prepare target cell infection: different viruses require shorter or longer intervals to express their proteins. Thus, you need to know the parameters for your system. During the penultimate hour of infection, the cells are labeled with 100 μCi $Na_2{}^{51}CrO_4$ per 3 × 10^6 cells. This is removed and cells are permitted to spontaneously release poorly bound isotope into fresh medium, for the last hour of incubation. Target cells are washed in fresh medium, counted, and dispensed into V-bottom 96-well plates at 1 × 10^4 cells/well in 50 μl.
   - For vesicular stomatitis virus, the "eclipse period", that is the time to release of progeny virions, can be as short as 4 h in some cell lines. We have found that a 30-min adsorption of virus in serum-free medium, followed by a 2-h incubation at permissive temperature (wild-type: 37°C; ts mutants: 31°C) is suitable. This contrasts with vaccinia viruses, which are used for the molecular genetic probing of the specificity of the T cells; the incubation period necessary depends on several factors, including the choice of promoter. If an early viral promotor is used, 4 to 6 h will suffice to allow enough vectored specific protein synthesis,[17] but an overnight infection is essential for gene expression driven by a late viral promoter to be sufficient. One consequence of long preincubation is that some viruses shut off host cell protein synthesis; this alone will diminish antigen presentation, since newly synthesized MHC class I is required, and this is also often seen for MHC class II-restricted antigen presentation.
   - At all times, uninfected, mock infected, or cells infected with irrelevant virus(es) are run as controls for virus-specific responses. When recombinant viruses are used, controls include recombinants carrying alternative gene(s) with the same promotor system.
   - It is also important to provide a multiplicity of infection, which will provide a uniform population. Thus, we recommend a moi of at least 3.
   - For class II MHC-restricted presentation to CD4 T cells, the exogenous route of presentation is generally very efficient. Viral antigens can be prepared as soluble protein, inactivated virus, defective interfering particles, cell lysates, etc.
   - Plan the organization of the plates, writing on the cover and on one identifying side *before* you put any cells in any plates. Include all target cell preparations. Include

all controls: replicates of target cells with medium for spontaneous release, and 5% Triton X-100 to measure maximal release.

2. Preparation of CTL effectors: remove cells from culture plates in approximately the numbers you will need. Count them and resuspend in fresh medium at 2×10^5/ml. Dispensing replicates (at least duplicates are strongly recommended) in 100 μl will produce an effector-to-target ratio of 20:1. This is very high for most assays of cultured clones, which are generally quite active. We recommend serial dilutions down to 0.3/1 target.

Note: *Never return long-term cultured cells or clones to wells, as the risk for inadvertent contamination is too great. Simply discard extra material.*

3. Prior to incubation at 37°C, we pre-spin the 96-well plates for 5 min at $50 \times g$ to gently bring target and effector cells together. Plates are incubated for 4 h, or more if your experience suggests longer is needed. Plates are then centrifuged at $100 \times g$ for 5 min and 100 μl aliquots of supernatant are removed for determination of release of the isotope using a gamma-counter (if available).
   - The pre-spin is optional. When inconvenient, allow a half hour more in the incubation period for gravity to compensate.
   - Liquid scintillation counters and a water-soluble scintillant in small vials will suffice, but are less efficient.
4. The activity of clones is determined by the formula:

$$\frac{100 \times \left[(\text{mean experimental dpm}) \pm (\text{mean spontaneous release dpm})\right]}{(\text{mean total releasable dpm}) \pm (\text{mean spontaneous release dpm})}$$

IV. Method: Preparation of Rat ConA Supernatant[14]

1. Using the spleens derived from 10 Lewis rats, approximately 250 g each, prepare a single cell suspension. Adjust the concentration of cells in culture medium supplemented with 5 μg ConA/ml to 5×10^6/ml, and seed 50 ml per 175 cm^2 culture flask. Incubate for 36 h at 37°C.
2. Decant the medium into centrifuge bottles, and remove cells. Pool supernatant medium into a 2 l beaker, noting the volume; place beaker on a magnetic stirrer. To both concentrate and to remove residual mitogen, using gloves and a mask, even if you are able to use a fume hood, slowly add enough $(NH_2)_2SO_4$ to reach the 50% saturated content. Remove precipitate. Add more $(NH_2)_2SO_4$ to reach the 75% saturated fraction and harvest the pellet. Resuspend the pellet in distilled water at 1% of the original volume of the medium. Dialyze the preparation against phosphate-buffered saline (PBS) extensively, filter through a 0.22 μ*M* sterile filter, and freeze in 1 ml aliquots.
   - We always determine the potency of the preparations using the IL-2 indicator cell line CTLL-2.
   - The preparations are usually used at 0.5% or 1% of culture medium to maintain the clones.

V. Method: Determination of MHC Restriction of T Cells

There are several complementary ways of mapping restriction. These include:

1. By using target/antigen-presenting cells that express one, but not other, MHC found on the stimulator cells. For instance, if you start with BALB/c mice, BALB/c-H-2^{dm2} cells lack the H-2 L^d molecule due to a spontaneous natural deletion on chromosome 17. Murine T cells from the dm2 donor would express H-2 K^d and H-2 D^d, but not the class II or the L^d molecules. A/J cells express a different K allele (K^k) but have both D^d and L^d. Thus, you could determine the class I restriction using murine (or human, porcine, etc.) genetics alone.
2. There is a wide library of transfected L cells available (as are other cells expressing isolated foreign MHC molecules; for instance, Reference 16). Using an L cell (originally derived from a C3H mouse, H-2^k) expressing H-2 L^d or I-A^d, for instance, you can unambiguously map recognition.

Note: *However, L cells are not the best antigen-presenting cells for all systems and are deficient in accessory molecules, so that some T cell properties are observed to be suboptimal (e.g., they are poor CTL targets when the avidity of the T cell could be enhanced with ICAM–LFA-1 interaction or B7–CD28 ligand binding).*

3. Monoclonal antibodies (mAbs) that recognize functional private domains of individual MHC molecules. Thus, using MKD-6, which recognizes (I-A^d) in the assay, it is possible to block specific recognition of I-A^d but not of I-E^d or L^d.

Note: *Using several distinct mAbs in the same assay, with just one and not ten targets/antigen-presenting cells, will provide the same genetic information as the other approaches described above. mAbs are readily available from ATCC (as cell lines) or commercially, as purified proteins.*

VI. Method: Preparation of T Cell Hybridomas Specific for Viral Antigens[15]

1. Follow the first two steps of the procedure for generating CTL clones, above.
2. Isolate T cell blasts, 3 d after the last stimulation, using a Ficoll-isopaque (or Ficoll/hypaque or Percoll) gradient. Wash to remove the density gradient material, count, and adjust to 1×10^7/ml.
3. Combine 2.5×10^6 azaguanine-resistant BW5147 cells with 1 ml T cells and centrifuge. Aspirate supernatant, resuspending the pelleted cells; immediately add culture medium supplemented with 40% polyethylene glycol. Incubate at 37°C with frequent gentle agitation for 15 min.

4. Wash and add culture medium to cells. Set up limit dilution cultures in 96-well, flat-bottom plates.
5. The following morning, add HAT-supplemented medium. Replace medium and screen wells for growth twice weekly.
6. When wells show signs of growth, expand into larger wells. Selective medium will no longer be necessary.
7. As with CTL clones, described above, freeze down as soon as possible, assay as soon as sufficient cells are available.
   - Hybridomas are, more frequently than not, class II-restricted, cytokine-secreting (either IL-2 or IL-4 plus variable other cytokines) cells; they rarely are cytolytic.
   - Hybridomas grow like weeds. They are not fastidious, but if they reach too high a density, will not be very useful.
   - It is not uncommon for hybridomas to spontaneously lose chromosomes. It may be necessary to subclone to retain antigen specificity.

VII. Method: Assay of Virus-Specific T Cell Hybridomas[15,18]

1. Mix at least three replicates 1×10^4 T cell hybridomas with an equal number of either control or virus-sensitized A20 cells in 200 μl complete medium per well in flat-bottom 96-well plates. Controls include one cell type alone, medium alone. Incubate overnight at 37°C.
   - You can inspect, visually, the culture wells before removing the supernatant. Hybridomas will undergo apoptosis, programmed cell death on stimulation, releasing their cytokines to the medium. This is a good predictor of the results of the experiment.
   - Plates can be frozen at this point for later assay, or cell-free supernatants can be harvested immediately for assay of the presence of cytokine.
2. Remove 100 μl supernatant and combine with the IL-2/IL-4 indicator cell line, 5×10^3 CTLL-2 cells per well in flat-bottom 96-well plates, for at least 6 h to overnight. Include a titration curve of a known source of IL-2, such as the rat ConA supernatant.

Note: *CTLL-2 cells are absolutely dependent on IL-2 and/or IL-4 for growth and survival. They must be passed in IL-2-supplemented medium and must be at log-phase growth for this assay. Generally they are passed twice weekly from about 1×10^4/ml in small TC flasks. Do not let the density get too high or they will die.*

3. Add 10 μl MTT (5 mg/ml in PBS, filtered) to each well, incubate at 37°C for 4 h.

Note: *You will see purple crystals form in wells where the CTLL-2 cells are viable. This is because MTT is metabolized in active mitochondria from the precursor to a colored product.*

4. Aspirate medium. We use a 27 gauge needle, bevel facing wall of well, to withdraw medium. Add 100 μl dimethyl sulfoxide (DMSO) to each well to dissolve cells and colored product. Read on an automatic plate reader using a 550-nm filter.
5. Calculate the relative concentration of IL-2 in your experimental samples by determining the IL-2 curve from your IL-2 standard and the CTLL-2 cells by the formula: OD 50% = 1 unit IL-2/ml = $[(OD_{max} - OD_{min}) \div 2] + OD_{min}$

VIII. Method: Construction and Propagation of Vaccinia Virus Vectors Incorporating Viral Genes of Choice or Fragments Thereof[17,19-22]

1. We have used vaccinia virus vectors engineered to express the entire glycoprotein gene, various constructs of the G gene, hybrid molecules, and fragments of it. We have used principally two promoters: the vaccinia early (P7.5) and the bacteriophage T7; the latter construct requires co-infection with T7 RNA polymerase-expressing vaccinia viruses.
   - Whenever a construct has been made, the expression of the protein we are interested in is determined by Western blot. For other genes, radioimmunoassay, enzyme-linked immunosorbent assay, immunofluorescence, or other techniques may be useful to show protein synthesis, subcellular location, etc.
2. New recombinant vaccinia viruses were constructed as follows. Recombinant plasmids were made: A *Bam*H1 fragment containing the vaccinia virus p7.5 early promoter was released from the plasmid pSC11 and ligated to the 5′ end of the VSV G gene (or fragment, recombinant, etc.) in order to obtain a plasmid containing the VSV G gene under the control of the p7.5 promoter. Plasmid contains interrupted TK gene for recombination into vaccinia. Plasmid also contains the ampicillin resistance gene for selection in *Escherichia coli*.
   - This can also serve as a source of material for further genetic manipulation. For instance, we replaced the C-terminal third of the G with a *Kpn*1-*Bam*H1-fragment of the "poison tail" pAR1473 mutant glycoprotein constructed initially by Rose and Bergmann.[23]
   - In decision making about construction of plasmids, some investigators may wish to use fusion proteins or may wish to use β-galactosidase expressing plasmids for screening blue plaques vs. white to detect recombinants, or alternatively luciferase, using commercial kits.
3. CV-1 cells are infected with the WR strain of vaccinia at a moi = 0.2 and then transfected (using $CaPO_4$ precipitation) with the plasmid described above. Plaques were picked, replaqued for purity, and cell lysates were screened for expression of G (or its fragments) by Western blot.
4. CV-1 cells are used for the propagation of vaccinia virus in our laboratory, as originally described by Moss and colleagues.[20] A multiplicity of infection of approximately 0.2 is used for a 48-h incubation. Inspect cells for CPE. Harvest monolayers with 0.25% trypsin-EDTA (ethylene diamine tetraacetic acid) solution (we use 75 ml per roller bottle) and place cells in 50-ml centrifuge tubes. Spin down, removing supernatant. Samples of the same preparation are pooled in 5 ml physiologic medium (PBS, HBSS,

RPMI, etc.; reserve a small aliquot for Western blot analysis of protein expression if under the control of the early promoter) for a three cycle freeze-thaw to release virus, which is highly cell-associated. Cellular debris is removed by centrifugation.

- This is because we want to have large stocks of a preparation so that for many experiments we are using the same material. It may be unnecessary for your purposes, if small amounts of virus are needed. We use roller bottles for the most efficient large preparation of virus.
- At this point, it is important to determine titer of the vaccinia virus preparation by plaque assay. We also use CV-1 cells for this. In order to assay the titer of a virus preparation, the solution should be trypsin-treated and sonicated briefly to disperse aggregates. Then the preparation should be serially diluted before adsorption of virus on monolayers of CV-1 cells. As time has gone on, we have reduced the monolayers from 60 mm tissue culture dishes to wells in 24-well plates.
- Depending on what you plan to do with the preparations, it may be important to purify the virus on sucrose gradients. This is especially important if you are going to distinguish exogenous protein from endogenous protein synthesis for antigen presentation.
- If you have selected a fusion protein, this can be purified by affinity chromatography to the foreign tag; however, you may lose the ability to express this protein in its "normal" manner inside vaccinia virus-infected cells (could alter half-life within cells), or "gain" novel site(s) of expression. (For example, if you construct a leader sequence on a normally cytoplasmic protein, you can force its entry into the secretory pathway in the endoplasmic reticulum-Golgi compartments. It is also possible to delete such a signal sequence and have a cytoplasmic rather than secretory product.)

5. You are now ready to use the virus for your experiments. Strategically, if you are trying to map region(s) recognized by the immune system in the protein whose gene you have engineered, a series of overlapping in-frame recombinants expressed in vaccinia vectors will permit localization of the epitopes. There are several ways of doing this; for instance, you can have staggered truncated molecules or you can mix and match alternative serotypes of the protein in the N- and C-terminal portions of the hybrid molecules.
   - mAb binding to proteins can be mapped to domains, whether neutralizing or non-neutralizing; CTL, T cell hybridoma, or proliferating epitopes can be mapped to infected or sensitized cells.
   - Functional domains, such as the CD4-binding portion of HIV gp120, or the regions conferring apical vs. basolateral expression of viral glycoproteins in polarized cells, can be mapped using these sorts of reagents.

IX. Method: Selection of Peptides for Synthesis

1. Using the analysis possible using the staggered recombinant hybrid viruses, above, you will be able to narrow important regions recognized by either T cells or antibody of your experimental viral antigen. You already know the restriction pattern of your T cells (method above), which narrows your search considerably. The next step would be to analyze the sequence using published motifs.[5] This should not be the initial step, as

only a small fraction of motif-bearing sequences are immunologically recognized. It will save you considerable expense of the synthetic laboratory charges.

2. In general, class II MHC antigens bind peptides as long as 20-mers, with somewhat ragged ends at either end permissible. In contrast, class I MHC heterodimers bind peptides of 8- to 10-mers, with stringent controls on the terminal residues. In real life, the peptides acquired by MHC molecules, in living cells, last the life span of the surface expression of the molecules. However, a very small percentage of these MHC molecules "breathe" and so, if cells are bathed in extraordinarily high concentrations (e.g., m*M* or μg) of peptides capable of binding them, MHC, a tiny but measureable fraction, will be bound. *The most sensitive detection method is cellular recognition.*
3. Antibodies recognize peptides that have more conformation than do T cells, which see peptides spread across the cleft of MHC. It is rare for mAbs to bind short peptides; they prefer native molecules or much longer peptides.

X. Method: Determination of TCR Expressed on T Cell Clone or Hybrid

When we started our studies on the clones a decade ago, the only way to determine the T cell receptor (TCR) used was to clone and sequence it. Now, we are fortunate to have two much simpler techniques available.

1. The molecular genetic approach would be to obtain a set of primers (available commercially) for each of the families of TCR of the species you are using. These would be used in reverse transcriptase-polymerase chain reaction (RT-PCR) to identify the relevant fragment from RNA obtained from the clone/hybridoma.

Note: *Controls would be critical, as a clone will express only one TCR, so most of the determinations will be negative. Few cells are required for this technique because of the amplification.*

2. For most, but not all, TCR, there are mAb available commercially which can recognize one family or subfamily of either α or β chains (most clones will not be γδ-expressing). This will tell you the important information. Stain cells with FITC- or PE-conjugated Fab′2 mAbs then evaluate in a flow cytometer.
   - You need access to a flow cytometer and the assistance of a skilled operator.
   - Positive controls should include MHC expression and/or CD4 or CD8. All clones and hybridomas will express CD3. Always include isotype controls and irrelevant mAb; using a panel of mAb to TCR, most will be negative. A clone can only express one.
   - More CTL clones are needed for this technique, depending on the size of your antibody panel, generally 1×10^5 cells for each reagent.
3. It may be important to sequence the CDR regions of the TCR if you are trying to distinguish fine sequence differences associated with antigen-recognition differences. This would require cloning the relevant regions from a series of clones having subtle recognition differences.

XI. Method: Adoptive Transfer of T Cell Clones and Infectious Challenge[9,10,24-26]

1. Clones growing in log phase, approximately 3 to 4 d after the last stimulation, are harvested from culture, and dead cells are removed using a Ficoll-hypaque gradient. The clones are adjusted to deliver graded doses (from 1×10^4 to 1×10^6 per recipient syngeneic host) in serum-free medium, and temporarily stored on ice.
   - You want to transfer approximately 0.2 ml medium to mice, as you do not want to interfere with the fluid volume of the hosts excessively.
   - Serum-free medium is essential, as fetal bovine serum is immunogenic.
   - Load the cells in tuberculin syringes and have on hand 27 gauge needles, as these are necessary for tail vein injections; smaller bore needles could damage the cells through the small orifice.
2. Infect mice and transfer cells to infected hosts via the tail vein. We anesthetize mice with halothane and apply a 10 µl aliquot of vesicular stomatitis virus (1×10^5 pfu) to the external opening of the nares on either one or both sides of the mouse's snout.
   - In designing experiments there are many issues to bear in mind. The first is the choice of experimental outcome(s) to be investigated. These could range from survival/morbidity observations (ruffled fur, activity, weight changes, grossly observable symptoms such as paralysis or behavioral changes, death), gross histological changes (influenza infection of mice readily results in lung consolidation secondary to inflammation), viral plaque assays using tissue homogenates, physiological staining (hematoxylin and eosin, for instance), immunohistochemistry (fixed vs. frozen tissue sections), or *in situ* hybridization (for virus, T cell cytokines, any other expressed RNA).
   - Determination of group size and of control groups is essential prior to the commencement of the experiment. Even using highly inbred animals of the same age and sex, there is biological variability in the responses. We never consider setting up a pathogenesis experiment without a group size of >8 for each time point included when determination of viral propagation is one end point. For observations of morbidity and mortality, groups of >10 are used. For immunohistochemistry, >3 mice per time point per perfusion condition are used. As there is variation from experiment to experiment, each study is replicated at least twice before we are confident of the data.
   - When infecting the nasal epithelium, if animals are awake, an aerosol is necessary (from a nebulizer). Mice that are just anesthetized will not sneeze the material out, and it is possible to infect the respiratory epithelium (influenza, respiratory syncytial, Sendai viruses) or the neuroepithelium (vesicular stomatitis virus, rabies, murine coronaviruses).
   - In our hands, if virus is applied to one side only, the virus remains on the ipsilateral side of the central nervous system until 4 to 5 d postinfection, when bilateral CNS infection is detectable.

Thus, a concerted effort of many people-years is required to generate these reagents and data. Typical data follow in the table.

TABLE 9.1
Analysis of the Fine Specificity and *In Vivo* Function of 7 CTL Clones and 4 T-T Hybridomas Specific for the Glycoprotein of Vesicular Stomatitis Virus and Restricted by Ia^d

Clone or Hybridoma	MHC Restriction	Peptide	T Cell Receptor Determined	Influence on Infection of Mice
9-F1	$I\text{-}E^d$	85–104	Vβ 8.1	Promotes recovery/survival
B1	$I\text{-}A^d$	337–368	Vβ 8.1	Increases mortality
B10	$I\text{-}A^d$	337–368	Vβ 8.1	Increases mortality
BALB/c No. 1	$I\text{-}E^d$	85–104	Vβ 10	Brain, 2 $\log_{10}$ less virus in homogenates
BALB/c No. 4	$I\text{-}E^d$	85–104	Vβ 14	Similar to media control; no effect
27H6	$I\text{-}E^d$	84–104	Vβ 5/12	1 $\log_{10}$ less virus in brain homogenates
G11	$I\text{-}E^d$	85–104	n.t.	Not tested, slow-growing line
3DV.155.14	$I\text{-}A^d$	n.d.	Vβ 3	No hybridomas were tested *in vivo*
3DV.9.18.24	$I\text{-}A^d$	n.d.	Vβ 3	No hybridomas were tested *in vivo*
3DV.315.11	$I\text{-}E^d$	85–104	Vβ 3	No hybridomas were tested *in vivo*
3DV.95.28	$I\text{-}E^d$	85–104	Vβ 3	No hybridomas were tested *in vivo*

XII. Conclusions

1. Molecular techniques and tools offer powerful reagents and strategies for determining the viral epitopes critical for host response and for viral pathogenesis. Although viruses can be larger, complex, protein-containing units, there are very few functional immunogenic epitopes in any protein recognized by the immune system of an infected host. The murine response to the glycoprotein completely dominates the class II-restricted T cell clones and hybridomas we have studied. The nucleoprotein dominates the class I-restricted response. The epitopes we have characterized for the clones and hybridomas *in vitro* are the same as those detected by the Hengartner-Zinkernagel group using *ex vivo* populations and sequential peptides for proliferation assays.[27]
2. The MHC restriction can be inclusive of all potential restricting heterodimers, as we have seen for the glycoprotein and class II molecules in the $H\text{-}2^d$ haplotype (Table 9.1), or can be extremely limited, as we and others have seen for the MHC class I restriction. In the $H\text{-}2^d$ haplotype, this is exclusively $H\text{-}2L^d$.[16]
3. There is no strict association of T cell receptor β genes employed despite a limited peptide and MHC restriction. For two peptides and two MHC heterodimers described here, at least five distinct T cell receptor β genes were used.
4. Despite superficial similarities or differences in MHC, peptide, and TCR use, the *in vivo* function of clones is not predictable. Some clones did nothing (BALB/c clone No. 4), others were pathogenic (B1, B10), while others were advantageous to the infected host (9F1, BALB/c clone No. 1, 27H6). The biologic effects are probably more influenced

by the cytokine profiles of the clones,[28] which can control inflammation and regulate the function of other cells.

Using monospecific T cells, it is possible to determine many of the immunologically important determinants of pathogenic viruses. Using different strategies, other host species or viruses can be similarly characterized; these strategies have to be based on the limitations of the systems to be investigated. Once epitopes are defined, uncloned populations of immune lymphocytes can be tested *ex vivo* to confirm and extend the reliability of the data. Further studies can be done to determine the precursor frequency of the reactive clones.

When epitopes have been determined, the relevancy of these epitopes in immunity or pathogenesis of disease can be determined. The development of vaccines depends on this knowledge, and also on the sites or regions where immunity is necessary. Having neutralizing antibody at a distant site (for instance, the peripheral circulation) will not help prevent an upper respiratory infection, where secretory immunity is preferable. The strategy for vaccine development also must include whether cellular or humoral immunity is desired.

Another consideration is the stability of the infectious virus. RNA viruses are notorious for variability, due to the lack of proofreading of the RNA-dependent polymerases. The immune response to epitopes negatively selects the population, and variants are therefore able to "escape" the best initial efforts of the hosts.

Acknowledgments

Without the long-term, generous support from the National Institutes of Health, AI1083 to CSR, this work would not have been possible. Much of this work was done in collaboration with Alice S. Huang (initially at Harvard Medical School, then at New York University), Stephanie Diment (formerly at New York University School of Medicine, now at John Wiley & Sons), and terrific postdoctoral fellows Michael J. Browning, Walter Keil, Ilia V. Plakhov, and a graduate student Michael A. Petrarca, who generated some of the essential reagents used in the studies. Additional experiments were done by rotating students at NYU, including JoAnne Bezault, Bruce Ng, Elizabeth Moranville, Boris Yusupov, and a technician, Lara M. Palevitz.

References

1. Moore, K. W., Vieira, P., Fiorentino, D. F., Trounstine, M. L., Khan, T. A., and Mosmann, T. R., *Science,* 248, 1230 (1990).
2. Racaniello, V. R. and Baltimore, D., *Science,* 214, 916 (1981).
3. McQueen, N. L., Nayak, D. P., Jones, L. V., and Compans, R. W., *Proc. Natl. Acad. Sci. U.S.A.,* 81, 395 (1984).
4. Fulginiti, V. A., Eller, J. J., Downie, A. E. et al., *JAMA,* 202, 1075 (1967).
5. Engelhard, V. H., *Annu. Rev. Immunol.,* 12, 181 (1994).
6. Doherty, P. C. and Zinkernagel, R. M., *Lancet,* 1, 1406 (1975).

7. Germain, R. N., in *Fundamental Immunology,* 3rd ed., Paul, W. E., Ed., Raven Press, New York, 1993, 629.
8. Reiss, C. S. and Aoki, C., *Rev. Med. Virol.,* 4, 121 (1994).
9. Plakhov, I. V., Arlund, E. E., Aoki, C., and Reiss, C. S., *Virology,* 209, 257 (1995).
10. Bi, Z., Komatsu, T., Barna, M., and Reiss, C. S., *J. Virol.,* 69, 6466 (1995).
11. Komatsu, T., Bi, Z., and Reiss, C. S., in preparation.
12. Levine, B., Hardwick, J., Trapp, B., Crawford, T., Bollinger, R., and Griffin, D., *Science,* 254, 856 (1991).
13. Braciale, T. J., Andrew, M. E., and Braciale, V. L., *J. Exp. Med.,* 153, 910 (1981).
14. Browning, M. J., Huang, A. S., and Reiss, C. S., *J. Immunol.,* 145, 985 (1990).
15. Petrarca, M. A., Reiss, C. S., Diamond, D. C., Boni, J., Burakoff, S. J., and Faller, D. V., *Microbial Pathogen,* 5, 319 (1988).
16. Reiss, C. S., Evans, G. A., Margulies, D. H., Seidman, J. G., and Burakoff, S. J., *Proc. Natl. Acad. Sci. U.S.A.,* 80, 2709 (1983).
17. Reiss, C. S., Gapud, C., and Keil, W., *Cell. Immunol.,* 139, 229 (1992).
18. Bartido, S. M., Diment, S., and Reiss, C. S., *Eur. J. Immunol.,* 25, in press (1995).
19. Mackett, M., Smith, G., and Moss, B., *J. Virol.,* 49, 851 (1984).
20. Chakrabarti, S., Breching, K., and Moss, B., *Mol. Cell. Biol.,* 5, 3403 (1985).
21. Keil, W. and Wagner, R., *Virology,* 170, 392 (1989).
22. Keil, W., Reiss, C. S., and Huang, A. S., unpublished.
23. Rose, J. K. and Bergmann, J. E., *Cell,* 34, 513 (1983).
24. Forger, J. M., III, et al., *J. Virol.,* 65, 4950 (1991).
25. Huneycutt, B. S. et al., *Brain Res.,* 635, 81 (1994).
26. Plakhov, I. V., Oien, S. A., Huang, A. S., and Reiss, C. S., in preparation.
27. Burkhardt, C. et al., *Viral Immunol.,* 7, 95 (1995).
28. Cao, B.-N., Huneycutt, B. S., Gapud, C. P., Arceci, R. J., and Reiss, C. S., *Cell. Immunol.,* 146, 147 (1993).

Section II

DNA Viruses

Chapter 10

Transient Genetic Approaches to Studying Multifactor Processes in Complex DNA Viruses

Gregory S. Pari, Andrea C. Iskenderian, and David G. Anders

Contents

0-8493-4412-3/96/$0.00+$.50

I. Introduction

Complex DNA viruses, with genomes that may express more than 200 gene products, encode collections of proteins that cooperate to perform or significantly modify processes normally executed by the host cell machinery, such as DNA synthesis and regulation of transcription, as well as other processes unique to viral amplification, such as capsid assembly and DNA packaging. These complex virus systems pose major challenges for analysis by classical genetic methods, because of the large number of mutants that must be produced and characterized to identify all of the components involved in such multifactor processes. Moreover, in many instances, classical genetic studies are further hindered by the limited host range and slow growth properties of the virus of interest. Likewise, direct biochemical studies of activities of interest in cell-free extracts made from virus-infected cells may be difficult or impossible, due to limited protein expression.

Human cytomegalovirus (HCMV), whose growth is restricted to cells of human origin, and which grows poorly or not at all in transformed cell lines, is a case in point;[1,2] even in fully permissive cells HCMV growth is slow and plaque assays, although feasible, require almost 3 weeks. For these reasons it has proven very difficult and generally impractical to do large-scale genetic analyses with HCMV.

In attempts to circumvent these roadblocks to defining critical genetic components of complex virus systems, methods to do "transient genetics" have been developed.[3] These methods transiently cointroduce genomic DNA fragments into appropriate host cells to reconstruct the process of interest, and in principle can be applied to any process for which a cell culture assay is available. Where comparative data are available, results obtained by transient genetic analyses have been shown to be consistent with those of classical genetic studies. This chapter describes the general strategy of these approaches, details the methods by which we have applied them to the problems of defining essential components of the HCMV DNA replication and transcription regulating machineries, and, finally, highlights some other applications and related approaches that illustrate the general applicability of transient genetics.

II. Strategy

The first step in the general strategy for using transient methods to dissect complex genetic systems, as outlined in the flow chart (Figure 10.1), is to fragment the genome by appropriate restriction enzyme treatment to produce a set of large and preferably overlapping DNA fragments. In general, genomic DNA fragments are cloned to assemble a library of characterized plasmid or cosmid recombinants. In at least one instance, however, uncloned DNA fragments were introduced directly into target cells to test the feasibility of the assay prior to cloning.[3] Next, the set of spanning fragments is introduced into carefully selected host cells by transfection or electroporation. Important properties of the host cells selected for use include transfectability and permissivity for the desired activity. Generally, but not always,[4] the selected cells will be permissive for virus infection. Following incubation for a suitable period, the transfected cells are harvested and assayed for the activity of interest. It may be necessary to test a variety of transfection protocols and cells in order to find conditions that will support complementation; indeed, finding the right combination may be the rate-limiting step in establishing a transient genetic system, and some potential pitfalls are discussed below. Once the desired activity is obtained, it is important to establish the fidelity of the complementation by examining the structure of the products, when possible.

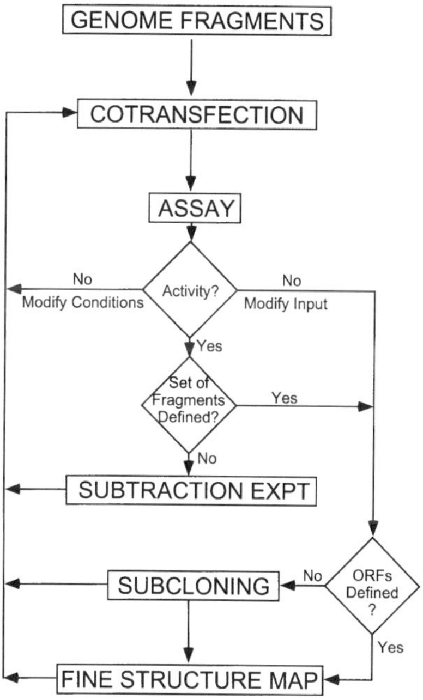

FIGURE 10.1
Flow chart describing the iterative strategy used in transient genetic studies for mapping cooperating genes.

After a useful transient assay is established, genes essential for complementation can be mapped by a three-step process. First, genomic segments containing essential genes are distinguished from dispensable segments in subtraction experiments, wherein each member of the complementing fragment sets is individually omitted from the transfecting mixture. Second, each essential fragment is dissected by subcloning to identify the essential coding regions present in each required fragment. Third, minimal coding regions are further refined by using the same assay to establish gene structure and study promoter requirements, in conjunction with DNA sequencing and transcript mapping studies. Expression of the essential proteins in complementing mixtures may require accessory trans-activating functions;[4-8] of course, in some cases, these may be overlapping sets.[9] Once defined, essential coding regions can be expressed under the control of a heterologous constitutive promoter to eliminate the need for accessory loci whose sole role is in trans-activating expression.[9,10] Finally, mutations can be generated in the coding and promoter regions of essential genes, and the resulting mutant proteins tested for function in the context of the transient assay.

III. Results and Discussion

Our efforts have sought to identify and then characterize the complete set of genetic components required to perform and regulate HCMV DNA synthesis. We first located the HCMV replicator, *ori*Lyt, using a standard transient method,[11,12] and subsequently used a transient complementation approach to establish the complete set of loci essential for DNA synthesis.[7,13] We found that 11 distinct loci, situated in 5 different cosmids, were needed for complementation of DNA replication. Subsequently, we have extended this approach to show that 4 of these 11 loci cooperate to enable early gene expression.[14] These experiments are discussed briefly here to illustrate our methods, and to demonstrate potential pitfalls.

A. Transient Complementation of HCMV DNA Replication

In developing a transient complementation assay for HCMV DNA replication, our strategy was to introduce large overlapping cloned segments of the HCMV genome into permissive human fibroblasts together with a plasmid containing HCMV *ori*Lyt. Available cosmid libraries[15] provided convenient sets of overlapping genomic fragments, and the *ori*Lyt-containing plasmid SP50[11] served as an assayable reporter of *ori*Lyt-mediated DNA synthesis; replication of pSP50 was measured by conversion of *dam*-methylated input DNA to *Dpn*I resistance. Replicated plasmids could then be detected as full-length molecules in Southern transfers. A particular challenge for this system was that the permissive human fibroblasts are difficult to transfect efficiently. Thus, initial experiments compared several transfection methods, including standard $CaPO_4$, DEAE-dextran and lipofection protocols, tested various combinations of available cosmids, plus or minus added pSVH,[16] which expresses the trans-activating proteins encoded by immediate-early regions 1 and 2 (MIE), and also tested several cell types.

For nearly a year, and dozens of transfections, these trials were unsuccessful. Although all of the transfection protocols were efficient enough for standard transfection experiments in which only *ori*Lyt was transfected and necessary trans-factors were supplied by viral infection, we found that only the BES-buffered, modified $CaPO_4$ procedure described by Chen and Okayama[17] yielded complementation.[13] Positive results, measured by conversion of *ori*Lyt to *Dpn*I resistance, first were obtained when the seven cosmids pCM1015, pCM1017, pCM1029, pCM1035, pCM1039, pCM1058, and pCM1052 were cotransfected with a roughly 20-fold molar excess of pSVH (Figure 10.2, lane 1). Thus, these seven partially overlapping cosmids expressed all of the proteins needed to direct detectable *ori*Lyt replication. Together, they comprise the entire HCMV strain AD169 genome (Figure 10.2), including all of the described homologs of HSV-1 replication genes. Without added pSVH the same set of cosmids reproducibly failed to complement DNA replication, even though pCM1058 contains a functioning MIE region. In this and other experiments only *ori*Lyt-containing plasmids were replicated.

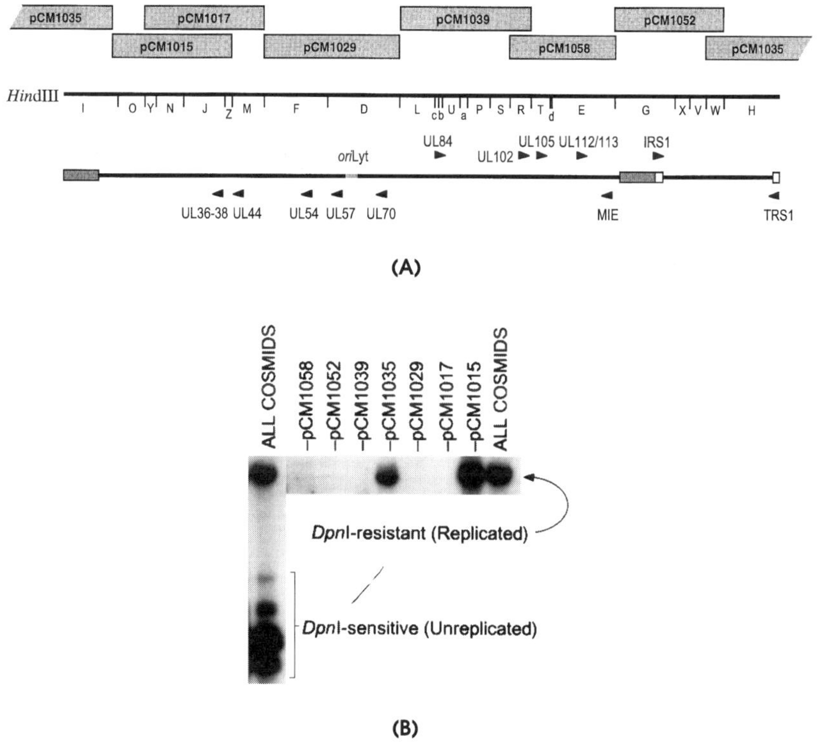

FIGURE 10.2
Cotransfected cosmids complement HCMV lytic-phase DNA replication. The top panel shows restriction and physical maps of the HCMV genome, indicates the relative locations of individual cosmids used in the experiments discussed in the text (grey boxes), and notes the positions of the 11 distinct loci found to be essential to complement lytic-phase DNA synthesis (filled arrows). The bottom panel presents results of a transient complementation subtraction experiment, in which transfection mixtures contained an *ori*Lyt-containing plasmid and pSVH, plus either all seven cosmids, or the complete set minus the indicated cosmid. The assay was done as described in Sections V and VI. Positions of *Dpn*I-resistant (replicated) and *Dpn*I-cut DNAs are noted. (Adapted from References 7 and 13. With permission of the American Society for Microbiology.)

While it is difficult to interpret these results quantitatively, we note that in typical experiments complemented replication produced a roughly 50- to 100-fold lower *Dpn*I-resistant signal than the parallel infected control (not shown). Several lines of evidence indicated that the observed replication was authentic replicator-mediated synthesis. First, in cells treated with phosphonoformic acid, a drug that selectively inhibits the viral DNA polymerase, no *Dpn*I-resistant product was detected at any time. Second, replication products of *ori*Lyt-containing plasmids in the transient complementation assay were tandem arrays of "endless" molecules, as are the products of viral DNA replication.[11,18] Finally, subsequent experiments showed directly that the viral DNA polymerase was essential for complementing DNA synthesis. Subtraction experiments showed that two of the cosmids were dispensable (Figure 10.2, lanes 2 to 8); thus, five cosmids contained all of the loci essential to HCMV lytic-phase DNA replication.

Next, we systematically generated subclones from each of the required cosmids, and tested their abilities to replace the corresponding cosmid in the transient complementation assay. Efforts to define the essential genes present in these loci have been greatly aided by the availability of the complete sequence of the HCMV strain AD169 genome.[19] Criteria that we have used for defining the structure of essential genes include: (1) elimination of other predicted coding regions from complementing plasmids by deletion; (2) determination that frameshift and/or nonsense mutations in the predicted essential coding regions abrogated complementation; (3) transcript data consistent with predictions of the sequence data; and (4) demonstration that expression of the deduced coding region under the control of a heterologous promoter allows complementation.

Three of the essential cosmids spanned more than one essential locus (Table 10.1). In all, 11 distinct loci were found to be required.[7,13] Eight of the 11 required loci contain a single significant open reading frame (ORF) and, based on somewhat limited transcript data, these probably express a single protein product via unspliced transcripts. Mutations in the coding regions of each of the predicted simple transcription units individually abrogated their ability to complement. The remaining three loci encompass complex transcription units that each express several proteins via differential splicing and promoter usage. Attempts to determine which of the protein products encoded within the complex transcription units are essential for DNA replication are ongoing. Six of the loci encode homologs, or probable homologs, of herpesvirus group-common replication genes that express proteins likely to constitute the replication fork (Table 10.1). Four of the remaining five loci encode regulatory proteins that cooperatively activate expression of the replication fork proteins.[14,20]

Results characterizing the locus spanning the UL101–102 ORFs serve to illustrate the utility of the transient complementation method for establishing essential gene structure. The smallest complementing fragment containing this locus extends from the *Asc*I site at nt 146384 to the *Hin*dIII site at 149645, and spans deduced ORF UL101 as well as UL102 (Figure 10.3). UL101 was first noted by Chee et al.[19] as a positional counterpart of HSV-1 UL9, an origin-binding protein that is essential for HSV-1 DNA replication.[6,21] However, the evidence now suggests that UL101 does not encode a protein essential for replication. Instead, an overlapping downstream ORF, referred to here as UL101X, is expressed continuously through UL102, and constitutes an upstream extension of the originally deduced UL102 coding region (Figure 10.3).

In the published sequence of HCMV strain AD169, UL102 initiates at nt 147127, terminates at 149140, and predicts a protein of 72 kDa. A frameshift mutation within

TABLE 10.1
Summary of Essential HCMV Loci and Herpesvirus Counterparts

		Herpesvirus Counterparts		
Cosmid	HCMV	HSV-1	EBV	Probable Function
pCM1017	UL44	UL42	BMRF1	*pol* accessory
	UL36–38	*a*	*b*	Trans-activation
pCM1029	UL54	UL30	BALF5	DNA polymerase
	UL57	UL29	BALF2	Single-stranded DNA binding
	UL70	UL52	BSLF1	Primase
pCM1058	UL105	UL5	BBLF4	DNA helicase
	UL102	UL8[c]	BBLF2/3[c]	Helicase-primase complex?
	UL112–113	*d*	*d*	Trans-activation
	IE1/IE2	*a*	*b*	Trans-activation
pCM1052	IRS1	*a*	*b*	Trans-activation
pCM1039	UL84	*d*	*d*	Unknown, early

[a] HSV-1 trans-activators are not required to complement DNA replication in transient assays.

[b] Three EBV trans-activators, Z, R, and M, are needed to complement *ori*Lyt replication in transient assay, although only one is essential to DNA synthesis; however, these genes are not obvious homologs of the HCMV trans-activators.

[c] Probable homologs displaying minimal sequence similarity.

[d] No HSV-1 or EBV counterparts of these HCMV early temporal-class proteins have been shown.

Adapted from Reference 7. With permission of the American Society for Microbiology.

UL102 abrogates complementation[7] consistent with a requirement for the UL102 protein. The deduced UL101 ORF initiates at nt 146392, and terminates at nt 146698. The overlapping ORF that we call UL101X is defined in the published sequence data by an ATG at nt 146522 and a UAG termination codon at 146744. Clones truncating the amino-terminal half of UL101 failed to replicate (Figure 10.3, *Eam*I-*Hin*dIII), yet it seemed unlikely that UL101 could be efficiently expressed from the *Asc*I-to-*Hin*dIII fragment, because only 7 nt 5′ with respect to the UL101 ATG remained in this construct, and thus it could not include an upstream UL101 promoter. In subsequent experiments we found that frameshift mutants made by Klenow treating the *Eam*1105I, *Xba*I, or *Bst*EII sites all failed to complement DNA synthesis (Figure 10.3). *Eam*1105I and *Xba*I mutations affect both UL101 and UL101X ORFs, whereas the *Bst*EII mutation leaves the UL101 ORF intact, altering only the UL101X frame. These results argue that the coding capacity of this region is essential to complement DNA replication, and also suggest that the UL101 ORF is not sufficient. We noted that UL101X lies in the same reading frame as UL102, and that the UL101X termination codon is the only stop codon in that frame between the UL101X ATG at nt 146522 and the UL102 ATG at 147127. A TESTCODE plot, based on Fickett's statistical measure of the "period three constant" used as a measure of protein coding likelihood,[22] predicts coding capacity beginning at the 5′ end of UL101X and extending uninterrupted into UL102. We sequenced the UL102 upstream region present in our complementing clones, and found that the published

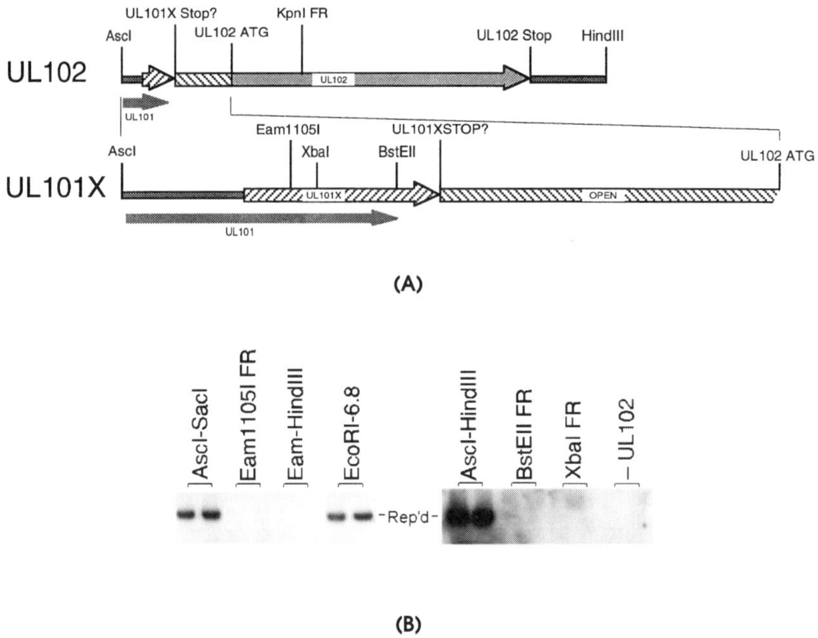

FIGURE 10.3
Transient complementation of DNA replication assays of UL101–102 region plasmids and frameshift mutants. The top panel shows a restriction map of the UL101/102 region of the HCMV strain AD169 genome, and an expansion of the UL101/UL101X segment. In the bottom panel, plasmids containing the indicated fragment or frameshift mutation were tested in the transient complementation assay, as described in Section VI. Only the portion of the Southern blot containing the replicated (*Dpn*I-resistant product) DNAs is shown. The "UL101X stop" codon identified in the published sequence was found to be absent in the HCMV strains AD169 and Towne DNA.

sequence contains a mutation; the strain AD169 genomic sequence does not encode a stop codon. Subsequently, the 5′ end of the UL102 transcript was mapped upstream of the UL101X start codon, and the sequence of an HCMV strain Towne cDNA clone was determined, confirming the absence of the stop codon.[23] Finally, we have shown that the sequence upstream of UL101X functions as a promoter.[14] We conclude, therefore, that the UL102 protein is expressed beginning at the UL101X ATG. In sum, these results provide a clear demonstration of the utility of this method for establishing essential gene structure by functional criteria.

B. Identification of Loci that Cooperate to Activate Gene Expression

Because initial experiments must rely on native promoter function to drive expression of essential functional proteins, it is possible to use these transient methods simultaneously to gain new insights into genetic requirements for their expression. In our example, the finding that five loci, in addition to those six encoding the replication fork proteins, were needed to complement HCMV DNA replication was surprising, because homologous proteins were not required to complement herpes simplex virus (HSV)-1 or Epstein-Barr virus DNA synthesis.[4,8,9] Moreover, none of these five "nonfork" loci showed evidence of similarity to the seventh essential HSV-1 replication component, the origin-binding initiator protein UL9 (Table 10.1).[19,21,24] We speculated that these five loci might cooperate to upregulate expression of replication proteins.[7] UL36–38, MIE, and IRS1/TRS1 proteins are expressed under immediate early conditions,[25,26] and had been shown to modulate expression of cellular and/or viral gene targets.[26-28] In contrast, the UL84 and UL112–113 loci both express early temporal class, nucleus-localized proteins with unknown functions.[29,30] The UL84 protein recently was shown to complex with the IE2 86-kDa protein,[31] but the significance of this association remains unknown. Studies of the UL112–113 locus characterized the transcripts, promoter, and encoded proteins and their expression, but had not elucidated its function.[30,32,33]

We directly tested the abilities of UL36–38, UL84, UL112–113, major immediate early, and IRS1 proteins to cooperatively regulate promoters for each of the replication fork proteins genes using an extension of the transient complementation assay.[14] For these experiments we cotransfected a set of five plasmids expressing, respectively, each of the "nonfork" candidate effector loci under the control of their native promoters, together with a replication gene promoter/luciferase reporter plasmid. We found that a lipofectamine reagent (Life Technologies, Gaithersburg, MD) transfection protocol produced levels of expression equivalent to or even greater than the $CaPO_4$ method.[14,34] Moreover, comparisons showed that the results obtained with these protocols were essentially identical. Because the lipofectamine transfections are technically easier to perform and, at least in our hands, more reproducible within and between experiments, the lipofectamine procedure has become our method of choice for gene expression studies in human fibroblasts. We have not, as yet, successfully employed lipofectamine for transient complementation of DNA replication.

Salient results obtained with the UL44 promoter are presented in Figure 10.4 and can be summarized as follows. When all of the five candidate effector plasmids were cotransfected with pAI1, luciferase expression was amplified more than 100-fold in comparison with no effectors ("ALL"). To determine which of these loci participate in activating the UL44 promoter, the candidate effectors were individually subtracted from the transfection mixture. Omitting either IRS1-, UL36–38-, UL112–113-, or MIE-expressing plasmids from these cotransfections reproducibly reduced UL44 promoter expression between two- and tenfold in comparison with the "ALL" effector transfections (MIE most dramatically, IRS1 least). In contrast, omitting UL84 from the transfection mixture always resulted in slightly increased luciferase expression. Although there was considerable variability in absolute levels of luciferase expression between experiments, the noted trends, when expressed as the percent of "ALL" activation, were clear and reproducible. Frameshift mutations in any of these loci eliminated these effects, arguing that the observed interactions are not due to promoter competition artifacts between the effector plasmids (not shown). Likewise, the activation was specific in that other loci, such as UL54 and UL70, did not have significant effects on reporter gene expression under the same transfection conditions. Thus, UL36–38, IRS1, and UL112–113 loci act together with the major immediate early proteins to activate expression from the UL44 promoter. Similar experiments demonstrated that interactions between the activating loci were significantly cooperative, and that promoters for each of the other replication fork proteins were likewise activated.

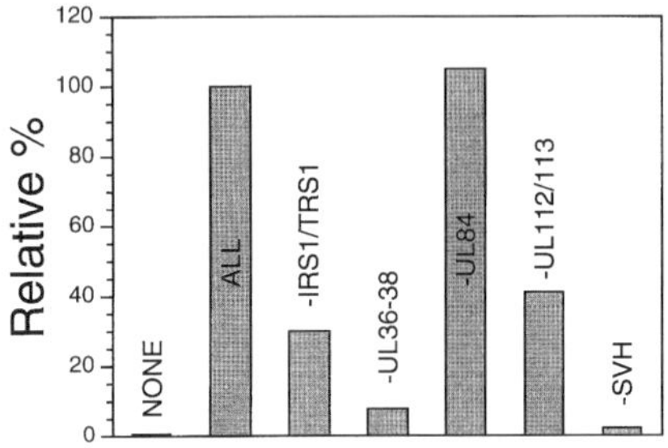

FIGURE 10.4
Multiple loci cooperate to activate HCMV early gene expression. Equimolar amounts of effector plasmids were cotransfected into HF cells along with a UL44 promoter-luciferase reporter plasmid (AI1), using the lipofectamine transfection protocol. The transfection mixtures contained 0.5 μg of the reporter plasmid and either: no effector DNA ("NONE"); 0.5 μg of an equimolar mixture of plasmids spanning the UL33–38 (ZP8), UL84 (ZP13), UL112–113 (ZP24), UL122–123 (SVH), IRS1 (ZP3), and TRS1 (XEXX6.1) loci ("ALL"); or the indicated plasmid was omitted and replaced by an equal mass of pBluescript DNA. Each well thus received a total of 1.0 μg of DNA. Luciferase activity was assayed 96 h after transfection, as described in Section VI. The percent activity relative to the "ALL" effectors mixture is plotted.

With respect to their role in transient complementation of HCMV DNA replication, these results are therefore consistent with a requirement for UL36–38, UL112–113, MIE, and IRS1/TRS1 to adequately express essential replication fork proteins in complementing DNA synthesis. Clearly, a significant reduction in expression of multiple early loci might produce too little accumulation of essential replication proteins to carry out DNA synthesis, especially if one or more are limiting. Nevertheless, these studies do not rule out direct involvement in initiating or performing DNA synthesis for any of these loci. If transcription or cognate transcriptional regulatory elements are critical components or regulators of DNA synthesis, as is the case in many other systems, including Epstein-Barr virus,[9,35] then each of these proteins becomes a candidate for involvement.

IV. Other Applications and Prospects

The transient genetic strategy was originally devised to identify the HSV genes required for lytic-phase DNA synthesis[3] Subsequently, it was used in other herpesvirus systems to find their replication genes, as well as regulatory functions needed to promote early and late gene expression.[4,7,9,13,14,26] More recently, similar approaches have been successfully applied to additional virus systems. For example, genes essential to carry out *Orygia pseudotsugata* nuclear polyhedrosis virus (NPV) DNA synthesis were located in this manner.[36,37] Most remarkably, 18 distinct genes that contribute to *Autographica californica* NPV late gene expression were identified.[5,38] In the NPV systems, late gene expression is linked to viral DNA replication, and is thought to involve a novel alpha-amanitin-resistant RNA polymerase activity. Thus, the entire cascade of events leading up to and including late gene expression has been reconstructed. Finally, once a set of genes that encode proteins comprising a functional unit, such as the DNA synthesis apparatus, has been defined, the essential genes can be overexpressed and the entire process reconstructed in foreign cell systems. Remarkably, it was demonstrated that origin-dependent HSV replication can be reconstituted in insect cells by coexpressing the essential replication proteins using recombinant baculoviruses.[39] Likewise, herpesvirus capsids could be assembled in insect cells by coexpressing the essential virus capsid and assembly proteins.[40] By using this extension of the transient genetic strategy, larger quantities of the proteins and products of interest can be produced to facilitate biochemical studies. Taken together, the above results argue that it should be possible to reconstitute *in vivo* any process for which a suitable assay can be devised using transient methods.

V. Materials

1. Dulbecco's modified Eagle medium supplemented with 10% fetal bovine serum (FBS) (10% DMEM).
2. CsCl (Ambion, Austin TX; molecular biology grade)-purified, double-banded DNA.
3. 2.5 *M* $CaCl_2$ (Sigma, St. Louis, MO; cell culture tested) filter sterilized through a 0.45-μm filter.
4. 2X BES-buffered saline (BBS): 50 m*M* *N*,*N*-bis(2-hydroxyethyl)-2-aminoethanesulfonic acid (CalBiochem, San Diego, CA), 280 m*M* NaCl, 1.5 m*M* Na_2HPO_4 (Sigma, St. Louis, MO). Adjust pH to 6.95 with 1 *N* NaOH. (See Notes.)
5. A 35°C 3% CO_2 humidified incubator.
6. Lipofectamine reagent (Life Technologies, Gaithersburg, MD).
7. Opti-MEM I (Life Technologies, Gaithersburg, MD).

VI. Procedures

A. $CaPO_4$ Cotransfection for Transient Complementation of HCMV DNA Synthesis

1. Plate approximately 5×10^5 human foreskin fibroblast (HF) cells on a 10-cm tissue culture dish 24 h before transfection. Cells should be no more than 50% to 70%

confluent for transient expression. Smaller plates (e.g., 6-cm) can be used with equivalent cell densities; in some cases this is sufficient, and easier when performing many transfections at one time. Note that the optimal cell density and number of cells to be plated may vary with the cell type employed, and must be determined experimentally.

2. Dilute the 2.5 *M* $CaCl_2$ solution to 0.25 *M* with sterile water.
3. Add 20 to 30 μg plasmid DNA per tube (Falcon No. 2058) or 10 to 20 μg for a 6-cm plate. Only high-quality plasmid DNA will work. Use only double-banded CsCl plasmid DNA. Carrier DNA is not necessary and actually will decrease efficiency. Also, linear DNA does not transfect well. To the DNA add 500 μl of 0.25 *M* $CaCl_2$. Then add 500 μl of 2X BBS. Mix well and incubate at room temperature for 10 to 20 min. At this point no precipitate should form. Use three different concentrations of DNA to help identify the DNA concentration necessary for optimal transfection. If 6-cm dishes are used this mixture can be split into two, and transfections can be done in duplicate. If duplicate transfections are not desired, then use half the volumes mentioned above and add the total volume to one 6-cm dish.
4. Add the calcium phosphate/DNA mixture to cells in a dropwise fashion, swirling the plate after each drop. Incubate the cells overnight in a 35°C 3% CO_2 incubator. The CO_2 concentration is critical. Measure the level with a Fyrite gas analyzer. Temperature is somewhat less critical; a 37°C incubator can be used.
5. Wash cells twice with 5 ml of phosphate-buffered saline (PBS), then add 10 ml of DMEM with 10% FBS. Incubation of cells from this point on is done in a 5% CO_2 37°C incubator.
6. For transient expression, harvest cells 48 to 72 h post-transfection. For the replication assay we usually harvest DNA 5 to 7 d post-transfection. Cells are lysed directly on the culture dish with 2% sodium dodecyl sulfate (SDS; United States Biochemical, Cleveland, OH), 10 m*M* Tris-HCl, pH 7.5, 10 m*M* EDTA (Sigma, St. Louis, MO), and transferred to a 1.5-ml microcentrifuge tube. Lysates are then treated with 500 μg per milliliter of proteinase (Boehringer Mannheim, Indianapolis, IN) for 2 h at 60°C, extracted once with phenol/chloroform/isoamyl alcohol (25:24:1), once with chloroform/isoamyl alcohol (24:1), ethanol precipitated, and resuspended in 100 μl of TE (10 mM Tris-HCl, pH 8.0, 1 mM EDTA) for analysis.
7. For Southern transfer, 25 μl of the redissolved DNA is cleaved in a volume of 50 μl with 20 U *Eco*RI and 10 U *Dpn*I (New England Biolabs, Beverly, MA) for 3 h at 37°C and electrophoresed through a 0.8% agarose gel, transferred to ZETA-probe nylon membrane (Bio-Rad, Hercules, CA), and hybridized with a random-primer ^{32}P-labeled pGEM7Zf(–) probe, according to manufacturer's instructions (Amersham, Arlington Heights, IL). Hybridizations are performed in a volume of 10 ml of buffer (1.5X SSPE [1X SPE is 0.18 *M* NaCl, 10 m*M* $NaPO_4$, and 1 m*M* EDTA (pH 7.5)], 7% w/v SDS, 10% polyethylene glycol (Sigma, St. Louis, MO) with 5 to 10 ng of labeled plasmid for 15 h at 65°C in a hybridization oven (Robbins Scientific, Sunnyville, CA). Blots are then exposed to X-ray film, X-OMAT AR (Kodak, Rochester, NY) with an intensifying screen (Dupont lighting plus), or equivalent, at –80°C for 15 h.

Note: *The pH of BBS is critical. We typically make several BBS solutions ranging in pH from 6.93 to 6.98, and test each to find the best preparation. Usually a visual inspection of cells after an overnight transfection will indicate* *which BBS mixture and DNA concentration works best. A coarse and clumpy precipitate will form when DNA concentrations are too low, a fine, almost invisible precipitate will form when DNA concentrations are too high. An even granular precipitate forms when the concentration is just right. This usually correlates to the highest level of gene expression and, consequently, replication.*

When performing cotransfections, vary the amount of any known effector plasmid in relation to the other plasmids in the mixture. The ratio of plasmids used in the mixture can be the difference between success and failure. We routinely find that a higher concentration of effector plasmid in the mix yields better results. Depending on the number of plasmids in the transfection mixture, vary the amount of DNA transfected at one time. Each plasmid can contain one or several essential genes.

B. Lipofectamine Cotransfections for Assays of Cooperative Activation

Our lipofectamine transfection protocol is adapted from the vendor's recommendations.[34]

1. Set up cells 18 to 24 h before transfection. We use 6-well plates, seeded at a density of 3×10^5 cells per well. Again, the appropriate cell density will probably vary with cell type.
2. Prepare DNA/lipid mixtures in sterile 12 × 75 mm tubes:

   Solution A: 100 μl OPTI-MEM I medium + 1 μg DNA (0.5 μg promoter construct + 0.1 μg of each transactivator).

   Solution B: 100 μl OPTI-MEM I medium + 6 μl Lipofectamine reagent.

   Combine solutions A and B, mix gently, and incubate at room temperature for 30 min.
3. While incubating transfection mixture, rinse cells (about 80% confluent) with 2 ml per well OPTI-MEM I, discard wash, add 0.8 ml per well OPTI-MEM I.
4. Add 200 μl per well of DNA/lipid mix to the cells and incubate in 5% CO_2 incubator at 37°C for 3 h.
5. After 3 h, add 1 ml per well DMEM + 20% FBS and incubate in 5% CO_2 incubator for 16 h.
6. Rinse cells once with 2 ml DMEM, discard wash, add 2 ml DMEM + 10% FBS.
7. Incubate in 5% CO_2 incubator for 96 h.
8. Harvest cells 96 h after transfection. We have been using the Promega Luciferase Assay System (Promega Corp., Madison, WI) per vendor's recommendations. Luminescence is measured in a liquid scintillation counter with coincidence turned off. A standard curve was constructed using purified firefly luciferase (Boehringer-Mannheim, Indianapolis, IN) to ensure that output was measured in the linear range.

Note: *Harvesting at 48 h after infection produced the same results for transient expression assays, but not for DNA replication assays.*

Acknowledgments

We thank the past and present members of our laboratories for their contributions to the application of these approaches to the problems that have interested us. We also gratefully acknowledge Tim Moran and Matt Shudt of the Wadsworth Center Molecular Genetics Core facility for oligonucleotide synthesis and DNA sequencing.

Experimental work described herein was supported by Public Health Service grants AI31249 and AI33416 to DGA.

References

1. Ho, M., *Cytomegalovirus Biology and Infection,* Plenum Press, New York, 1991.
2. Mocarski, E. S. J., in *The Human Herpesviruses,* Roizman, B., Whitley, R. J., and Lopez, C., Eds., Raven Press, New York, 1993, 173.
3. Challberg, M. D., *Proc. Natl. Acad. Sci. U.S.A.,* 83, 9094 (1986).
4. Fixman, E. D., Hayward, G. S., and Hayward, S. D., *J. Virol.,* 66, 5030 (1992).
5. Todd, J. W., Passarelli, A. L., and Miller, L. K., *J. Virol.,* 69, 968 (1995).
6. Wu, C. A., Nelson, N. J., McGeoch, D. J., and Challberg, M. D., *J. Virol.,* 62, 435 (1988).
7. Pari, G. S. and Anders, D. G., *J. Virol.,* 67, 6979 (1993).
8. Olivo, P. D., Nelson, N. J., and Challberg, M. D., *J. Virol.,* 63, 196 (1989).
9. Fixman, E. D., Hayward, G. S., and Hayward, S. D., *J. Virol.,* 69, 2998 (1995).
10. Heilbronn, R. and zur Hausen, H., *J. Virol.,* 63, 3683 (1989).
11. Anders, D. G. and Punturieri, S. M., *J. Virol.,* 65, 931 (1991).
12. Anders, D. G., Kacica, M. A., Pari, G. S., and Punturieri, S. M., *J. Virol.,* 66, 3373 (1992).
13. Pari, G. S., Kacica, M. A., and Anders, D. G., *J. Virol.,* 67, 2575 (1993).
14. Iskenderian, A. C., Huang, L., Reilly, A., Stenberg, R. M., and Anders, D. G., *J. Virol.,* 70, 383 (1996).
15. Fleckenstein, B., Muller, I., and Collins, J., *Gene,* 18, 39 (1982).
16. Stenberg, R. M. et al., *J. Virol.,* 64, 1556 (1990).
17. Chen, C. and Okayama, H., *Mol. Cell. Biol.,* 7, 2745 (1987).
18. Challberg, M. D. and Kelly, T. J., *Annu. Rev. Biochem.,* 58, 671 (1989).
19. Chee, M. S. et al., *Curr. Top. Microbiol. Immunol.,* 154, 125 (1990).
20. Kerry, J. A. et al., *J. Virol.,* 70, 373 (1996).
21. Elias, P. and Lehman, I. R., *Proc. Natl. Acad. Sci. U.S.A.,* 85, 2959 (1988).
22. Fickett, J. W., *Nucleic Acids Res.,* 10, 5303 (1982).
23. Smith, J. A. and Pari, G. S., *J. Virol.,* 69, 1734, (1995).
24. Elias, P., O'Donnell, M. E., Mocarski, E. S., and Lehman, I. R., *Proc. Natl. Acad. Sci. U.S.A.,* 83, 6322 (1986).
25. Tenney, D. J. and Colberg-Poley, A. M., *Virology,* 182, 199 (1991).
26. Stasiak, P. C. and Mocarski, E. S., *J. Virol.,* 66, 1050 (1992).
27. Pizzorno, M. C., O'Hare, P., Sha, L., LaFemina, R. L., and Hayward, G. S., *J. Virol.,* 62, 1167 (1988).

28. Colberg-Poley, A. M., Santomenna, L. D., Harlow, P. P., Benfield, P. A., and Tenney, D. J., *J. Virol.,* 66, 95 (1992).
29. He, Y. S., Xu, L., and Huang, E. S., *J. Virol.,* 66, 1098 (1992).
30. Wright, D. A., Staprans, S. I., and Spector, D. H., *J. Virol.,* 62, 331 (1988).
31. Samaniego, L. A., Tevethia, M. J., and Spector, D. J., *J. Virol.,* 68, 720 (1994).
32. Staprans, S. I. and Spector, D. H., *J. Virol.,* 57, 591 (1986).
33. Iwayama, S. et al., *J. Gen. Virol.,* 75, 3309 (1994).
34. Harms, J. S. and Splitter, G. A., *FOCUS,* 17, 34 (1995).
35. Schepers, A., Pich, D., and Hammerschmidt, W., *EMBO J.,* 12, 3921 (1993).
36. Kool, M., Ahrens, C. H., Goldbach, R. W., Rohrmann, G. F., and Vlak, J. M., *Proc. Natl. Acad. Sci. U.S.A.,* 91, 11212 (1994).
37. Ahrens, C. H. and Rohrmann, G. F., *Virology,* 207, 417 (1995).
38. Lu, A. and Miller, L. K., *J. Virol.,* 69, 975 (1995).
39. Stow, N. D., *J. Gen. Virol.,* 73, 313 (1992).
40. Thomsen, D. R., Roof, L. L., and Homa, F. L., *J. Virol.,* 68, 2442 (1994).

Chapter 11

Epstein-Barr Virus: Gene Expression and Regulation

Nirupama Deshmane Sista and Joseph S. Pagano

Contents

0-8493-4412-3/96/$0.00+$.50

I. Introduction

Epstein-Barr virus (EBV) is the causative agent of infectious mononucleosis and oral hairy leukoplakia. It also causes polyclonal B-lymphocytic proliferative conditions in immunosuppressed patients who are recipients of organ transplants or have AIDS. EBV infection is universally linked to nasopharyngeal carcinoma and is also associated with parotid carcinoma.[1] The virus predisposes to the development of Burkitt's lymphoma in endemic regions. The molecular and cellular pathogenesis of EBV infection helps to explain these diverse virus–disease associations. EBV exists in the cell as a latent infection or can produce a cytolytic virus-productive state. Infection of lymphocytes is predominantly latent where the EBV exists as an episome (Figure 11.1), whereas infection of the mucosal epithelial cells of the oropharynx is characterized by cytolysis and viral replication.[2] Viral reactivation from the latent state can be induced by a variety of different agents, such as phorbol esters, sodium butyrate, calcium ionophores, and anti-immunoglobulins, all of which activate Z, the EBV immediate-early gene product of the BZLF-1 open reading frame (ORF), which is sufficient to trigger disruption of viral latency in EBV-infected cells (reviewed in Reference 3). By binding to AP1-like sites in the upstream regions of genes, Z trans-activates several EBV early promoters, including the promoter for the DNA polymerase gene.[4] This starts a cascade of events that ultimately results in cytolysis.

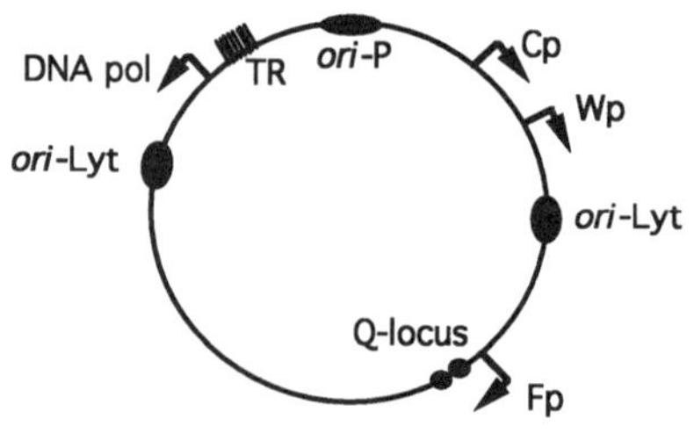

FIGURE 11.1
Schematic presentation of the Epstein-Barr virus episome. The locations of the pol promoter as well as the latent promoters Cp, Wp, and Fp and the origins of replication in the latent cycle (*ori*P) and cytolytic cycle (*ori*lyt) are shown.

In this chapter, we review how commonly used molecular methods are practically applied to analyze the different processes that result in reactivation of the virus from the latent phase. These studies provide the framework for this chapter. We commence by describing cell lines used to study EBV gene expression, how to induce viral reactivation, and the markers for its analysis. Next are reviewed the methods for study of transcription of EBV genes, starting with analysis and mapping of the mRNAs for the EBV DNA polymerase gene. The promoter for this gene provides an example of how regulation of its expression is produced by a combination of viral and cellular factors, and analysis of interactions between both the transcription factors and cis-acting elements as well as protein–protein interactions are described. An unusual aspect of the expression of the EBV polymerase gene is post-transcriptional regulation conferred by 3′ processing of its mRNAs. Amplification of the 3′ ends of mRNA and *in vitro* cleavage and polyadenylation assays to identify cis-acting components of polyadenylation signals are also briefly reviewed. We end with a description of assays of the function of the EBV DNA polymerase protein, which is a target for antiviral drugs, and detection of its product, EBV genomes.

II. Cell Lines

Analysis of gene expression involves the use of several EBV-positive lymphoid cell lines, some of which are derived from Burkitt's lymphomas. EBV is completely latent in several such cell lines, such as Raji, Jijoye, and Daudi, which can therefore be used for study of latent gene expression. Since the latent viral genome can be induced to replicate, such cell lines can also be used for study of viral reactivation. EBV is in a productive state in less than 10% of the cells in lines such as B95-8, P3HR1, and Akata; with induction 40% or more of the cell population may produce virus, depending on the cell line and method of induction. These cells are routinely grown in RPMI 1640 with 10% (v/v) fetal calf serum (FCS) and antibiotics at 37°C. EBV-negative Burkitt's lymphoma cell lines (such as BJAB, Louckes) are often used as controls. All cell lines are maintained in a 5% CO_2 environment.

III. Induction of Virus Replication

The EBV cytolytic cycle can be induced by a variety of agents. Cells can be grown in growth medium containing reduced serum concentration (3 to 5%) or can be incubated with phorbol 12-myristate 13-acetate (PMA, 30 ng/ml, also called 12-*o*-tetradecanoyl-phorbol-13-acetate, TPA) and sodium butyrate (5 m*M*).[5,6] The latently infected Raji cells contain the EBV genome in an episomal form and can be superinfected with P3HR1 virus to induce reactivation and recombinants of the exogenous and endogenous viral genomes.[7] Akata cells have surface IgG receptors, and viral replication can be very efficiently induced by treatment with 100 μg of anti-human immunoglobulin G (IgG) per milliliter of medium (Sigma, St. Louis, MO).[8] More recently, transfection of DNA of an expression clone for Z into EBV-positive cells has been sufficient to induce reactivation[9] in up to 10% of cells. The protocol used for the chemical induction of Raji cells is detailed:

1. Grow 100 ml of Raji cells in 3 to 4 d.
2. Centrifuge at 1200 rpm for 10 min in two 50-ml conical tubes.
3. Resuspend the cell pellets in 5 ml of the regular growth medium: RPMI 1640, 10% FCS, penicillin/streptomycin.
4. Pipette the resuspended cells into one 75 cm^2 tissue culture flask.
5. Dilute to 100 ml with fresh growth medium.
6. Add the TPA.

Note: *Be careful when working with TPA. This chemical is extremely carcinogenic. The stock TPA is diluted in 50:50 acetone-methanol at a concentration of 1 mg/ml. Since the acetone-methanol evaporates quickly, and the volume of stock TPA is so small, pipette the 3 μl into 1 to 2 ml of growth medium and then add to 100 ml of cells. This prevents loss of TPA by evaporation and ensures safer handling of the chemical. When pipetting TPA, use disposable pipettes. Also, TPA is light-sensitive, and exposure to direct light is to be avoided. Dispose of all equipment coming in contact with the TPA by autoclaving and discarding. If glass is used, autoclave immediately after use.*

7. TPA-treated Raji cells are then maintained in RPMI 1640 medium until harvested.
8. Centrifuge at 1200 rpm for 10 min in two 50-ml conical tubes. Be sure to autoclave the supernatant fluids.
9. Wash cells with phosphate-buffered saline solution (PBS) and then harvest.

IV. Monitoring of Virus Production

Induced EBV-positive cells are usually harvested between 4 and 48 h, depending on the type of mRNA and protein to be extracted. Early proteins are usually detected 48 h after induction. Virus production is monitored by indirect immunofluorescence for early and viral capsid antigens (EA/VCA) using sera from patients with nasopharyngeal carcinoma (NPC) that typically have high titers of antibodies against EBV replicative antigens.[10] Fluorescein-isothiocyanate(FITC)-conjugated goat anti-human antibody is used as the secondary antibody for immunofluorescence. The detailed protocol for detection of EBV antigens is as follows:

1. Induced Raji cells are resuspended in PBS so that the cell concentration is equal to 5×10^6 cells/ml.
2. Pipette 20 to 25 μl of the resuspended cells onto each well of a clean slide.
3. Air-dry until all liquid has evaporated.
4. Fix cells in 10% acetone (–20°C) for 10 min.

5. Air-dry slides. At this point they are ready for use or storage at –70°C up to 1 year.
6. Dilute EA/VCA-positive human serum 1:800 or use a monoclonal antibody to EA-D and store at –20°C.
7. Apply diluted serum to the cells as follows:
   a. Warm slides to room temperature.
   b. Pipette 15 to 20 μl of diluted sera onto the wells of the slide; use EA/VCA-positive and -negative control sera.
   c. Place the slides in a moist chamber and incubate for 45 min to 1 h at 37°C. After incubating, wash twice in PBS and once in distilled water for 10 min per washing.
   d. Air-dry slides.
8. Prepare FITC anti-human IgG:
   a. The FITC anti-IgG is aliquoted into 0.1 ml portions and stored at –70°C until needed.
   b. Dilute the FITC anti-IgG to 1:30 using PBS. Add the 0.1 ml aliquot to 2.9 ml PBS.
9. Pipette 15 to 20 μl of the diluted anti-IgG onto the wells of the dried slides.
10. Incubate slides at 37°C for 30 min in a moist chamber.
11. Wash four times in PBS. During the distilled water washing, add 3 to 4 drops of Evan's Blue, 1% in H_2O; stain for 10 min.
12. Air-dry, mount, and study under the fluorescence microscope.

V. Immunoblotting

An alternate method to monitor induction of EBV-positive cell lines is the detection of EA-D protein by Western blots.[11] This is a relatively easy and fast method that gives very reproducible results. EBV-positive cells that have been induced are harvested after 24 h and washed twice with 1X PBS and resuspended in 200 μl of ELB^+ buffer (250 m*M* NaCl, 0.1% Nonidet-P-40, 50 m*M* *N*-2-hydroxyethylpiperazine-*N*-2-ethanesulfonic acid [HEPES], pH 7.0, 5 m*M* ethylene diamine tetraacetic acid [EDTA], pH 8.0, 0.5 m*M* dithiothreitol [DTT], and 1 m*M* phenyl methyl sulfonyl fluoride [PMSF]). Cells are then lysed by sonication for 10 sec, and the lysate is cleared by centrifugation for 10 min at 4°C. Equal amounts of samples are resuspended in Laemmli sample buffer, boiled, and loaded onto a sodium dodecyl sulfate (SDS)-8% polyacrylamide electrophoresis gel (PAGE). Following electrophoresis, the gel is transferred to polyvinylidene difluoride membrane (Schleicher and Schuell, Keene, NH) for 2 h at 140 mA and then immunoblotted with EA-D monoclonal antibody (Dupont, Boston, MA, No. 9240) and visualized by the chemiluminescence method, as specified by the manufacturer (Amersham, Arlington Heights, IL). In Figure 11.2, latently infected Raji cells (lane 1) are induced by cotransfection with a Z-expression clone (lane 2) and increasing concentrations of an expression clone for another EBV immediate-early gene product, RAZ (lanes 3 to 5). Detection of EA-D by Western blotting indicates reactivation of the latent EBV. RAZ can inhibit viral reactivation produced by Z as marked by the expression of the early antigen.

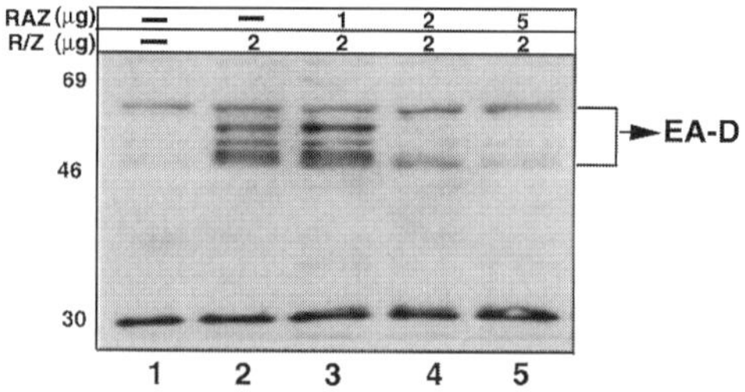

FIGURE 11.2
Inhibition of viral reactivation by the EBV transdominant repressor protein, RAZ. Latently EBV-infected Raji cells were cotransfected with a R/Z genomic clone and increasing concentrations of RAZ. Reactivation was marked by the expression of the EBV EA-D protein, detected by Western blotting and probed with EA-D-specific monoclonal antibody (Dupont). (From Furnari, F. B. et al., *J. Virol.*, 68, 1827, 1994. With permission.)

Most investigations of EBV gene regulation have centered on maintenance of latency and disruption of this state by the expression of immediate-early genes encoding Z and the gene product of the BRLF1 ORF, R. The immediate-early proteins, Z and R, then trans-activate the promoters for several early genes, sometimes synergistically. The promoter and the 5′ regulatory sequences of EBV genes have to be identified before undertaking analysis of gene regulation. In order to map the 5′ end involved in transcription initiation of genes, RNA from induced EBV-positive cells is isolated and used in Northern blot analyses.

VI. RNA Isolation and Northern Analysis

Total cellular RNA can be isolated from cell lines by resuspending the pelleted cells in 4 *M* guanidine isothiocyanate.[12] An alternative method is to lyse the cells with a mild detergent, such as buffer containing 50 m*M* Tris-hydrochloride, pH 7.4, 150 m*M* NaCl, 20 m*M* EDTA, 2% (v/v) *N*-lauryl sarcosine, and 200 μg/ml of proteinase K. The lysate is incubated at 30°C and then vortexed thoroughly to reduce the viscosity. The lysate is then mixed with CsCl (1 g/ml of lysate) and underlaid with 2.5 ml of 5.7 *M* CsCl. After 18 h at 27,000 rpm in a Beckman SW40 rotor (20°C), the RNA is recovered as a DNA-free pellet, dissolved in 200 μl of 300 m*M* sodium acetate (pH 5.4), and precipitated with 3 volumes of ethanol.[13] Polyadenylated mRNA can then be enriched by oligo(dT)-cellulose chromatography. For Northern analysis, 20 μg of total cellular RNA or 5 μg of poly(A^+) RNA are denatured for 15 min at 65°C or temperatures up to 90°C in the presence of formaldehyde and resolved on a 1% agarose gel containing 0.66 *M* formaldehyde, 20 m*M* 3-(*N*-morpholino) propanesulfonic acid, 5 m*M* sodium acetate, and 1 m*M* EDTA. Following electrophoresis, RNA can be blotted on to nitrocellulose filters, fixed by heat, and prehybridized for 2 h at 59°C in prehybridization solution containing 50% formamide, 10X SSC (1X SSC is 0.15 *M* NaCl, 0.015 *M* sodium citrate), 50 m*M* sodium phosphate (pH 7.0), 2.5X Denhardt's solution, and 200 μg of salmon sperm DNA per milliliter.

Hybridization is performed in the same solution at 59°C. Filters are then washed at 65°C in 0.1X SSC-0.1% sodium dodecyl sulfate and treated with RNAse A.

The major problem with RNA extraction is the low abundance of some RNA, as well as the sensitivity to the highly abundant and ubiquitous RNAses in the cell. Many RNAse inhibitors, such as RNAsin, are available and are used in extractions. The EBV DNA polymerase message was detected by Northern blotting, as described above, only after successful induction of viral reactivation in a substantial percentage of cells, as monitored by early antigen (EA) immunofluorescence staining.[4] In order to identify the regulatory sequences for the *pol* gene, the transcription start site had to be mapped. In order to examine promoters for this and other EBV genes, the 5′ start site has first to be identified by techniques such as S1 nuclease analysis or RNAse protection and, more recently, by a method that amplifies 5′ cDNA ends, called rapid amplification of cDNA ends (RACE).

VII. S1-Nuclease and Primer-Extension Analyses

For S1-nuclease protection assays of viral RNA to identify the pol RNA start site, mRNA was prepared as described from TPA-induced and uninduced B95-8, P3HR1, and Raji cells. S1 probes are specifically designed to protect the 5′ end of RNA so that it will not be cleaved by the single-strand-specific S1 nuclease. S1 probes were prepared by subcloning EBV fragment SphI 153,176 to StuI 156,695 into phagemid pBS+ (Stratagene, La Jolla, CA) and generating single-stranded DNA template by using helper phage M13K07. A ^{32}P-end-labeled primer (156,801 to 157,118) is annealed to the template and extended with Klenow enzyme. After digestion with SphI, the strands are separated on a denaturing polyacrylamide-urea gel, and the appropriate size band is recovered. Five micrograms of poly(A)$^+$ RNA is hybridized for 12 h with 5×10^4 cpm of the end-labeled probe (Figure 11.3) in 30 µl 40 m*M* PIPES (pH 6.4), 1 m*M* EDTA (pH 8.0), 0.4 *M* NaCl, and 80% formamide at 45°C. S1 digestion is carried out by addition of 800 U S1 nuclease in 300 µl 0.28 *M* NaCl, 0.05 *M* sodium acetate (pH 4.5), 4.5 m*M* $ZnSO_4$, and 20 mg/ml single-stranded calf thymus DNA. Reactions are terminated by addition of 80 ml of a mixture containing 4 m*M* ammonium acetate, 50 m*M* EDTA (pH 8.0). Products are resolved on 7 *M* urea, 6% polyacrylamide gel and visualized by autoradiography. Since RNA samples may migrate differently based on the secondary structure, radioactively labeled pol RNA that was transcribed *in vitro* using the SP6/T7 promoters is used as size marker.

For primer-extension analysis, 5×10^4 cpm of ^{32}P-end-labeled primer are annealed as described above, precipitated, and resuspended in 30 µl reverse transcriptase buffer containing 50 m*M* Tris (pH 7.6), 60 m*M* KCl, 10 m*M* $MgCl_2$, and 1 m*M* dNTPs. Reactions are carried out with 30 U AMV reverse transcriptase at 42°C for 1 h, and products are resolved as described for S1 analysis.

Generally, primer-extension analyses used alone are inadequate for mapping transcriptional start sites, especially in G-rich genomes, since extension may stall

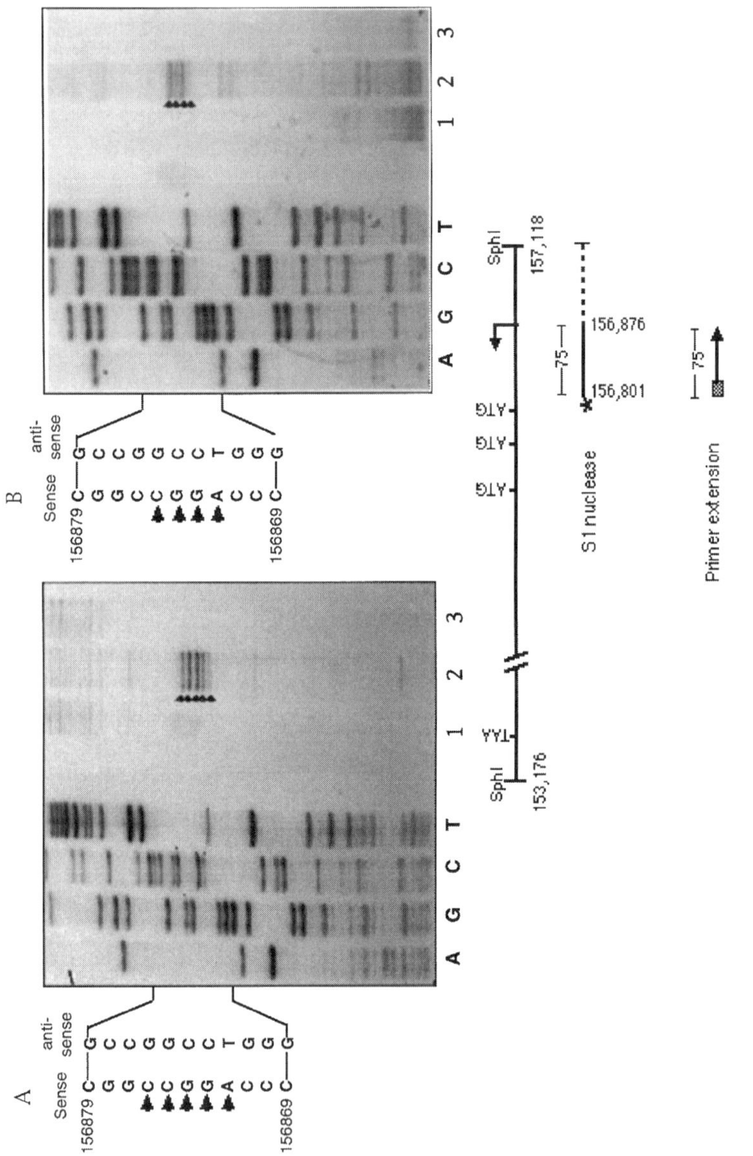
A
B
Sense
anti-
sense
156879
156869
A
G
C
T
1
2
3
SphI
157,118
153,176
ATG
TAA
156,876
156,801
75
S1 nuclease
Primer extension

FIGURE 11.3

at stretches of Gs. The combination of S-1 nuclease and primer extension analyses provides reliable positioning that can also be confirmed by mapping of cDNAs.

VIII. Promoter Analysis

The DNA sequences surrounding the 5′ start site are analyzed for promoter activity by cloning the DNA into reporter constructs. Most EBV promoters have been cloned upstream of the chloramphenicol acetyl transferase (CAT) gene. The pBS.CAT was constructed by isolating the 1.7 kb *Bgl*II-*Bam*HI CAT gene of pCAT3M[4] and cloning it into the *Bam*HI site of pBS+ (Stratagene, La Jolla, CA). Subsequently, pPolCAT was constructed by isolating a 1.29 kb *Bal*I restriction fragment that included the identified start site and cloning it into the *Hinc*II site directly upstream of CAT.[4]

IX. Transfections and CAT Assays

Plasmid DNA is amplified from *Escherichia coli* and purified over two sequential cesium chloride gradients. For each sample, 10 μg of reporter and 10 μg trans-activator plasmid DNA are electroporated into 10^7 cells at 1,500 V[14] with the University of Wisconsin Zapper electroporation unit. Cells were suspended in 10 ml of RPMI 1640 medium supplemented with 10% fetal calf serum and incubated for 48 h at 37°C in 5% CO_2. Extracts of transfected cells were prepared by washing cell pellets twice in PBS, suspending them in 200 μl of 0.25 *M* Tris-HCl (pH 7.5), and freeze-thawing four times. Reactions are carried out by incubating each sample with acetyl coenzyme A and [^{14}C]-chloramphenicol at 37°C for 1 h.[15] Reaction products are separated by thin-layer chromatography, visualized by autoradiography, and quantitated either by scintillation counting or by scanning with a phosphoimager. Percent acetylation is calculated as the ratio of acetylated reaction products to the entire sample. A sufficient number of assays has to be done to provide reliable results with relatively low standard deviation. Transfection efficiency and purity of DNA preparations are key factors. The background CAT activity produced by the vector alone is a negative control.

FIGURE 11.3
S1-nuclease and primer-extension mapping of the 5′ end of EBV DNA polymerase mRNA. (A) For S1 analysis, ^{32}P-end-labeled probe was annealed to 5 mg poly(A)+ RNA isolated from B95-8 uninduced (lane 1), induced (lane 2), or Jurkat cell (lane 3). (B) For primer-extension analysis, ^{32}P-end-labeled oligonucleotide was annealed to 5 mg poly(A)+ RNA isolated from B95-8 cells, uninduced (lane 1), induced (lane 2), or Jurkat cells (lane 3). Hybridized mixtures were either treated with S1-nuclease or reverse-transcribed with AMV-RT and products were resolved as described. Sequence ladders were generated using the same oligonucleotide primer and single-stranded DNA from a genomic pol subclone to determine precisely the site of transcription initiation. S1 probe (156,801 to 157,118) and oligonucleotide primer (156,801 to 156,840) used in mapping are as diagrammed. (From Furnari, F. B. et al., *J. Virol.*, 66, 2837, 1992. With permission.)

X. Mutational Analyses of Promoters

Once the promoter region is identified, and the correct start site for transcription is confirmed, the cis-acting elements are identified by creating deletions in the promoter DNA sequences, and the promoter activity of such deletion constructs is assayed. Protein-binding assays may also suggest the different transcription factors that may bind to the promoter DNA sequence and activate or repress expression. Site-directed mutations are generated, and the effect on promoter activity conferred by specific point mutations is analyzed.

XI. Analysis of the EBV Z Trans-Activator: DNA Binding Assays

All cellular pathways that induce viral reactivation converge to a single event: the activation of the BZLF-1 protein, Z. Z, an immediate-early viral transcription factor that is a member of the bZip family of proteins, initiates a cascade of events starting with the activation of early viral promoters involved in replication of viral DNA and then late gene expression. The promoters for EBV DNA polymerase (pol) and BMRF1 (EA-D), a cofactor required for replication, are trans-activated by Z. Z is a 34-kDa nuclear protein that has similarity to c-fos protein and binds to AP1-binding sites (TGAGTCA) and similar DNA sequences, called Z-response elements (ZRE). Mobility shift assays are commonly used to demonstrate the DNA-binding capacity of proteins. The assays are based on the fact that DNA-protein complexes migrate through nondenaturing polyacrylamide gels at a slower rate than free DNA. Although the basic principle involves the incubation of radiolabeled DNA probes with extracts containing the DNA-binding protein and then subjecting the probe with the protein bound to it to gel electrophoresis, each protein has specific requirements for binding to specific probes. Z protein can be obtained in several ways. The Z ORF has been cloned downstream of several strong promoters, for example, the CMV IE promoter, in expression clones. *In vitro*-transcribed and translated constructs in rabbit reticulocyte extracts or wheat germ extracts can be used as a protein source for binding assays. Alternatively, the Z ORF has also been cloned downstream of the glutathione-*S*-transferase gene; the fusion protein obtained works well in binding assays with or without removal of the GST portion of the fusion protein. Finally Z can be obtained in extracts of EBV-infected cells. The *in vitro* transcription and translation of Z and the conditions for binding of Z to ZRE are outlined below.

A. *In Vitro* Transcription and Translation of Z Protein

The Sp64Z clone is linearized with *Eco*RI, transcribed with SP6 RNA polymerase, and the resulting mRNA is translated in mRNA-dependent wheat germ or rabbit reticulocyte extracts. The translation product (3 μl) was mixed with a ^{32}P-labeled

ZRE-containing EBV-DNA probe from the *Bam*HI A fragment (coordinates 156,860 to 157,006) in 20 μ of 100 m*M* KCl, 20 m*M* HEPES, pH 7.3), 10% glycerol, 0.2 m*M* fluoride, and 2 μg of poly(dl-dC). Binding reactions were carried out for 20 min at 27°C and loaded onto 4% nondenaturing acrylamide gels, which were electrophoresed in 0.25X Tris-borate-EDTA, dried, and visualized by autoradiography.

An example of Z-binding to the pol promoter is shown in Figure 11.4. Oligonucleotides containing known DNA sequences that bind to Z were used as cold competitors to verify that binding to the probe was specific. The affinity of a given protein to a known binding site may vary and is reflected in competition studies. For example, binding of an E2F-1-like protein from cell extracts that recognizes a DNA sequence in one of the EBV latency promoters used for transcription of EBNA-1,[16] Fp, can be competed with very well by the E2F sites from the adenovirus E2 promoter, but not with the E2F sites from the DHFR promoter.

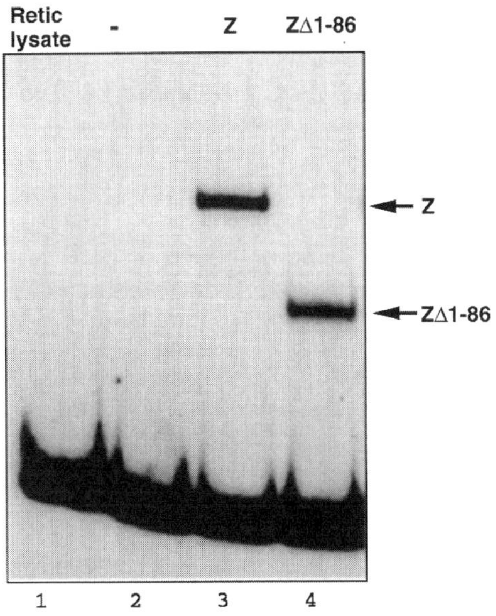

FIGURE 11.4
Electrophoretic mobility shift assays with *in vitro*-transcribed and translated EBV activator protein, Z. A ^{32}P-end-labeled ZRE-containing probe was incubated with rabbit reticulocyte lysate (lane 1), or lysates programmed with RNA for a protein that does not bind to ZRE (lane 2), Z RNA (lane 3), or ZΔ1-86 RNA (lane 4). A shown in lane 3, Z as well as ZΔ1-86 proteins can bind to Z. (From Furnari, F. B. et al., *J. Virol.,* 68, 1827, 1994. With permission.)

In order to analyze and confirm the proteins that bind to promoter sequences, cellular extracts are used as a source of protein in binding reactions. For example, Z from cellular extracts of productively EBV-infected cells binds to ZRE-containing probes under the same conditions used with *in vitro*-made protein except that 5 μg of the protein are used in binding reactions. When the source is cell extract, antibodies are used to identify the proteins that bind to promoter probes. Specific monoclonal or polyclonal antibody is incubated with the cellular extracts for 10 to

15 min, either at 4°C or at room temperature, before or after addition of the probe. Addition of the antibody may result in either slower mobility of the protein-bound probe (supershift) or may disrupt complex formation (blocking). Supershifts of complexes bound to probes may not be specific and therefore may be misleading. To circumvent that problem, antibodies to specific peptides in the protein are used. A specific supershift with such an antibody can be reversed with the addition of that peptide. As negative controls, nonspecific antibodies are used. To determine the size of the proteins present in the complex, proteins bound to the radioactive probe are cross-linked by ultraviolet light and analyzed by SDS-PAGE.[17] The proteins are radioactive because they are now cross-linked to the probe, migrate according to their molecular weight, and can be visualized by autoradiography.

In order to determine the actual bases that are contacted by the protein, protein-DNA complexes are subjected to footprinting using DNAse[17] or chemical methods such as copper-phenanthroline.[18] Another method, known as *in vivo* foot-printing, offers greater fidelity in identification of factors that bind to promoters in the cell and may also be used in studies of tissue-specific and temporal regulation.

B. Protein–Protein Interactions

Viral proteins that bind to the viral promoters and regulate gene expression can be identified with relative ease because of their suspected role in viral regulation. However, many cellular proteins also interact with promoter regions and add another level of complexity. Cell proteins may bind directly to the promoter, interact with other proteins, or both.

Although one level of regulation of EBV gene expression is achieved by binding of Z to DNA sequences in the promoters of certain genes, another level involves a direct interaction between Z and cellular and viral proteins. Cellular factors interact with Z and may modulate viral reactivation. Several years ago, we and others had shown that retinoic acid can repress TPA-mediated induction of the EBV cytolytic cycle.[19] However, the molecular basis of this phenomenon was not understood at the time. Recently, we have demonstrated that, at the molecular level, retinoic acid receptors have an inhibitory effect on Z action.[20,20a] The two proteins directly interact, and each inhibits binding of the other protein to its cognate binding site. Other cellular proteins interact with Z, such as the tumor-suppressor gene, p53, and the NFkB component, p65.[21,22] Z also interacts with the TFIID protein[23] and with the viral protein, RAZ.[11] Protein–protein interactions have been studied by affinity chromatography, in which proteins are purified, one protein is attached to a matrix, and the other protein is exposed to the attached protein. This is a cumbersome and time-consuming process that requires considerable skill to purify proteins without disrupting conformation and structure. Bacterial GST-fusion proteins take advantage of some aspects of affinity chromatography. As mentioned earlier, the protein of interest, in this case Z, was cloned downstream of the GST ORF. The resulting GST-Z fusion protein can be bound to glutathione-coated Sepharose beads. The GST-Z-coated beads, which serve as an affinity column, will bind to Z itself, as well as the proteins that bind to Z. Such specific affinity columns can then be exposed to cell extracts or specific *in vitro*-transcribed and translated proteins that may be made radioactive by labeling with ^{35}S-methionine. We used wild-type Z as well as mutants of Z fused to GST to determine the ability of the cellular protein, RXRα, a retinoic acid receptor, to bind to Z. The detailed protocol follows:

1. Grow *E. coli* cultures containing the pGEX vector or pGEX-Z clones to logarithmic phase, using an overnight culture as inoculum.
2. Induce bacterial cultures (50 ml) with isopropylthiogalactoside (IPTG) (4 m*M*) for 3 h.
3. Pellet and resuspend in 5 ml PBS, sonicate three times for 10 sec and clear the bacterial debris by centrifugation at 12,000 *g*.
4. For each condition, incubate 100 µl of the cell lysate in 1 ml PBS for 20 min at room temperature with 50 µl glutathione-Sepharose beads (Sigma, St. Louis, MO).
5. The GST proteins were purified by three cycles of centrifugation and washing in 1 ml aliquots of buffer containing: 10 m*M* Tris, pH 8.0, 40 m*M* KCl, 0.05% NP-40, 6% glycerol, 1 m*M* DTT, 0.2 µg poly dI/dC.
6. For dimerization experiments, ^{35}S-labeled RARα protein (10,000 cpm) was incubated with 50 µg of purified GST-fusion protein bound to beads in the reaction buffer used above for 1 h at 37°C.
7. GST-Z-coated beads that retained proteins were washed at least five times in the above buffer, pelleted, and boiled in Laemmli buffer.
8. Bound proteins were resolved in 10% SDS-PAE and visualized by autoradiography.

With this protocol, a direct interaction between RXRα and Z could be demonstrated. The advantage of this system is that it is fairly rapid, reproducible, and simple. The disadvantage is that some GST fusion proteins may not retain the protein conformation of the wild type and therefore may not mimic what happens *in vivo*. Furthermore, not all GST fusion proteins are soluble and therefore cannot be used. Also some GST-fusion proteins are sensitive to degradation when expressed in *E. coli*. Nonetheless, affinity chromatography with GST-fusion proteins has advanced the study of protein–protein interactions greatly, especially when coupled with the use of introduced mutations.

XII. Coprecipitation of Proteins

A confirmatory and alternative method to study the interaction between EBV and cellular proteins or between viral proteins involves immunoprecipitation of one protein from a cell extract followed by Western blotting of the gel and probing with antibody to the other protein. This approach detected interaction of proteins *in vivo* and is a more definitive method to show such interactions. For example, the interaction between p53 and Z was demonstrated by coprecipitating both the proteins from EBV-positive cells.[22] The cells were pelleted, washed with cold PBS, resuspended in 1 ml of buffer (50 m*M* Tris, 150 m*M* NaCl, 0.5 m*M* EDTA, 0.5% NP-40 [pH 8.0] supplemented with 1 m*M* PMSF) and then freeze-thawed two times. The cell lysate was then sonicated three times, pelleted again, and equal amounts of protein were incubated overnight with Z-specific antibody, BZ1. Extracts were then immunoprecipitated with protein A-Sepharose beads (Sigma, St. Louis, MO), washed at 4°C in buffer containing 20 m*M* HEPES (pH 7.7), 75 m*M* KCl, 0.1 m*M* EDTA, 25 m*M* $MgCl_2$, 10 m*M* DTT, and 0.15% NP-40. The immunoprecipitates were separated by 10% SDS-PAGE, transferred to Immobilon paper, and analyzed by Western blot using anti-p53 antibody.

The blot was blocked with 5% bovine serum albumin in PBS overnight at 4°C, rinsed three times with PBS, and then incubated at room temperature with monoclonal anti-p53 for 1 to 2 h with gentle shaking. The blot was then washed with PBS plus 0.05% Tween 20 using six changes over 1 h with vigorous shaking. The secondary antibody (a light-chain specific anti-mouse antibody linked to horseradish peroxidase was used so that the heavy chain which comigrates with the p53 will not interfere as a background band) was diluted in 5% nonfat dried milk in PBS + 0.05% Tween 20 and incubated at room temperature with gentle shaking for 1 h, washed as before, and briefly rinsed two or three times with TBS. The proteins were detected using enhanced chemiluminescence (ECL) performed as recommended by the manufacturer (Amersham, Arlington Heights, IL). The presence of the p53 protein indicated that Z and p53 formed a complex in the cells that could be precipitated by the BZ1 antibody. The same experiment was also done in reverse, where the cell extract was first immunoprecipitated with anti-p53 antibody and then probed by BZ1 in a Western blot.

It is not always possible to coprecipitate proteins that interact with each other in the cell. If an antibody is targeted to the domain that is required for interaction, then it will not be able to immunoprecipitate that protein complex. It is also possible that proteins in a complex may have an altered conformation, making them unrecognizable by the antibody. Spurious results are common in coprecipitation and meticulous controls are mandatory, such as preclearing with preimmune sera and the use of nonspecific sera as negative controls.

XII. Regulation at the 3′ End of Genes

The genetic switch between latency and the productive phase of the virus is finely tuned and controlled at several levels. The transcriptional activation of EBV early genes is regulated by the immediate-early proteins, Z and R, and can be modulated by several cellular proteins, some of which may bind to the promoter sequences, and others of which may interfere with Z action. The messages are alternatively spliced, adding another level of control. The EBV DNA polymerase message is predicted to have strong secondary structure, especially in the 5′ untranslated region, which may impose some restraints on translation of the protein, adding yet another level of control. The herpes simplex virus (HSV) polymerase message also has pronounced secondary structure and does impose translational controls.[24] However, we have demonstrated a unique aspect of post-transcriptional processing of the pol mRNA. Most higher eukaryotes exhibit, as a signal for mRNA cleavage and polyadenylation, a AAUAAA sequence or variations thereof, located 10 to 30 nucleotides upstream of the poly-A addition site. This sequence is crucial for accurate, efficient cleavage and polyadenylation of pre-mRNA *in vivo* and *in vitro*.[25] In addition, there is sequence referred to as the "G/U cluster", downstream of the poly-A addition site. The 3.7 kb polymerase message from the prototype EBV-producing cell line (B95-8) is perhaps unique in that it does not have the canonical AAUAAA polyadenylation signal or its variants, yet it is precisely cleaved and polyadenylated. In most other EBV strains, however, a larger 5.1 kb processed polymerase mRNA is detected. This major pol mRNA form has an unusual poly-A signal (UAUAAA) which has been rarely detected in eukaryotic mRNA (hepatitis B virus and Fig wart virus). These unique features of the pol mRNAs are of great interest because the 3′ untranslated region (UTR) may play a crucial role in the regulation of expression of this gene. The 3′ end of the mRNA was identified by a combination of S-1 nuclease digestion and primer extension analysis, as described before. In order to confirm identification of the 3′ ends of the EBV pol mRNAs, cDNAs were sequenced from a λgt10 library generated from TPA-induced B95-8 cells.

XIV. Construction of λgt10 cDNA Library

A λgt10 library was prepared from poly(A)$^+$ RNA isolated from B95-8 cells treated for 48 h with TPA and sodium butyrate by a modification of the method of Gubler and Hoffman.[26] *Eco*RI adapters (Promega, Madison, WI) were ligated to double-stranded size-selected (>0.5 kb) cDNA, ligated to *Eco*RI-digested λgt10 DNA, and packaged with Gigapack extracts (Stratagene, La Jolla, CA). Bacteriophage were plated on the C600 hfl$^-$ strain of *Escherichia coli*, and 3′ coterminal pol cDNAs were recovered by hybridization with the ^{32}P-end-labeled oligonucleotide probe.

XV. 3′ Rapid Amplification of cDNA Ends (RACE)

3′ RACE was performed as described by Frohman et al.[27] using a gene-specific amplimer within the EBV pol ORF (coordinates, 153904 to 153924) and an oligo-dT primer (GGACTCGAGTCGACATCGAT$_{(17)}$) containing restriction sites for *Sal*I, *Xho*I, and *Cla*I at the 5′ end. Polymerase chain reaction (PCR) products were resolved on a 1% agarose gel, blotted onto nitrocellulose filters, fixed by heat, and prehybridized for 1 h at 37°C in prehybridization solution containing 6X SSC, 5X Denhardt's solution, 0.15% sodium pyrophosphate, 100 mg/ml salmon sperm DNA, and 0.1% SDS. Hybridization was performed in the same solution at 37°C with a ^{32}P-end-labeled pol-specific oligonucleotide probe (coordinates, 153794 to 153803). Filters were washed at 50°C in 0.1X SSC-0.1% SDS. Positive PCR bands were cloned into pBS+ (Stratagene, La Jolla, CA), sequenced, and compared with published B95-8 sequence. In B95-8 cells, these experiments yielded identical cDNAs terminating exactly 480 bp downstream of the ORF, in a region of the genome remarkable for its lack of an apparent polyadenylation signal. Analysis of the 3′ ends of other EBV strains indicated that pol mRNA from these strains terminated about 1.4 kb downstream of the B95-8 strain, in a region that is deleted in the B95-8 virus. Since both processed mRNA termini lack canonical polyadenylation signals, *in vitro* polyadenylation assays are being used in our laboratory by Cathy Silver Key to identify, through mutational analyses, the signals and other cis-acting elements that may contribute to signaling.

XVI. *In Vitro* Polyadenylation Assays

In vitro transcription: templates for RNA synthesis were constructed by cloning a 313 bp region of the 3′ UTR of the DNA polymerase gene from the B95-8 genome or a 304 bp region from the Raji genome into pBS- and linearizing the clones with *Eco*RI. The template was transcribed as discussed previously, but in the presence of radiolabeled uridine triphosphate (UTP) and capped by addition of 5 μl of 5 μ*M* GpppG to the transcription reaction.

XVII. Purification of the Template for Cleavage and Polyadenylation Assays

The template has to be very clean in order to be used for the cleavage reaction. The following steps are optional and may be needed for some templates and not others; this protocol was first obtained from the laboratory of Clements[28] and modified in this laboratory.[27a]

1. Add 2.5 μl DNAse. Incubate 10 min, at 37°C.
2. Increase the reaction volume to 200 μl and extract with an equal volume of Tris-buffered phenol-chloroform. Extract with an equal volume of chloroform. This step should remove the majority of counts of unincorporated UTP. At this stage, monitor radioactivity before and after every step.
3. Remove supernatant fluid to a fresh Eppendorf tube and add 20 μl 6 *M* NH_4Ac and 2 μl tRNA (20 mg/ml) and then 220 μl isopropanol. Place on dry ice at least 10 min. It is best to leave the mixture in isopropanol at –20°C overnight to precipitate efficiently.
4. Spin at 4°C for 10 min, remove supernatant fluid and wash pellet with 70% ethanol. Most of the counts should be in the pellet and only a few in the supernatant fluid.
5. Monitor counts and resuspend the pellet in Tris or water to obtain 200 cpm/μl.
6. The purity of the template can be checked using a 6% polyacrylamide sequencing gel.
7. Resolve the template on a 6% gel and expose the film for 5 to 10 min. Develop the film, orient gel to film and, using a scalpel, cut out the clean template band, which should be at the top of the lane. The gel slice should contain a large portion of the counts.

8. Put gel slice into a 0.5-ml Eppendorf tube with a hole pierced in the bottom. Remove the lid and place this tube in a 1.5-ml tube and spin for 2 to 3 min until the gel slice is sheared through the small Eppendorf into the large one.
9. Add 200 μl of gel-purifying solution (enough to cover the gel slice) and leave overnight or longer. (Gel-purifying solution: 400 m*M* NaCl, 50 m*M* Tris, pH 7.5, 0.1% SDS, and 0.1 μg Proteinase K.)
10. Transfer gel and solution to a small Spinex tube and spin in a microfuge for 5 to 10 min, to separate gel and supernatant fluid. Most of the counts should be in the supernatant fluid.
11. Phenol-chloroform extract the precursor and precipitate as before. Resuspend the required amount of distilled water and check on 6% sequencing gel.

The radioactive RNA template is then incubated with nuclear extracts from EBV-negative cells, such as Hela, or EBV-positive lymphoid cells or D98-HR-1, a hybrid cell line to study RNA cleavage and polyadenylation. To determine the cleavage site, the reaction is carried out in the presence of Cordycepin, an ATP analog, which terminates polyadenylation after the first A is added (Figure 11.5). To determine polyadenylation efficiency, the reaction is carried out in the presence of ATP. Incubation with Cordycepin theoretically yields a discrete band (actually a small cluster is detected), whereas incubation with ATP will yield a ladder of much larger products which indicate increasing numbers of A added to the cleaved template.

Cleavage reaction mixture:

25 m*M* Cordycepin	1 μl
50 m*M* creatine phosphate	2.5 μl
10 m*M* Tris, pH 7.6	6.5 μl
25% polyethylene glycol (PEG)	3.0 μl

The reaction mixture is incubated in a total volume of 25 μl with the nuclear extract for at least 1 h at 30°C. The reaction products are resolved on 7% urea acrylamide gels, dried, and exposed to film.

XVIII. Analysis of EBV Polymerase Activity

The multifaceted regulation of EBV DNA polymerase gene expression underscores the crucial role of the polymerase in the replication of the viral DNA in the productive phase. The existence of an EBV DNA pol was first suspected by Miller et al.[29] The enzyme has a molecular weight of 110 kDa when extracted from P3HR1 cells[30] and in superinfected Raji cells. The EBV gene product, BMRF1 (EA-D), copurified with EBV pol and appears to stabilize enzymatic activity.[31] The two proteins colocalized to the nucleus.[32] EBV DNA polymerase has some similarity to cellular α-polymerase and yet can be differentiated from it by the salt-stimulatory properties of the herpesvirus enzyme.[33,34] This key protein serves as a target for antiviral drugs. Most of the antiviral drugs are nucleoside analogs that inhibit DNA polymerase activity, which can be used to screen for antiviral activity.

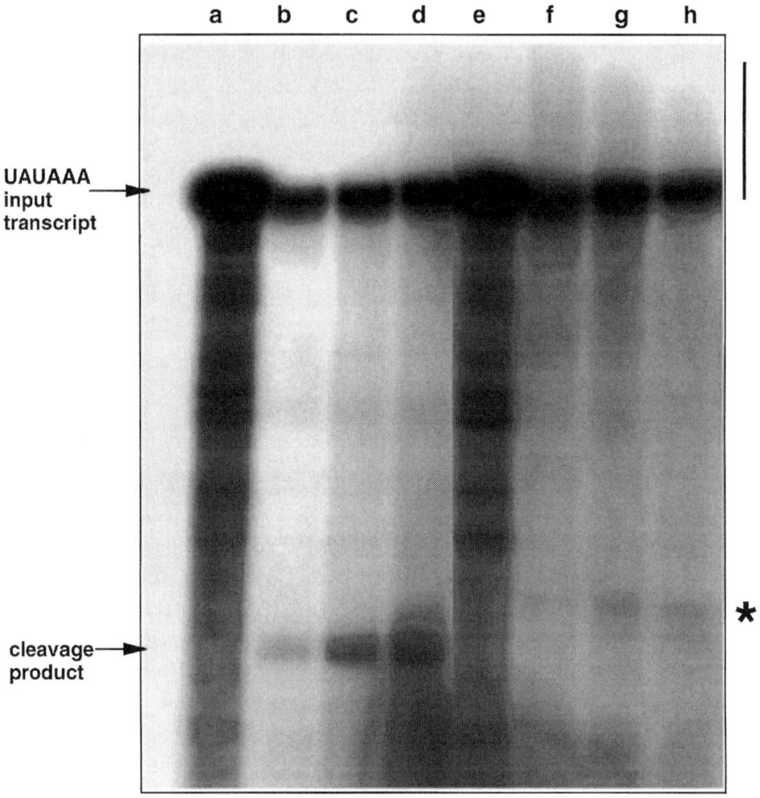

FIGURE 11.5
Cleavage and polyadenylation of EBV polymerase RNA. Arrows indicate the cleavage products generated in D98/HR1 nuclear extracts. Vertical lines indicate polyadenylation products. The asterisk denotes oligoadenylation or the addition of less than 9 adenosine residues onto the UAUAAA Raji transcript cleaved product. A time course cleavage assay is shown in lanes b (30 min), c (1 h), and d (2 h). Polyadenylation of the transcript is shown in lanes f (30 min), g (1 h), and h (2 h). Lanes a and e show cleavage and polyadenylation assays performed in the absence of nuclear extract plus cordycepin or adenosine triphosphate, respectively.

The EBV DNA polymerase has an intrinsic 3′ to 5′ exonuclease activity and is sensitive to drugs such as aphidicolin and phosphonoacetic acid (PAA) and also to nucleoside analogs such as acyclovir.[35,36] Dr. Paul Zajac in our laboratory has cloned the EBV pol ORF in a vector that can be expressed in the vaccinia virus system. This system is very convenient for mutagenesis as well as assay of the enzyme and can reduce the time required to analyze expressed products from several days to less than 24 h.

The EBV DNA polymerase activity is routinely assayed in a reaction mixture (25 μl) that contains 50 m*M* Tris-HCl (pH 8.0), 10% glycerol, 6 m*M* $MgCl_2$, 100 μg bovine serum albumin per ml, 80 μg of activated calf thymus DNA per ml, 1 m*M* DTT, dATP, dCTP, and dTTP at 25 μ*M* each, and [^{3}H]dGTP at 10 μ*M*

(900 cpm/pmol).[37] The reaction is started by addition of the enzyme source, and incubated for 20 min at 35°C. Reactions are stopped by addition of 30 μl of sonicated salmon sperm DNA (2 mg/ml) containing 0.1 *M* sodium pyrophosphate and quenched with 0.6 ml of cold 10% trichloroacetic acid containing 0.1 *M* sodium pyrophosphate. Incorporation of labeled nucleotides into acid-insoluble material is measured in a liquid scintillation counter.

XIX. Analysis of Viral Genomes

EBV has two well-established modes of replication, one utilizing the virus-encoded DNA polymerase to produce the linear double-stranded, ~180 kbp DNA genome, which is encapsulated in virions, and the other utilizing host DNA polymerase to replicate the circular form of the same genome, the EBV episome, which is found only intracellularly within the chromatin. The episome is the molecular basis for latent EBV infection. Integrated forms of the EBV genome have also been detected in cell lines but rarely in EBV-infected tumor or other tissues, which contain episomes. The episome, which is present in low, constant copy number in each cell, replicates once every cell cycle. It has been characterized extensively since its discovery.[38]

Viral reactivation is characterized by synthesis of linear genomes and viral replication. Two origins of DNA replication (*ori*Lyt) are used in productive infection.[39-41] The viral DNA polymerase-mediated replication proceeds via linear concatamers as replicative intermediates, perhaps by a rolling-circle mechanism. Cleavage of concatamers generates a ladder of restriction fragments in linear DNA which are heterogeneous in size by increments of 500 bp caused by differing copy numbers of the terminal repeats. These fragments are eliminated or greatly increased after induction of replication and are greatly reduced by treatment with acyclovir. Therefore, percentage reactivation is directly proportional to the amounts of linear DNA induced and can be quantitated.[42]

XX. Analysis of Genomic Termini

Termini analysis is carried out by digesting EBV intracellular or virion DNA from EBV-positive cells with *Bam*HI, electrophoresis through a 0.6% gel, and transferring to nitrocellulose. The right-terminal *Bam*HI fragments can be identified by hybridization with a ^{32}P-labeled probe synthesized from a 1.9 kb *Xho*I fragment cloned into the pGEM2 vector. This fragment represents unique DNA adjacent to the terminal repeats at the right end of the genome. The left-terminal restriction enzyme fragments can be identified by using the ^{32}P-labeled *Eco*RI-*Bam*HI portion of the *Eco*RI fragment, representing unique DNA adjacent to the terminal repeats at the left end of the genome.[42]

XXI. Conclusion

Analysis of the Epstein Barr virus depends upon standard methods as it is reactivated from the latent state and enters a cytolytic phase. Many of these methods can also be used to study latent gene expression.[16,43,44] Other methods and techniques that are used to study the virus, such as *in situ* hybridization,[45] analysis of potential viral candidate proteins required for transformation using transgenic and SCID mice,[46] and transformation assays[47,48] to appraise the transforming potential of virus isolates, and typing and characterization of EBV strains have been well described elsewhere.[53]

EBV infection evokes many immune responses as well, and a large number of immunologic assays have been employed in their study.[49] Finally the molecular pathogenesis of EBV infection in normal and diseased tissues has been much studied and integrates molecular and histopathologic techniques with the virology of the Epstein-Barr virus.[50-52]

Acknowledgments

We thank Chunnan Liu, Val Zacny, Matt Davenport, and Cathy Silver Key for their help with the figures. A portion of this work was supported by U.S. Public Service Grants 2-PO1-Ca-19014-17A1 and RO1 CA 56695-02 from the National Cancer Institute.

References

1. Raab-Traub, N., Rajadurai, P., Flynn, K., and Lanier, A. P., *J. Virol.*, 65, 7032 (1991).
2. Pagano, J. S., in *Textbook of Internal Medicine*, Kelley, W. N., Ed., J. B. Lippincott, Philadelphia, 1991, 1499.
3. Miller, G., in *Virology*, Fields, B. N. and Knipe, D. M., Eds., Raven Press, New York, 1990, 1921.
4. Furnari, F. B., Adams, M. D., and Pagano, J. S., *J. Virol.*, 66, 2837 (1992).
5. Luka, J., Kallin, B., and Klein, G., *J. Virol.*, 65, 2728 (1979).
6. zur Hausen, H., O'Neil, F., and Freese, U., *Nature*, 272, 373 (1978).
7. Henle, W. et al., *Science*, 169, 188 (1970).
8. Takada, K., *Int. J. Cancer*, 33, 27 (1984).
9. Takada, K., Shimizu, S., Sakuma, S., and Ono, Y., *J. Virol.*, 57, 1016 (1986).
10. Lin, J. C., Sista, D., Besencon, F., Kamine, J., and Pagano, J. S., *J. Virol.*, 65, 2728 (1991).
11. Furnari, F. B., Zacny, V., Quinlivan, E. B., Kenney, S., and Pagano, J. S., *J. Virol.*, 68, 1828 (1994).
12. Furnari, F. B., Adams, M. D., and Pagano, J. S., *Proc. Natl. Acad. Sci. U.S.A.*, 90, 378 (1993).
13. Hitt, M. M. et al., *EMBO J.*, 8, 2639 (1989).
14. Toneguzzo, F., Hayday, A. C., and Keating, A., *Mol. Cell. Biol.*, 6, 703 (1986).

15. Gorman, C. M., Moffat, L. F., and Howard, B. H., *Mol. Cell. Biol.,* 2, 1044 (1982).
16. Sung, N. S., Wilson, J., Davenport, M., Sista, N. D., and Pagano, J. S., *Mol. Cell. Biol.,* 14, 7144 (1994).
17. Ausubel, F. M., Kingston, R. E., Moore, D. D., Seidman, J. G., Smith, J. A., and Struhl, K., Eds., in *Current Protocols in Molecular Biology,* Sec. 12.4, John Wiley & Sons, New York, 1990.
18. Kuwabara, M. D. and Sigman, D. S., *Biochemistry,* 26, 7234 (1987).
19. Lin, J.-C., Smith, M. C., and Pagano, J. S., *Virology,* 111, 294 (1981).
20. Sista, N. D., Pagano, J. S., Liao, W., and Kenney, S., *Proc. Natl. Acad. Sci. U.S.A.,* 90, 3894 (1993).
20a. Sista, N. D., Sampson, K., and Pagano, J. S., *Nucl. Acids. Res.,* 23, 1729 (1995).
21. Gutsch, D. E. et al., *Mol. Cell. Biol.,* 14, 1939 (1994).
22. Zhang, Q., Gutsch, D., and Kenney, S., *Mol. Cell. Biol.,* 14, 1929 (1994).
23. Lieberman, P. M. and Berk, A. J., *Genes Dev.,* 5, 2441 (1991).
24. Yager, D. R. and Coen, D. M., *J. Virol.,* 62, 2007 (1988).
25. Birnstiel, M. L., Busslinger, M., and Strub, K., *Cell,* 41, 349 (1985).
26. Gubler, U. and Hoffman, B. J., *Gene,* 25, 263 (1983).
27. Frohman, M. A., Dush, M. K., and Martin, G. R., *Proc. Natl. Acad. Sci. U.S.A.,* 85, 8998 (1988).
27a. Key, S. C. S. and Pagano, J. S., *Mol. Cell Biol.,* Submitted (1996).
28. McLauchlan, J., Phelan, A., Loney, C., Sandri-Goldin, R. M., and Clements, J. B., *J. Virol.,* 66, 6939 (1992).
29. Miller, R. L., Glaser, R., and Rapp, F., *Virology,* 76, 494 (1977).
30. Kallin, B. et al., *J. Virol.,* 54, 561 (1985).
31. Li, J.-S. et al., *J. Virol.,* 61, 2947 (1987).
32. Kiehl, A. and Dorsky, D. I., *Virology,* 184, 330 (1991).
33. Larder, B. A., Kemp, S., and Darby, G., *EMBO J.,* 6, 169 (1987).
34. Wang, T. S.-F., *Annu. Rev. Biochem.,* 60, 513 (1991).
35. Tsurumi, T. et al., *J. Virol.,* 67, 4651 (1993).
36. Tsurumi, T., Daikoku, T., Kurachi, R., and Nishiyama, Y., *J. Virol.,* 67, 7648 (1993).
37. Datta, A. K., Feighny, R. J., and Pagano, J. S., *J. Biol. Chem.,* 255, 5120 (1980).
38. Pagano, J. S., in *Extrachromosomal DNA,* Cummings, D. J., Ed., Academic Press, New York, 1978, 235.
39. Yates, J., Warren, N., Reisman, D., and Sugden, B., *Proc. Natl. Acad. Sci. U.S.A.,* 81, 3806 (1984).
40. Reisman, D., Yates, J., and Sugden, B., *Mol. Cell. Biol.,* 5, 1822 (1985).
41. Hammerschmidt, W. and Sugden, B., *Cell,* 55, 427 (1988).
42. Raab-Traub, N. and Flynn, K., *Cell,* 47, 883 (1986).
43. Sista, N. D., Barry, C. M., and Pagano, J. S., manuscript in preparation (1994).
44. Sung, N. S., Kenney, S., Gutsch, D., and Pagano, J. S., *J. Virol.,* 65, 2164 (1991).
45. Gulley, M. L. and Raab-Traub, N., *Arch. Pathol. Lab. Med.,* 117, 1115 (1993).
46. Cannon, M. J., Pisa, P., Fox, R. I., and Cooper, N. R., *J. Clin. Invest.,* 85, 1333 (1990).
47. Lin, J. C., Zhang, Z. X., Chou, T. C., Sim, I., and Pagano, J. S., *J. Infect. Dis.,* 159, 248 (1989).
48. Pagano, J. S., in *Antiviral Chemotherapy,* Jeffries, D. J. and DeClerq, E., Eds., John Wiley & Sons, New York, 1995, 155.

49. Rickinson, A. B. et al., *Cancer. Surv.,* 13, 53 (1992).
50. Sixbey, J. W. and Pagano, J. S., in *Current Clinical Topics of Infectious Diseases,* Vol. 5, Remington, J. and Swartz, M., Eds., McGraw-Hill, New York, 1984, 146.
51. Sixbey, J., Nedrud, J., Raab-Traub, N., Hanes, R., and Pagano, J., *N. Engl. J. Med.,* 310, 1225 (1984).
52. Sixbey, J. W., Lemon, S. M., and Pagano, J. S., *Lancet,* 2, 1122 (1986).
53. Walling, D. M., Edmiston, S. N., Sixbey, J. W., Abdel-Hamid, M., Resnick, L., Raab-Traub, N., *Proc. Natl. Acad. Sci. U.S.A.,* 89, 6560 (1992).

Chapter 12

Functional Analysis of the Herpes Simplex Virus DNA Polymerase

Paul Digard, William Bebrin, Connie Chow, and Donald M. Coen

Contents

I. Introduction

Since its development in the 1970s, micrococcal nuclease-treated rabbit reticulocyte lysate[1] has been used extensively as a system for studying the process of translation. However, because of the ease with which synthetic mRNA can be produced from cloned DNA by bacteriophage RNA polymerase-mediated *in vitro* transcription, it has also been widely used as a rapid and convenient method for producing small

0-8493-4412-3/96/$0.00+$.50

amounts of biologically active protein. Moreover, given the ease with which plasmid-borne genes can be manipulated, *in vitro* translation in rabbit reticulocyte lysate (RRL) provides an attractive system for performing mutagenic analyses of protein function. What follows is a description of the techniques our laboratory has found useful for studying the two subunits of the herpes simplex virus DNA polymerase.[2-4] We also describe a transient genetic assay that allows phenotypic assessment of the mutants in the context of an infected cell, permitting a correlation between the *in vitro* biochemical activities of a protein and its function *in vivo*. Although applied here to a specific problem, similar approaches have been used to investigate a variety of viral and nonviral proteins. We hope, therefore, that a discussion of the techniques used and problems encountered will be of general interest.

Herpes simplex virus (HSV) provides an excellent model system for the study of DNA replication in eukaryotic cells, in part because it encodes many, if not all, of the proteins that are directly required for replication of the virus genome.[5] The virus is also a ubiquitous human pathogen whose control relies primarily on antiviral drugs, the most successful of which are directed against the catalytic site of the DNA polymerase.[6] Therefore, the function of the enzyme is of considerable interest. The polymerase consists of two subunits, a large catalytic subunit encoded by the HSV *UL30* gene (Pol), and a smaller subunit encoded by the *UL42* gene (UL42). Although Pol is capable of synthesizing DNA by itself, one function of UL42 is to stimulate the ability of Pol to polymerize long stretches of DNA via an increase in processivity.[7,8] The fact that UL42 has an intrinsic affinity for double-stranded DNA (dsDNA) suggests the attractive hypothesis that it tethers Pol to the template and, thereby, increases processivity by discouraging dissociation of the enzyme from the primer terminus.[8]

Therefore, our objective was to investigate the interactions between Pol, UL42, and dsDNA, and thereby test their involvement in processive DNA synthesis. The experimental approach was to introduce mutations into the Pol and UL42 genes, express the altered polypeptides, and assay them for the various biochemical functions. The methods used to introduce mutations into the Pol and UL42 genes, and to analyze the protein–protein interaction between the two polypeptides are described in detail elsewhere.[9] Here, the approaches used to investigate DNA-binding activity and enzymatic activities will be detailed.

II. Expression of HSV Pol and UL42 *in Vitro*

Uniformly sized capped mRNAs encoding Pol, UL42, or derivatives thereof were generated by *in vitro* transcription of the appropriate plasmid (linearized downstream of the open reading frame by restriction enzyme digest) in the presence of a molar excess of cap analog (New England Biolabs, Beverly, MA) according to standard procedures.[10] Individual or mixed transcripts at a final concentration of 1 to 10 ng/μl were translated in rabbit reticulocyte lysate, according to the manufacturer's instructions (Promega, Madison, WI), with the exception that [^{35}S]-methionine was included at a concentration of 2 μCi/μl, and the reactions were incubated at 37°C instead of the recommended 30°C. The higher temperature was found to increase the yield of larger polypeptides such as Pol (135 kDa). The addition of larger quantities of RNA will result in more protein synthesis, but often at the expense of full-length product.[11] Recent systems for coupled *in vitro* transcription and translation (ie., Reference 12, or the Promega T.N.T. system) may also produce satisfactory results.

For experiments analyzing the combined function of Pol and UL42, generally the mutant polypeptide was translated alone, and the wild-type partner polypeptide supplied by the post-translational addition of exogenously overexpressed and purified protein.[3,4] If a source of purified protein is available, this approach has the practical advantage that it makes it easier to control the relative amounts of the two polypeptides, since achieving equal levels of *in vitro* translation from two transcripts that may translate with quite different efficiencies can be difficult. In our studies, Pol and UL42 were purified from insect cells infected with the appropriate recombinant baculoviruses,[8,13] although any expression system from which active protein can be produced will suffice. However, most experiments should still be feasible even if both the polypeptides of interest are produced in reticulocyte lysate, and indeed, some systems actually require cotranslation of the proteins (e.g., Reference 14).

III. Analysis of DNA Binding Activity

Analysis of *in vitro* translated non-sequence-specific DNA binding proteins is complicated by the large quantities of endogenous nucleic acid binding proteins present in RRL. Whereas sequence-specific binding of proteins to a target sequence can conveniently be examined by following shifts in the electrophoretic mobility of an appropriate radiolabeled target sequence, sequence-independent binding generally has to be measured by following radiolabel present in the polypeptide. Therefore, to analyze the interaction of UL42 with DNA, we examined the *in vitro* translated polypeptides by DNA cellulose chromatography, using the protocol described below.

Ten microliter aliquots of RRL containing the polypeptide of interest were diluted into 100 µl of TM buffer (20 m*M* Tris-Cl, pH 7.6, 5 m*M* $MgCl_2$), containing 50 m*M* NaCl, and incubated with 2 µl of a mixture of RNAse A and a nonspecific RNAse from *Aspergillus oryzae* (RNAce-it; Stratagene, La Jolla, CA) at room temperature for 20 min to destroy free RNA. Samples were then applied to a 0.5 ml bed volume dsDNA-cellulose (Sigma, St. Louis, MO) column equilibrated in the same buffer. The column was washed twice with 0.5 ml of TM buffer plus 50 m*M* NaCl, and then sequentially eluted with 2 column volumes each of TM buffer containing 100 m*M*, 250 m*M*, 500 m*M*, and 1 *M* NaCl. Two volumes of ethanol and 40 µg of carrier bovine serum albumin were added to the resulting 0.5 ml fractions, and after 1 h at –20°C, precipitated protein was collected by microcentrifugation, resuspended in 50 µl of sodium dodecyl sulfate-polyacrylamide gel electrophoresis (SDS-PAGE) sample buffer, and analyzed by gel electrophoresis.

The results obtained when a panel of UL42 mutants were analyzed by the above method is shown in Figure 12.1. The elution profile of the wild-type polypeptide is shown in panel A; the majority of the protein was retained by the column at 50 m*M* NaCl, but was eluted by 250 m*M* NaCl. Control experiments using cellulose alone as a column matrix revealed no significant retention of the polypeptide (data not shown), suggesting that the assay reflects a genuine interaction with DNA. In contrast, several mutant polypeptides, including ones with internal deletions (panel G, Δ20–131; panel H, Δ161–223), or a large carboxy-terminal deletion (panel D, ΔC307) appeared mainly in the wash fractions, suggesting that they had lost the ability to bind DNA. Other UL42 mutants, especially those containing mutations near the N-terminus of the protein, were still retained by the DNA cellulose column at low salt concentrations, but displayed an altered elution profile, with substantial amounts of protein eluting at 1 *M* salt (panel E, ΔN10; panel J, I-20). Control experiments showed that these polypeptides did not interact with cellulose alone, suggesting that the altered elution pattern from dsDNA cellulose reflected an altered interaction with DNA. However, the behavior of these N-terminal mutants illustrates a major weakness of the column method of analysis; it does not provide direct information on the strength of the protein–DNA interaction, therefore precluding any conclusion about whether the altered elution profiles resulted from an increase or decrease in affinity. Also, because of the extremely high effective DNA concentration, this method of column analysis may be a rather insensitive method for detecting the phenotypes of mutants that are only partially impaired for DNA binding. An alternative method of analysis has been described by Tenney and colleagues, in which radiolabeled *in vitro* translated UL42 is mixed with unlabeled plasmid DNA, and electrophoresed on an agarose gel.[15] Under the conditions employed, UL42 migrated as a diffuse smear in the absence of DNA, but in discrete bands colocalizing with the plasmid when bound to DNA. However, this approach may not be applicable to all proteins, depending on their charge and intrinsic electrophoretic migration properties. We also note that both approaches produced similar data about the regions of UL42 necessary for interaction with DNA,[4,15] suggesting that the two methods are of broadly similar veracity and sensitivity.

IV. Analysis of DNA Polymerase Activity

The original demonstration that RRL-expressed HSV Pol possessed enzymatic activity used the incorporation of radioactive nucleotide triphosphates (NTPs) into an activated (nicked and gapped) calf thymus DNA template as a measure of polymerase function.[16] However, the sensitivity of this assay is limited by the relatively low ratio of Pol activity to background incorporation of radiolabel (not necessarily into DNA) by RRL, with an overall stimulation of around sevenfold for wild-type Pol. In addition, such an assay provides no information on the size of DNA product synthesized, and so is of little use for investigating the effects of a processivity factor on polymerase activity. However, previous work has shown that Pol alone synthesizes relatively short products on a singly primed single-stranded DNA (ssDNA) template when at a low polymerase-to-template ratio, but will synthesize much longer products in the presence of UL42, and that this can be taken as a measure of the

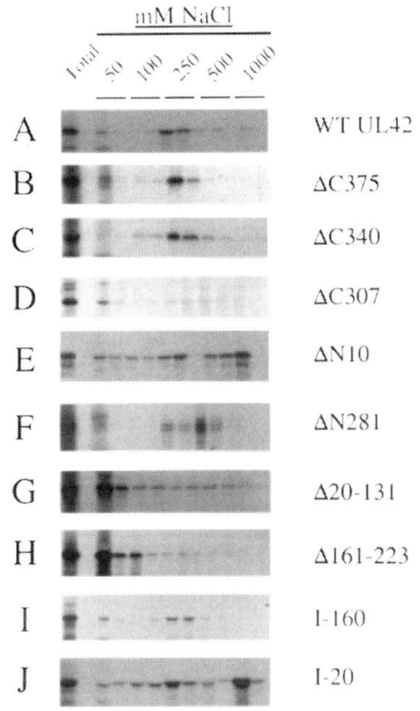

FIGURE 12.1
DNA binding assay: [^{35}S]-methionine-labeled UL42 polypeptides expressed in RRL were applied to DNA cellulose columns and eluted with buffer containing the indicated concentration of NaCl. Fractions were collected and analyzed for protein content by SDS-PAGE and autoradiography. An equivalent unfractionated amount of each sample was also analyzed (Total). WT, wild-type. (From Digard, P., Chow, C., Pirrit, L., and Coen, D. M., *J. Virol.*, 67, 1159 (1993). With permission.)

processivity of the enzyme.[7,8] We therefore tested for DNA polymerase activity from RRL containing *in vitro* translated Pol in the presence and absence of UL42 on primed ssDNA templates, and monitored newly synthesized DNA by electrophoresis on alkaline agarose gels, as described below.

A. Preparation of DNA Templates

Singly primed M13 template was prepared by hybridizing 25 μg of M13mp18 positive strand DNA (Pharmacia, Piscataway, NJ) with a twofold molar excess of M13 single-strand primer (Pharmacia) in 500 μl of 100 m*M* NaCl, 50 m*M* Tris-Cl (pH 7.6), 1 m*M* EDTA. The mixture was heated to 90°C for 5 min, and then cooled slowly to room temperature. Excess primer was removed by filtration through a Centricon-100 spin-filter (Amicon, Beverly, MA) according to the manufacturer's instructions. Primed ϕX174 DNA (New England Biolabs, Beverly, MA) was prepared by the same procedure, except that the oligonucleotide 5′-GGCGCATAAC-GATACCACTGACC was used as the primer. An alternative template, that we presently prefer, which works well with the HSV polymerase (although not necessarily with other polymerases) can be made by hybridizing poly(dA) with oligo(dT)

[10:1 w/w]. This template (at 10 μg/ml) is a very poor substrate for Pol alone, but is used much more efficiently by a Pol–UL42 complex,[15] and this efficiency is also reflected in the size of product synthesized.[17,18]

B. Polymerase Activity Assay

Aliquots (5 μl) of RRL were diluted in 50 μl of buffer containing 100 m*M* $(NH_4)_2SO_4$, 20 m*M* Tris-Cl (pH 7.5), 3 m*M* $MgCl_2$, 0.1 m*M* EDTA, 0.5 m*M* dithiothreitol, 4% glycerol, 40 μg/ml of bovine serum albumin, 60 μ*M* dATP, dGTP, and dTTP, 10 μ*M* [α-^{32}P]dCTP (2 μCi), and 50 fmol of a singly primed M13 or ϕX174 DNA template or 0.5 μg of a poly(dA):oligo(dT) template prepared as described above. For certain reactions, 80 fmol of purified UL42 was added. After 30 min incubation at 37°C, 5 μl samples were taken from each reaction, added to an equal volume of SDS-PAGE sample buffer, and analyzed for radiolabeled protein by gel electrophoresis and autoradiography. DNA synthesis in the remainder of the reactions was terminated by the addition of an equal volume of 1% SDS, 10 m*M* EDTA, 10 m*M* Tris-Cl (pH 8), 200 μg/ml proteinase K, and digestion for a further 1 h at 37°C. The reactions were extracted once with phenol/chloroform (1:1 v/v), ethanol precipitated in the presence of 1 *M* ammonium acetate, and resuspended in 50 μl of 50 m*M* NaOH, 2.5 m*M* EDTA, 25% glycerol, 0.025% bromocresol green. After fractionation on a 1.3% alkaline agarose gel,[19] newly synthesized DNA was detected by autoradiography.

An example of such an experiment is shown in Figure 12.2, where it can be seen that Pol alone, either purified from cells infected with a recombinant baculovirus (lane 1), or expressed in RRL (lane 5), possessed enzymatic activity, but only synthesized relatively short products from a uniquely primed M13 template. However, on addition of 80 fmol of purified UL42, Pol from either source synthesized significantly longer products, including some full-length M13 (lanes 2 and 6). The polymerase activity of two Pol mutants (expressed in RRL) is also shown; a mutant containing a four-amino acid insertion at position 1131 (lanes 7 and 8) behaved similarly to the wild-type Pol, while one lacking the extreme C-terminus (F-1208; lanes 9 and 10) which is unable to bind UL42,[3] failed to synthesize longer DNA products in the presence of UL42. Under the reaction conditions employed, no endogenous DNA polymerase activity was detected in RRL, irrespective of the presence or absence of UL42 (lanes 3 and 4). However, this reflects the assay conditions rather than a true lack of polymerase activity in RRL, as endogenous polymerase activity can be detected in the presence of moderate amounts of NaCl, and the absence of $(NH_4)_2SO_4$ (data not shown). In this respect it is fortuitous that the salt optima of the HSV Pol and Pol + UL42 complex[20] permits detection of their activity under conditions that result in negligible background. Nevertheless, we note that several other viral polymerases have been expressed in active form in RRL, including not only DNA polymerases from other herpes viruses such as Epstein-Barr virus[21,22] and pseudorabies virus,[23] but more disparate enzymes, such as reverse transcriptase from duck hepatitis B virus,[24] and the RNA-dependent RNA polymerase from poliovirus.[25] Therefore, endogenous polymerase activity in RRL does not appear to be a major problem for analysis of heterologous polymerase function.

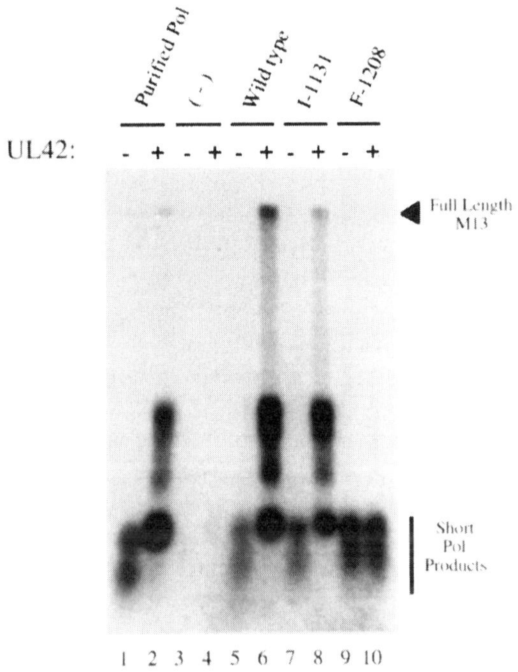

FIGURE 12.2
DNA polymerase activity assay. Purified Pol (lanes 1 and 2), or RRL programmed with water (lanes 3 and 4), or mRNAs encoding the indicated Pol proteins (lanes 5–10) were tested for polymerase activity on a singly primed M13 single-stranded DNA template in either the presence (+) or absence (–) of 80 fmol of purified UL42. Newly synthesized single-stranded DNA was monitored by autoradiography after fractionation on a 1.3% alkaline agarose gel. The migration of full-length M13 DNA is indicated. (From Digard, P., Bebrin, W. R., Weisshart, K., and Coen, D. M., *J. Virol.*, 67, 398 (1993). With permission.)

V. Transient Genetic Complementation Analysis of Polymerase Subunits

Genetic analysis of herpes simplex viral mutants has established that both subunits of the DNA polymerase are essential for viral growth, and has provided much evidence regarding the regions of Pol involved in substrate interaction (References 5 and 6, and references therein). This approach has utilized naturally occurring and artificially constructed mutants, but the notable effort required to isolate and characterize such viruses limits the numbers of mutations that can reasonably be studied. Therefore, as we wished to examine the *in vivo* phenotypes of a large number of Pol and UL42 mutants that had been partially characterized for their biochemical functions in RRL, we chose to use a transient complementation assay based on mutant viruses lacking functional copies of either the *Pol* or *UL42* genes. These viruses are unable to replicate unless a viable copy of the gene is supplied *in trans,* either by a complementing cell line, or by transfection of the host cell with a plasmid containing the gene.[26,27] Accordingly, if Vero cells are transfected with plasmids encoding Pol or UL42 (under the control of their normal HSV promoters), and then superinfected with the corresponding disrupted virus, the overwhelming majority of the resulting progeny will most likely have arisen because of complementation of the defect by the plasmid encoded gene. Low levels of recombinant virus are difficult to eliminate unless there is no sequence overlap between the virus and the host-cell resident complementing gene. The progeny arising from complementation will still lack a functional copy of the polymerase subunit gene, and are therefore titered on a cell line that supplies the appropriate polypeptide. Recombinant virus that has reacquired a functional gene can be titered on Vero cells and the background subtracted from the titer obtained from the complementing cell line. Comparison of the virus titers obtained by transfection of the wild-type and mutant genes therefore provides a measure of the ability of the mutant to support virus growth, allowing correlations to be drawn between specific *in vitro* biochemical functions and the role they play in virus replication.[28,29] Detailed protocols for this method of analysis are described below.

African green monkey kidney cells (Vero) were propagated at 37°C in the presence of 5% CO_2 in modified Dulbecco medium (DME) supplemented with 5% newborn calf serum.[30] Cells were then trypsinized and plated onto 35-mm tissue culture dishes at a density sufficient to give approximately 80% confluence (although a range of cell densities appeared to give reasonable transfection efficiency). After 2 to 4 h incubation to allow the cells to attach, medium was removed and the cells washed twice with serum-free DME (SFM). Three hundred nanograms of plasmid DNA (an amount shown to be saturating by dose-response experiments, so as to minimize the effects of small errors in DNA quantification) in 0.5 ml SFM containing 80 μg/ml DEAE-dextran was added to the cells and incubated for 12 min at room temperature. Three milliliters of DME with 5% newborn calf serum was then added and the cells incubated at 37°C in the presence of 5% CO_2 for 3 to 4 h. Fresh medium was the added and the cells incubated for 16 h. Next, the cells were washed once with DME and infected for 2 h with the appropriate mutant herpes simplex virus at multiplicity of infection of 2. For analysis of *pol*-containing plasmids, the virus HP66 was used, which contains a 2.2 kb deletion of the *pol* gene replaced by a *LacZ* gene.[26] For analysis of *UL42*-containing plasmids, the virus CgalΔ42 (generously supplied by Drs. P. Johnson and D. Parris) was used, which similarly contains a large deletion and a *LacZ* insertion within the *UL42* gene.[27] To remove extracellular virus, the cells were washed twice with serum-free medium before and after treatment for 2 min with 3 ml of 0.1 *M* glycine, pH 3[29] and fresh serum-containing medium added. Twenty-four hours later, virus was harvested as described previously,[31] and titered on the appropriate complementing cell line. *Pol*-negative viruses were titered on DP6 cells,[26] a stable cell line derived from Vero cells that contains the HSV *pol* gene and is permissive for the growth of the *pol* null mutant HP66. *UL42*-negative viruses were titered on V9 cells, which similarly contain a resident copy of the *UL42* gene.[27] Background arising from residual inoculum was assessed by titering the progeny from wells transfected with an irrelevant plasmid. All progeny were also titered on Vero cells to quantify viruses that had reacquired a functional polymerase subunit gene by recombination, either with the complementing cell line the virus stock was prepared from, or with a plasmid-borne gene during the transient assay. Although the former case represents an infrequent event, the presence of a single "jackpot" replication competent virus in the superinfecting inoculum will give large numbers of progeny, irrespective of the ability of the test plasmid to complement a defective virus. Recombination between null mutant virus and a plasmid containing a functional copy of the gene can also produce progeny able to form plaques on Vero cells, but this is also a relatively infrequent event. The viral yield due to complementation by the plasmid was then calculated by subtracting the background titers of recombinant and residual input virus from the viral titer on the complementing cells. Results were finally presented as a percentage of the complementation seen with a plasmid containing the wild-type gene, and were averaged from at least three separate experiments.

Table 12.1 shows sample results obtained from two experiments analyzing transient complementation of a *pol* null mutant by various plasmid-borne *pol* genes. After infection of Vero cells with the *pol* null mutant virus, no progeny were found from untransfected wells in either experiment, confirming that the mutant virus is unable to replicate in the absence of a functional *pol* gene. However, wells transfected with a plasmid containing a wild-type copy of the *Pol* gene gave rise to over 1000 plaque forming units (pfu) per six-well dish, indicating that the plasmid-borne gene was able to complement virus replication *in trans*. In contrast, two mutant polymerase genes (F-1208 and *n*-1216, both encoding polypeptides lacking amino acids from the extreme C-terminus of the protein) were unable to significantly complement growth of the defective virus. No progeny able to grow on the complementing cell line were found in either experiment from wells transfected with the F-1208 plasmid. A few plaques were found to arise from the *n*-1216 construct in the first experiment, but similar numbers of these progeny were also able to grow on Vero cells, indicating that they had reacquired a functional *pol* gene, most likely by recombination with the host cell-resident copy in the DP6 cell line used to grow stocks of the mutant virus. Although the F-1208 and *n*-1216 mutants possessed normal polymerase activity when assayed alone as described above, each was impaired for binding of UL42 in a co-immunoprecipitation assay and for stimulation of long-chain DNA synthesis by UL42, thus indicating that the protein–protein interaction is necessary for virus replication.[3] *Pol* mutants intermediate in their ability to support viral growth were also found. In the first experiment, the mutant I-1208, containing a 12 bp insertion at codon 1208 of the *pol* gene, gave rise to around a tenth of the progeny seen from complementation with a wild-type copy of the gene. However, substantial numbers of these progeny were able to replicate on Vero cells, suggesting that they had reacquired a functional copy of the *pol* gene. Therefore, it could not safely be concluded from this experiment that the I-1208 mutant possessed the ability to substantially complement virus replication. However, in the second experiment the I-1208 mutant again gave rise to around a tenth of the progeny seen from complementation with a wild-type copy of the gene, but none of these viruses were able to replicate on Vero cells. Similar results were obtained from other experiments (data not shown), allowing the conclusion that the I-1208 mutant did substantially complement replication of the *pol* null virus. Therefore, the assay possesses sufficient sensitivity to detect mutants that are phenotypically impaired for growth, as well as more defective examples.

TABLE 12.1
Ability of Selected Pol Mutants to Complement a *pol* Null Mutant Virus in a Transfection Assay

Experiment	*trans* Pol Gene	Titer on DP6 Cells	Titer on Vero Cells	% Complementation[a]
1	WT	1400	0	100
1	I-1208	130	53	5.5
1	F-1208	0	0	<0.07
1	*n*-1216	8	5	0.2
1	No plasmid	0	0	<0.07
2	WT	1900	0	100
2	I-1208	240	0	13
2	F-1208	0	0	<0.05
2	*n*-1216	0	0	<0.05
2	No plasmid	0	0	<0.05

[a] Values are calculated as a percentage relative to the complementation seen with the wild-type gene.

To conclude, *in vitro* translation in RRL is a convenient method for small-scale synthesis of polypeptides, which are often biologically active. In conjunction with the ease with which mutations can be introduced into plasmid-encoded genes, this method can provide an outstanding system for mutagenic analysis of protein function. When used to analyze polypeptides from a system amenable to genetic analysis such as herpes viruses, the approaches outlined above combine to form a powerful means of assessing the functions of the polypeptides *in vivo*.

Acknowledgments

The authors thank J. Brown for help with manuscript and figure preparation, and laboratory colleagues for helpful discussions. Grant support from the National Institutes of Health (R01 AI19838 and U01 AI26077 to DMC) and the Medical Research Council (G9232370PB to PD) is gratefully acknowledged.

References

1. Pelham, H. R. B. and Jackson, R. J., *Eur. J. Biochem.,* 105, 445 (1976).
2. Digard, P. and Coen, D. M., *J. Biol. Chem.,* 265, 17393 (1989).
3. Digard, P., Bebrin, W., Weisshart, K., and Coen, D. M., *J. Virol.,* 67, 398 (1993).
4. Digard, P., Chow, C., Pirrit, L., and Coen, D. M., *J. Virol.,* 67, 1159 (1993).
5. Challberg, M. D. and Kelley, T. J., *Annu. Rev. Biochem.,* 58, 671 (1989).
6. Coen, D. M., *Sem. Virol.,* 3, 3 (1992).
7. Hernandez, T. R. and Lehman, I. R., *J. Biol. Chem.,* 265, 11227 (1989).

8. Gottlieb, J., Marcy, A. I., Coen, D. M., and Challberg, M. D., *J. Virol.*, 64, 5976 (1990).
9. Digard, P. and Coen, D. M., *Methods Enzymol.*, 262, 303 (1995).
10. Melton, D. A., Krieg, P. A., Rebagliati, M. R., Maniatis, T., Zinn, K., and Green, M. R., *Nucleic Acids Res.*, 12, 7035 (1984).
11. Dasso, M. C. and Jackson, R. J., *Nucleic Acids Res.*, 17, 3129 (1989).
12. Craig, D., Howell, M. T., Gibbs, C. L., Hunt, T., and Jackson, R. J., *Nucleic Acids Res.*, 20, 4987 (1992).
13. Marcy, A. I., Olivo, P. D., Challberg, M. D., and Coen, D. M., *Nucleic Acids Res.*, 18, 1207 (1990).
14. Masters, P. S. and Banerjee, A. K., *J. Virol.*, 62, 2651 (1988).
15. Tenney, D. J., Hurlburt, W. W., Bifano, M., Stevens, J. T., Micheletti, P. A., Hamatake, R. K., and Cordingley, M. G., *J. Virol.*, 67, 1959 (1993).
16. Dorsky, D. I. and Crumpacker, C. S., *J. Virol.*, 62, 3224 (1988).
17. Hamatake, R. K., Bifano, M., Tenney, D. J., Hurlburt, W. W., and Cordingley, M. G., *J. Gen. Virol.*, 74, 2181 (1993).
18. Digard, P., Williams, K., Hensley, P., Brooks, I. S., Dahl, C. E., and Coen, D. M., *Proc. Natl. Acad. Sci. U.S.A.*, 92, 1456 (1995).
19. Maniatis, T., Fritsch, E. F., and Sambrook, J., *Molecular Cloning: A Laboratory Manual,* Cold Spring Harbor Laboratory Press, Plainview, New York, 1989.
20. Hart, G. J. and Boehme, R. E., *FEBS Lett.*, 305, 97 (1992).
21. Kiehl, A. and Dorsky, D. I., *Virology,* 184, 330 (1991).
22. Lin, J.-C., Sista, N. D., Besencon, F., Kamine, J., and Pagano, J. S., *J. Virol.*, 65, 2728 (1991).
23. Berthomme, H., Monahan, S. J., Parris, D. S., Jacquemont, B., and Epstein, A. L., *J. Virol.*, 69, 2811 (1995).
24. Wang, G.-H. and Seeger, C., *Cell,* 71, 663 (1992).
25. Barton, D. J. and Flanegan, J. B., *J. Virol.*, 67, 822 (1993).
26. Marcy, A. I., Yager, D. R., and Coen, D. M., *J. Virol.*, 64, 2208 (1990).
27. Johnson, P. A., Best, M. G., Friedmann, T., and Parris, D. S., *J. Virol.*, 65, 700 (1991).
28. Quinlan, M. P. and Knipe, D. M., *J. Virol.*, 54, 619 (1985).
29. Cai, W., Person, S., DebRoy, C., and Gu, B., *J. Mol. Biol.*, 201, 575 (1988).
30. Weller, S. K., Lee, K. J., Sabourin, D. J., and Schaffer, P. A., *Virology,* 130, 290 (1983).
31. Coen, D. M., Fleming, H. E., Leslie, L. K., and Retondo, M. J., *J. Virol.*, 53, 477 (1985).

Chapter 13

Molecular Genetics of Adeno-Associated Virus

Barrie J. Carter

Contents

I. AAV Biology

Adeno-associated viruses (AAV) are small, DNA-containing viruses that belong to the family Parvoviridae within the genus *Dependovirus*. AAV have been isolated

0-8493-4412-3/96/$0.00+$.50

directly from humans and a variety of other animals, including mammalian and avian species. AAV is a defective parvovirus that grows only in cells in which certain functions are provided by a coinfecting helper virus. AAV helper viruses include adenoviruses and some herpes viruses. The nature of the helper function is not known, but appears to be an indirect effect of the helper virus, which renders the cell permissive for AAV replication. More detailed summaries of the AAV life cycle and AAV genetics can be found in several recent reviews.[1-3]

AAV has a very broad host range *in vitro* without any obvious species or tissue specificity, and will replicate in cell lines of human simian or rodent origin provided an appropriate helper is present. It is not yet clearly established if AAV exhibits any tissue specificity *in vivo*. At least five serotypes, AAV1, 2, 3, 4, and 5, have been distinguished. AAV2 and 3 have been most frequently isolated from humans. Virtually all molecular genetic studies of AAV have used AAV2 and this is the only serotype that will be discussed in this chapter.

Infection of cells with AAV in the absence of helper functions may result in integration of AAV into the host cell genome. The integrated AAV provirus may be rescued and replicated to yield a burst of infectious progeny AAV particles if the cells are superinfected with a helper virus such as adenovirus. Recent results have suggested that AAV may exhibit preference for integration at a site on human chromosome 19 but the generality and mechanism of this phenomenon has not been elucidated fully. Although AAV has been isolated from humans, it has not been associated with the cause of any disease. AAV is not a transforming or oncogenic virus and integration into human cell chromosomes of stable cell lines does not cause any significant alteration in the growth properties or morphological characteristics of the cells.

AAV particles are about 20 nm in diameter with icosahedral symmetry and are nonenveloped, stable to heat, and resistant to ether, chloroform, and anionic detergents. The particles are comprised of a protein capsid having three capsid proteins VP1, VP2, and VP3 and enclosing a DNA genome. The AAV DNA genome is a linear single-stranded DNA molecule with a molecular weight of about 1.5×10^6 Da and 4681 nucleotides long. Strands of either complementary sense, "plus" or "minus" strands, are packaged into individual particles, but each particle has only one DNA molecule. Equal numbers of AAV particles contain either a plus or minus strand. Duplex or single-strand copies of AAV genomes inserted into bacterial plasmids are infectious when transfected into adenovirus-infected cells and this has allowed the study of AAV genetics and the development of AAV vectors.

In permissive cells in which the appropriate helper functions are expressed by coinfecting helper virus, AAV undergoes a productive replication cycle. Infectious AAV particles absorb to the cells and enter the cell nucleus in a process that is not well characterized but which does not require, and is not affected by, the presence of helper virus. Replication, assembly, and accumulation of infectious AAV particles occurs in the cell nucleus. The infecting parental AAV single-strand genome is converted to a parental duplex replicating form (RF) by a self-priming mechanism which takes advantage of the ability of the inverted terminal repeat (ITR) to form a hairpin structure which provides a self-priming, base-paired 3′ hydroxyl group to initiate replication. This process probably utilizes a cellular DNA polymerase and it is unclear if any helper function is required for this process. This parental RF molecule is then amplified to form a large pool of progeny RF molecules in a process which not only requires the helper functions but also the action of the AAV *rep* gene products, Rep78 and Rep68. From this pool of duplex RF molecules, progeny single-strand (ss) DNA genomes are generated by a strand displacement mechanism and packaged into preformed empty AAV capsids composed of VP1, VP2, and VP3 proteins. Accumulation of progeny ssDNA requires the *rep* proteins Rep52 and Rep40 as well as VP2 and VP3. VP1 appears to be required to generate stable, infectious AAV particles.

The AAV2 genome (Figure 13.1) has one copy of the 145-nucleotide long ITR at each end and a unique sequence region of about 4470 nucleotides long[4] that contains two main open reading frames for the *rep* and *cap* genes. The unique region contains three transcription promoters p5, p19, and p40 that are used to express the *rep* and *cap* genes, which provide trans-acting functions for replication and encapsulation of the viral genome, respectively. The *rep* gene open reading frame is expressed from p5 and p19 and both unspliced and spliced mRNAs to yield four Rep proteins, Rep78, Rep68, Rep52, and Rep40, which all contain a common internal region but differ with respect to their amino- and carboxyl-terminal regions. The proteins VP1, VP2, and VP3 all share a common overlapping sequence, but differ in that VP1 and VP2 contain additional amino-terminal sequence. All three are coded from the same *cap* gene reading frame expressed from a spliced 2.3 kb mRNA transcribed from the p40 promoter. VP2 and VP3 are generated from the same mRNA by use of alternate initiation codons. VP1 is coded from a minor mRNA using a 3′ donor site that is 30 nucleotides upstream from the 3′ donor for the major mRNA that encodes VP2 and VP3.

II. AAV Genetics

Genetic analysis of AAV presents several unique problems. The defectiveness of AAV and need for a coinfecting, lytically replicating helper virus renders a conventional genetic analysis of AAV impractical. Also, the expression of overlapping proteins (Rep or Cap) from single open reading frames complicates mutational analysis. However, a molecular approach to AAV genetics became possible when duplex copies of AAV genomes were molecularly cloned into bacterial plasmids.[5,6] Transfection of such clones into mammalian cells that are also infected with an appropriate helper virus, such as adenovirus, results in excision and replication of the AAV genome free of any plasmid sequence and generation of a yield of infectious progeny AAV particles. Standard molecular biology techniques applied to mutate the genome sequence of molecular clones of AAV followed by transfection into helper virus infected cells permitted establishment of a basic genetic analysis and description of several mutant phenotypes of AAV.[7,8]

A complementation analysis of AAV mutants by transfecting two plasmids has been hampered in these transfection assays by recombination between mutant plasmid

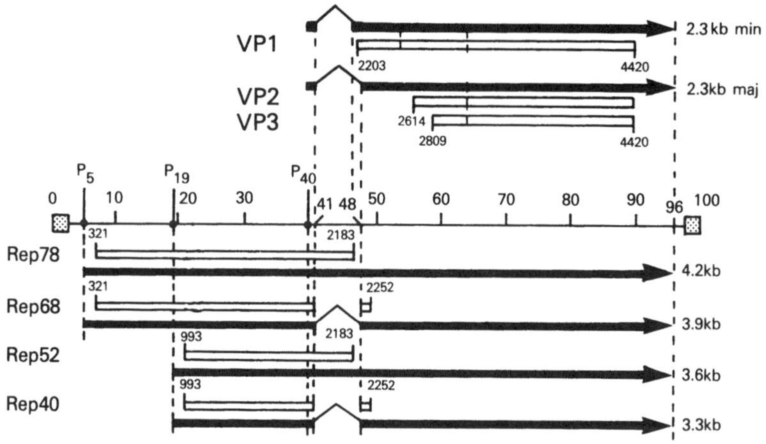

FIGURE 13.1
Structure of the AAV genome. The AAV2 genome is shown as a single bar with a 100 map units scale (1 unit = approximately 47 nucleotides). Stippled boxes indicate inverted terminal repeats (ITRs, replication origins) and solid circles indicate transcription promoters (p5, p19, and p40). The poly(A) site is at map position 96. RNAs from AAV promoters are shown as heavy arrows with the introns indicated by the caret. The coding regions for the four rep proteins (Rep78, Rep68, Rep52, and Rep40) and for the viral capsid proteins (VP1, VP2, and VP3) are shown with open boxes. (Redrawn with permission from Smuda, J. W. and Carter, B. J., *Virology,* 184, 310, 1991.)

clones.[9] However, in one case such a complementation assay was useful to demonstrate a dominant negative phenotype.[10] Expression of mutant AAV proteins, especially rep, from transfected plasmids has allowed analysis of the biochemical properties of the protein. Although this biochemical genetics is extremely valuable, it is beyond the scope of this chapter and is summarized here only briefly. This chapter focuses on the ability to analyze AAV mutations by reconstructing them into AAV genomes that can be used to generate pure populations of AAV particles in order to study properties of any mutation in the biological context of an AAV infection. This has proved difficult to do for AAV, but one approach using conditional lethal mutants is described in detail. This illustrates a general approach and assays that will be useful in performing AAV genetics as additional cell lines that can complement AAV gene functions become available.

III. Mutant Phenotypes

Analysis of various mutant AAV plasmids has demonstrated several mutant phenotypes. The only clearly defined cis-acting phenotype is Ori$^-$, whereas at least four trans-acting phenotypes, Rep$^-$, Ssd$^-$, Lip$^-$/Inf$^-$, and Cap$^-$ have been defined.[1,2,7,8,11] Mutations in the ITR exhibited a DNA replication-negative phenotype (Ori$^-$) which cannot be complemented in trans. The ITR sequences are required in cis and are sufficient to provide a functional origin of replication *(ori)*. The ITRs appear also sufficient to provide signals required for packaging into AAV particles *(pac)*, for integration into the cell genome *(int)*, and for excision and rescue from host cell chromosomes or from recombinant plasmids *(res)*. It is unclear if mutational analysis is able to separate pac, int, and res functions from ori functions.[2] It is important to note that mutational analysis of the ITRs is complicated by the need to place the same mutation in each ITR to avoid a gene conversion process that can repair mutations in one ITR by using the sequence in the ITR at the other end of the genome.[12,13]

Only Rep78 and Rep68 are required for AAV DNA replication. Mutations in Rep78 or Rep68 (Rep^-) lead to a complete DNA replication defect because these proteins are required for replication and for rescue of AAV genomes that are integrated into plasmids or the host cell chromosome.[7,8] Rep68 and Rep78 possess several enzyme activities required for AAV DNA replication, including binding to the ITR, a strand-specific, site-specific endonuclease activity, and a DNA helicase activity.[14-18] These activities combine to provide for hairpin resolution, which is a combined process that converts a covalently closed hairpin conformation of the ITR to the extended linear duplex conformation and is required at each round of AAV DNA replication. Rep52 and Rep40 have none of these properties,[19] but appear to be needed for generation of progeny AAV genomes and for assembly of infectious particles. A mutation that eliminated Rep52 and Rep40 by converting the first AUG codon in the reading frame for these proteins to a glycine codon exhibited a partial replication defective phenotype (Ssd^-) in which duplex RF DNA was normal but no ssDNA and no infectious progeny particles were generated. VP1, VP2, and VP3 are all required for capsid production.[11] Mutations that eliminate all three proteins (Cap^-) prevent accumulation of progeny ssDNA, whereas mutations in the VP1 amino terminus (Lip^-, Inf^-) permit ssDNA accumulation but prevent assembly of stable, infectious particles.[7,8] This implies that VP2 and VP3 may be sufficient to generate the performed capsid but that VP1 is required to stabilize the structure.

All of the Rep^-, Cap^-, and Lip^- mutants can be complemented in trans. The Rep proteins, primarily Rep78 and Rep68, exhibit several pleiotropic regulatory activities, including positive and negative regulation of AAV genes and expression from some heterologous promoters, as well as inhibitory effects on cell growth. The AAV p5 promoter is negatively autoregulated by Rep78 or Rep68 in the absence of helper functions but this may be partly overcome by helper virus infection. Because of the inhibitory effects of expression of *rep* on cell growth, constitutive expression of *rep* in cell lines has been hampered and this has slowed the development of cell lines able to complement *rep* functions.

One interesting class of Rep^- mutants exhibited a dominant-negative phenotype and could completely block replication even from a cotransfected wild-type AAV genome.[10] One such mutant generated by mutation (K340H) of a lysine in the ATP binding domain of the Rep protein produced mutant Rep 68/78 proteins that were not negatively autoregulated and thus were overproduced. Biochemical analysis showed that the mutant proteins bind to the ITR but could not mediate the site-specific nicking and hairpin resolution reactions.[24]

IV. Biochemical Properties of AAV Proteins

The AAV ITR forms a hairpin structure and AAV rep proteins interact with this structure in a series of reactions that are essential to AAV DNA replication and probably play an important role in AAV integration and rescue and perhaps packaging. These reactions include binding to the hairpin, probably as a multi-Rep protein complex, then mediating a concerted series of events including partial unwinding of the hairpin, followed by a strand-specific endonuclease cleavage at a specific site (the terminal resolution site), covalent binding to the 5′ side of the site, followed by helicase unwinding of the hairpin to allow cellular DNA polymerase to complete repair replication and convert the ITR from a hairpin to an extended linear duplex. The biochemical properties of AAV rep proteins can be assayed in a series of elegant *in vitro* assays that measure each of these reactions.[14,15,17-19]

A variety of methods have been utilized to generate mutant AAV rep proteins, including expression in insect cells using baculovirus vectors,[16] transfection into human 293 cells of plasmids containing rep gene reading frames expressed from a heterologous promoter such as the HIV long terminal repeat (LTR) promoter,[16] expression from *Escherichia coli,* or by *in vitro* translation.[26] This has permitted significant analysis of functional domains of the rep gene and in some cases has allowed correlation of biochemical functions with genetic phenotypes. For instance, the dominant negative Rep$^-$ phenotype of the K340H mutant *in vivo*[10] was recapitulated *in vitro* by showing that it inhibited the *in vitro* terminal resolution reactions.[16,24] For a more detailed description of this biochemical genetic approach to analysis of AAV, the references cited in this section should be consulted.

V. Conditional Lethal Mutants

Full genetic analysis of AAV requires generation of infectious virus particles containing mutant genomes in order to analyze the relevance of any mutation to the *in vivo* biological functions of AAV. This goal has been obtained most readily in genetic studies of viruses by generation of conditional lethal mutants such as temperature-sensitive, host-range, or suppressible nonsense mutations. For AAV, temperature-sensitive mutants have not been isolated because there is no simple infectious center assay or plaque assay. Host-range mutants have not been isolated readily because in cell culture this is generally regulated by the helper virus rather than AAV. However, as complementing cell lines begin to be developed that are capable of expressing *rep* or *cap* genes, then the ability to complement a much wider range of AAV mutants will be possible.

Suppressible nonsense mutants were constructed in AAV by taking advantage of an inducible amber suppression system developed for a mammalian cell line. Sedivy et al.[27] constructed a monkey cell line, BSCSupD12, having a human amber suppressor tRNA^{ser} gene contained in an SV40 replicon. The human amber suppressor tRNA^{ser} gene was derived by modification of the anti-codon region of a human tRNA^{ser} to recognize the amber termination codon UAG. These SupD12 cells also contain a temperature-sensitive SV40 T-antigen gene. Upon shift-down from nonpermissive (39.5°C) to permissive (33°C) temperature, the temperature-sensitive SV40 T-antigen becomes functional and causes amplification of the suppressor tRNA gene to enable efficient suppression of amber codons. The control, nonpermissive cell line, BSC Sup0 contains only the temperature-sensitive SV40 T-antigen gene.

Amber mutants of AAV were generated by site-specific mutagenesis in AAV plasmids to convert serine codons to amber termination codons.[28,29] Individual mutations were made in serine residues at nucleotide 1033 in the *rep* gene (rep*am*) and in the *cap* gene at nucleotides 2527 (VP1*am*) or nucleotide 3001 (cap*am*). The rep*am* mutation terminates all four *rep* reading frames. The VP1*am* mutation terminates only VP1, whereas the cap*am* mutation terminates all three capsid proteins VP1, VP2, and VP3. Transfection of AAV plasmids, prep*am*, pVP1*am*, pcap*am*, containing any one of these amber mutations into BSC SupD12 cells at 33°C allowed the generation of stocks of virus which contained only mutant genomes (Tables 13.1 and 13.2). When these particles were used to infect nonpermissive cells (BSC Sup0 or human HeLa cells), the mutant phenotype was expressed, whereas at the permissive temperature, 33°C, in BSC SupD12, the nonsense mutation was efficiently suppressed.

TABLE 13.1
INFECTIOUS TITER OF WILD-TYPE AND REP*AM* AAV STOCKS

		AAV2 Infectivity Assayed in[b]	
Virus Stock[a]	Grown on	SupD12	Sup0
wt	SupD12	4.0×10^8	1.5×10^8
wt	Sup0	3.2×10^8	1.4×10^8
rep*am*	SupD12	4.0×10^7	$<1.0 \times 10^3$
rep*am*	Sup0	0	0

[a] Virus stocks were prepared by transfecting 1 μg of the wt or rep*am* plasmid onto SupD12 or Sup0 cell at 33°C. Lysates were prepared at 90 h after transfection.

[b] Infectivity of the viral stocks was assayed at 33°C in SupD12 or Sup0 cells using the indirect immunofluorescence procedure.[33] Titers are expressed as AAV2 infectious units per milliliter.

The titer of viral stocks derived by transfection of plasmids in SupD12 cells at 33°C was assayed by the indirect immunofluorescence assay (see below) using antibodies against either AAV capsid protein or AAV Rep protein in SupD12 and Sup0 cells at 33°C. The virus stocks were then used to infect permissive (SupD12) or nonpermissive (Sup0, HeLa) cells. The efficiency of amber suppression was about 10% both for the initial generation of virus following transfection or for infection with mutant virus particles.

The level of reversion or leakiness of the amber mutants was extremely low, as judged by the relative titer on permissive and nonpermissive cells. Thus, for the rep*am* mutant, transfection of the mutant plasmid into Sup0 cells yielded no infectious

TABLE 13.2
Yields of Virus Following Growth on Three Cell Lines

Inoculum	SupD12	Sup0	HeLa
wt DNA	5.8×10^8	4×10^8	n.d.[a]
VP1*am* DNA	6.0×10^7	5×10^2	n.d.
Cap*am* DNA	2.7×10^7	0	n.d.
AAV2*wt*	5.5×10^9	4.2×10^9	7.5×10^9
VP1*am*	3.5×10^8	1.1×10^3	6.4×10^3
Cap*am*	2.3×10^8	0	0

Note: 10^6 SupD12 or Sup0 cells growing at 33°C or HeLa cells growing at 37°C were transfected with 10 μg plasmid DNA or infected with AAV virus stocks at a m.o.i. (multiplicity of infection) of 5 infectious units per cell. All cells were also infected with adenovirus type 5 at a multiplicity of 5 pfu per cell. Viral lysates were prepared after 96 h at 33°C or 50 h at 37°C and assayed for AAV infectivity on HeLa cells using AAV Rep antibody in an indirect immunofluorescence assay. All titers are listed as the yield per 10^6 cells.

[a] n.d., not done.

mutant virus particles and the proportion of wild-type revertants in rep*am* stocks produced on SupD12 cells was less than 10^{-5} of the mutant titer (Table 13.1, Figure 13.2). The cap*am* mutation appeared to be completely non-leaky and no reversion was detected in mutant stocks (Table 13.2). This suggests that stoichiometric amounts of all three capsid proteins are required for particle production. The VP1*am* mutant appeared to be more leaky and gave a low level of production in nonpermissive cells (Table 13.2). However, this appeared to reflect biochemical leakiness rather than reversion, since continued passage of VP1*am* in nonpermissive cells did not lead to virus stocks with improved infectivity for nonpermissive relative to permissive cells.

The phenotypes of the several amber mutants are illustrated in Figures 13.2 and 13.3. Figure 13.3 illustrates a standard AAV viral DNA replication assay that can be performed in adenovirus-infected cells that are either infected with AAV particles or transfected with molecular clones of AAV DNA. At an appropriate time after infection, cells are harvested, viral DNA is selectively extracted using a modified Hirt procedure (described below), and analyzed by agarose gel electrophoresis followed by Southern blotting with an AAV DNA-specific probe. The three main species of AAV DNA observed are the replicating monomeric duplex (RFm) and dimeric duplex (RFd) species and the progeny single strand (ss).

Figure 13.3 shows that the rep*am* mutant exhibited a Rep^- phenotype. In nonpermissive cells (Sup0) no viral DNA replication occurred, as indicated by the absence of duplex RF DNA and the consequent absence of progeny ssDNA, whereas in SupD12 cells the normal replicating and progeny DNA species were observed. This experiment also shows the level of suppression of the amber mutation in SupD12 cells was above 10% (compare rep*am* to wt on SupD12) and that reversion was less than 10^{-5} (compare rep*am* to wt on Sup0).

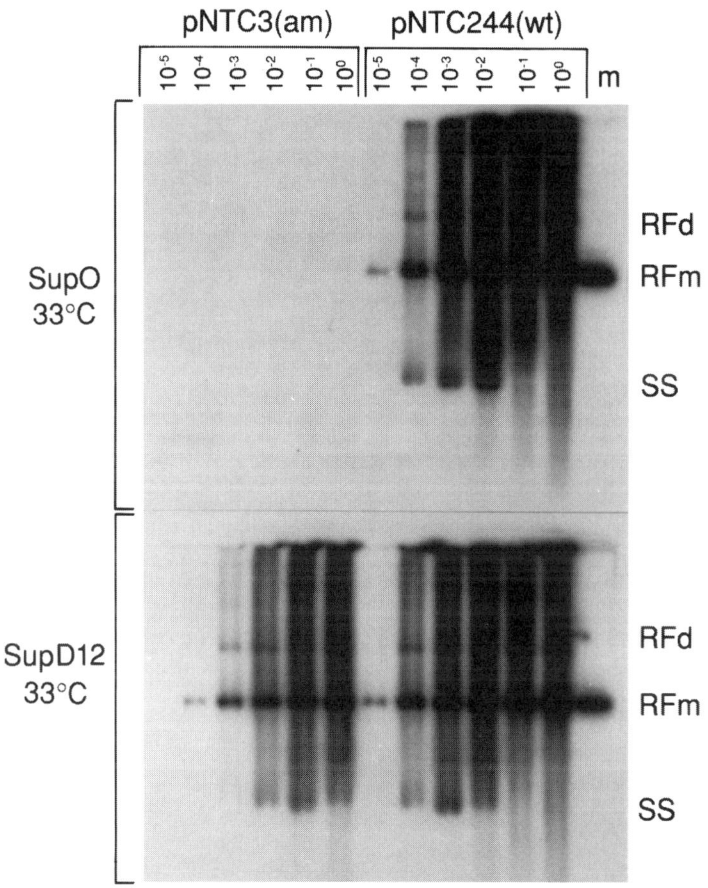

FIGURE 13.2
Phenotype of rep*am* virus. Stocks of rep*am* or wt AAV were generated by transfecting the rep*am* plasmid (pNTC3) or wt plasmid (pNTC244) into adenovirus-infected SupD12 cells at 33°C. Viral lysates were prepared and serial dilutions (10^0 to 10^{-5}) were used to infect adenovirus-infected Sup0 or SupD12 cells at 33°C. At 72 h after infection, viral DNA replicating forms were analyzed by Hirt extraction of cell lysates followed by agarose gel electrophoresis and Southern blotting with an AAV2 ^{32}P-DNA probe. Dimeric duplex (RFd), replicating monomeric duplex (RFm), and progeny single strand (ss), intracellular AAV DNA species; M, a molecular weight marker of duplex AAV DNA. (Reprinted with permission from Chejanovsky, N. and Carter, B. J., *Virology,* 171, 239, 1989).

Figure 13.3 shows an analysis to establish that the cap*am* and VP1*am* mutants exhibit Cap⁻ and Lip⁻ phenotypes, respectively. Infection of nonpermissive (HeLa) cells with cap*am* resulted in production of normal levels of RF DNA but no accumulation (i.e., no packaging) of single-strand progeny DNA (Figure 13.3A) and no accumulation of AAV capsid proteins, as measured in an immunoblot assay (Figure 13.3). This is consistent with the Cap⁻ phenotype defined above.

The VP1*am* mutant generated normal amounts of RF and ssDNA (Figure 13.3A) and normal levels of VP2 and VP3, but no VP1 (Figure 13.3B). Although normal

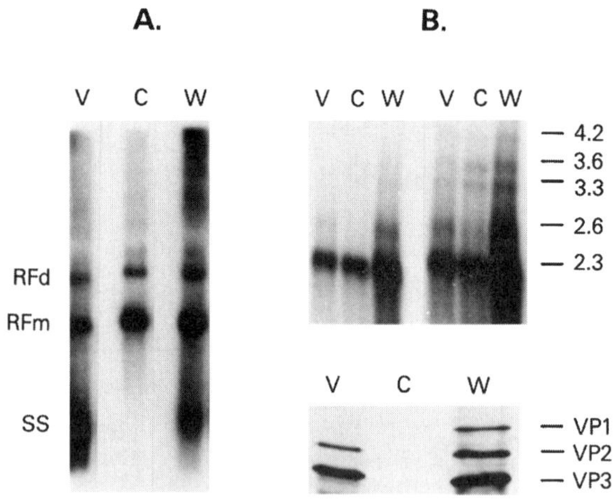

FIGURE 13.3
Phenotype of AAV amber mutants on nonpermissive HeLa cells. Cultures of HeLa cells (10^6 cells per 10-cm dish) growing at 37°C were infected with adenovirus (5 pfu/cell) and AAV VP1*am* (V), AAV Cap*am* (C), or AAV*wt* (W), at 10 infectious units per cell. Individual cultures were taken for analysis of AAV DNA, RNA, or capsid protein synthesis. (A) AAV DNA synthesis analyzed by Southern blotting. (B) Analysis of AAV RNA synthesis (upper) and AAV capsid protein synthesis (lower). For the RNA synthesis, two autoradiographic exposures of 2 h (left) and 4 h (right) are shown. The sizes (in kb) of the individual RNA species are indicated at the right. (Reprinted with permission from Smuda, J. W. and Carter, B. J., *Virology,* 184, 310, 1991.)

levels of progeny ssDNA accumulated, there were no infectious particles (Table 13.2). This is consistent with the previously defined Lip⁻ phenotype and presumably indicates that VP1 is required for generation of stable particles. In contrast, the *Cap* phenotype indicates that both VP2 and VP3 are required for accumulation of ssDNA. The additional RNA analysis by Northern blot analysis (Figure 13.3B) confirms that absence of the capsid proteins in the mutant infected cells was due to failure to suppress the translational defect and not due to absence of mRNA.

These experiments with the amber mutants[28,29] define the three major phenotypes characterized clearly to date and illustrate the basic set of assays for analysis of *in vivo* AAV genetics. Summaries of procedures for these assays are described below and this general approach should also be applicable to assays of other classes of AAV mutants, especially as *Rep* and *Cap* complementing systems become available.

VI. Complementing Cell Lines for *Rep* and *Cap*

The amber mutant approach is limited to the use of an amber termination codon and thus far has only been used to mutate serine residues to amber codons so that suppression results in insertion of the wild-type (serine) residue. It could be extended to substitution of other residues by serine. An alternate approach would be to generate cell lines that are capable of supplying AAV rep or cap functions so that additional mutants might be analyzed. However, because of the properties of rep protein, the generation of such lines has been hampered, although HeLa and 293 cells that constitutively expressed low levels of Rep78 or Rep52 were described.[30]

Recently a 293-derived cell line containing an inducible rep gene has been described.[31] This cell line contains a *rep* gene cassette in which the p5 promoter was replaced by the murine metallothionein promoter, which is inducible by cations such as zinc or cadmium. Upon induction, this cell line complemented Rep^- mutants of AAV to generate stocks of mutant virus with titers of 10^7 to 10^8 per milliliter of crude lysate. This cell line was also capable of packaging a Rep^- Cap^- AAV vector containing a reporter gene, provided that a plasmid expressing the cap gene was also transfected.

Another group[32] derived two HeLa cell lines that expressed a *rep* gene in which the p5 promoter was replaced by the mouse mammary tumor virus promoter, which is inducible by glucocorticoids. These cell lines could also complement replication of a Rep^- AAV mutant, but did not support packaging of the mutant to generate infectious particles even though the three AAV capsid proteins were expressed by the Rep^- mutant. This packaging defect could be overcome by additionally transfecting a rep gene plasmid. Apparently, these cells had some unidentified defect in the expression of the inducible rep gene that is formally analogous to the Ssd^- phenotype.

Because of the interest in developing AAV as a vector for gene therapy it is probable that cell lines will become available that will allow packaging of pure populations of AAV mutants. However, recombination will be an important issue for determining if such systems can be used for production of mutant AAV genomes, as opposed to packaging of gene therapy vectors. In AAV vectors most of the AAV coding sequence is removed, and the complementing system can be designed without homology between the complementing gene and the vector. For analysis of mutants useful for genetic studies of AAV it is likely that there will be extensive homology between the mutant and the complementing gene. In a two-plasmid complementing system there is extensive recombination between the plasmids if there is overlapping homologous sequence and even when there is no overlapping homology, recombination may occur at a detectable frequency.[9]

VII. Methods

A. Cells

HeLa cells or human 293 cells are obtained from the American Type Culture Collection (ATCC) and grown in Dulbecco's modified Eagle's medium (DMEM) with 5% fetal calf serum (FBS) at 37°C in 5% CO_2.

The monkey cell lines BSCSup0 and BSCSupD12[27] are obtained from J. Sedivy and P. Sharp (Massachusetts Institute of Technology, Boston, MA) and grown in 10-cm dishes at 39.5°C in DMEM with 10% fetal calf serum in 5% CO_2. For routine passage, the cells were split 1:50 about every 6 d. The antibiotic geneticin (700 μg/ml total concentration) is included in the medium in alternate passages and care must be taken to avoid exposure of the cell cultures to temperatures below 39.5°C. For experiments, one 10-cm dish of confluent cells is split 1:4 into 6-cm dishes and immediately placed at 39.5°C. The following day the cells that are about 70% confluent (10^6 cells per dish) are infected with human adenovirus type 5 (five plaque forming units, pfu, per cell) and transfected with AAV plasmids using a $CaPO_4$ procedure or infected with AAV particles. Four hours after transfection, the medium is changed and the cells maintained at 39.5°C or shifted to 33°C. For preparation of viral lysates, cells maintained at 39.5°C are harvested at 50 h after transfection or infection, while those held at 33°C are harvested at 4 d.

For analysis of AAV DNA replication, 39.5°C cultures were harvested at 24 h after virus infection or 40 h after DNA transfection, and 33°C cultures were harvested at 40 h after infection or 70 h after transfection. For assay of viral proteins or infectious progeny virus, cells were harvested at 48 h after infection or transfection at 39.5°C or 72 h after infection or transfection at 33°C. Viral lysates from transfected cultures are concentrated 10-fold.

B. Viruses

Human adenovirus type 5 and AAV serotype 2 are obtained from ATCC and can be grown by infection of human 293 cells or HeLa cells. AAV2 stocks can also be generated by transfection of the plasmid pAV2 into adenovirus-infected 293 cells. This plasmid is also available from ATCC.

Growth or assay of AAV is performed by infecting cells that are also infected with adenovirus 5 at a multiplicity of about 5 pfu per cell. Because adenovirus stocks are a necessary requirement for studies with AAV, the plaque assay for adenovirus infectivity is also described below.

C. Generation of AAV Mutants

For generation of AAV mutants, a variety of molecular clones of infectious AAV genomes are available.[5,6,13,28] However, a convenient starting point is the plasmid pAV2,[6] which is obtainable from ATCC. pAV2 contains a full-length clone of AAV2 inserted into a pBR322-derived backbone using *Bgl*II linkers. This allows for convenient excision of the intact AAV2 genome and ready transfer to other plasmid backbones.

D. Assay of Viral Infectivity

This is measured using a fluorescent focus assay.[33] HeLa cells are grown in a monolayer on microscope slides containing eight chambers (Lab-Tek trays) at 37°C. Each chamber contains 5×10^5 cells. The cells were infected with the helper adenovirus 5 (pfu/cell) and appropriate dilutions of the AAV preparation. Virus dilutions are made in phosphate-buffered saline (PBS) with 2% FBS and absorption is allowed to proceed for 1 h in a volume of 0.1 or 0.2 ml/chamber. The inoculum is then removed, fresh medium (0.4 ml) containing 5% FBS is added, and the cultures are incubated at 37°C in a CO_2 incubator. Cells are fixed at 30 h after infection by removing the medium and immersing in 95% methanol, 5% PBS at room temperature for 10 min, then air dried and stored at –20°C until stained as below.

The number of cell nuclei containing AAV antigen is determined using an indirect immunofluorescent assay. The fixed cells are rinsed with PBS then incubated with anti-AAV2 rabbit IgG for 45 min at 37°C. The cells are washed thoroughly with PBS and incubated with FITC-conjugated goat anti-rabbit IgG for 45 min at 37°C. The cells are then washed again with PBS and covered with one drop of PBS-glycerol (2:3) and a coverslip. The proportion of cells exhibiting nuclear fluorescence is determined using a Zeiss microscope equipped with UV-epifluorescent optics.

With adenovirus as the helper, the maximum proportion of infected cell nuclei fluorescent for AAV is usually about 50 to 65%. Therefore, according to the Poisson equation, the amount of the virus dilution that gives 63% of this maximal number of infected cell fluorescent nuclei is taken as being equivalent to a multiplicity of one infectious unit of AAV per cell.

Assays of amber mutants use SupD12 or Sup0 cells as appropriate. Also, either antibodies directed at the AAV *rep* protein or the AAV capsid protein can be used for assays.

E. Plaque Assay of Adenovirus Infectivity

1. Human 293 cells are grown in DMEM and 10% FBS at 37°C and 5% CO_2 in 6-well dishes. Plate 10^6 293 cells per well in 5 ml medium at 37°C.
2. Next day, make adenovirus dilutions in cold PBS containing 2% FBS.
3. Remove medium from cells and add virus inoculum in 1.0 to 1.5 ml of PBS-2% FBS. Absorb 1 h at room temperature.
4. Remove inoculum and add slowly 2 ml of agarose overlay medium comprised of 1:1 mix of solution A (1.8% Seakem LE Agarose) and solution B (2X modified Eagle medium [MEM] without phenol red, plus 15% FBS and antibiotics). To make the overlay mix, melt solution A (agarose) and hold at 45°C. Hold solution B at 37°C. Then mix equal parts A + B and immediately overlay.
5. Incubate at 37°C and add another 2 ml per well of overlay at day 4.
6. Plaques are usually visible at 5 to 7 d. For counting the plaques, overlay with 1 ml overlay mix containing 0.1% neutral red and score the next day.
7. 2X MEM (without phenol red) is obtained from Gibco, Life Sciences, Gaithersburg, MD. Neutral red 0.5% is obtained from Quality Biological Inc., Gaithersburg, MD.

F. Preparation of Viral Lysates

1. Wild-type virus lysates grown at permissive temperature usually can be harvested at 40 to 50 h after infection. For production of virus at 33°C, harvesting is usually done

at about 72 to 90 h after infection. The easiest guide usually is to look for the adenovirus cytopathic effect. For cells in monolayer culture, when they are all rounded up due to the adenovirus infection it is appropriate to harvest them.

2. For routine lysates it is usually sufficient to harvest the cells and resuspend them in 1/10 volume of cell growth medium.
3. The cells should then be subjected to three to six cycles of freeze-thaw between dry-ice ethanol and 37°C baths.
4. Cellular debris can be removed by low-speed centrifugation.
5. The lysate is heated at 60°C for 15 min to inactivate adenovirus infectivity.
6. Store lysate at –70°C.
7. If it is desired to use purified stocks of virus, the following procedure can be used.

G. Purification of AAV Particles

1. Resuspend infected cells in 10 m*M* Tris, pH 8.0, at concentration of up to 2×10^8 cells per 5 ml, and subject to three to six cycles of freeze-thaw between dry-ice ethanol and 37°C.
2. Add 1/10 volume of 10% sodium deoxycholate and 1/10 volume of trypsin (0.25%) that is normally used for splitting cells. Incubate at 37°C for 30 min then chill on ice.
3. Adjust with CsCl to a density of 1.40 gm^3. Add 6.48 g CsCl/12.0 ml of suspension. Check density on refractometer. (CsCl at 1.40 g cm^3 has a refractive index of approximately 1.3715.)
4. Transfer to 5 ml ultracentrifuge tube for swinging bucket rotor (e.g., SW50, SW65) and overlay with at least 0.25 ml mineral oil. Centrifuge at 35,000 rpm at 4°C for about 24 h.
5. Collect fractions by puncturing hole at bottom. You should see a protein pellicle at the top of the gradient, the band of adenovirus at a density of about 1.35 g cm^3 (above the AAV), and if there is enough AAV you may see the visible band located at a density of approximately 1.40 g cm^3. If the AAV band is not easily visible it can be located precisely by infectivity assay or by a hybridization dot-blot assay. The yield from 2×10^8 cells should be visible. Note also that there may be a DNA pellicle at the very bottom of the tube, so pierce the hole slightly to one side of this.
6. If desired, the AAV virus may be rebanded one or two additional times in CsCl.
7. Most of the adenovirus can be removed by the cycles of banding in CsCl but the heat inactivation step as above is still required to inactivate residual adenovirus.
8. Virus can be stored at 4°C as a concentrated suspension in the CsCl or it can be dialyzed against an appropriate buffer such as 10 m*M* Tris, pH 7.4, 1 m*M* $MgCl_2$, 10% glycerol and stored frozen at –70°C. Working stocks can be diluted into cell growth medium with serum and stored in aliquots at –20°C.
9. The number of AAV particles can be determined spectrophotometrically by diluting stocks into buffer containing 0.5% sodium dodecyl sulfate (SDS) to prevent particle aggregation and light scattering and by using the numbers below.
10. Useful AAV numbers: AAV2 DNA is 4681 nucleotides long with a molecular weight of 1.544×10^6. The molecular weight of 1 AAV particle is about 5.4×10^6. Purified

wild-type AAV particles have an $A_{260:280}$ absorbance of about 1.38 and 1 A 260/ml is equivalent to about 1.5×10^{13} particles. The particle:infectivity ratio for highly purified AAV stocks is about 40.

H. Analysis of AAV DNA Replication

AAV DNA replication can be analyzed using a modified Hirt procedure. This procedure ensures that ssDNA is extracted from virus particles and does not anneal to duplex molecules.[35] The following procedure is for an experiment performed in 6-well dishes.

1. Remove medium and pellet any cells that detached in medium by low-speed centrifugation.
2. Add 0.8 ml lysis buffer (10 m*M* Tris, pH 8.0, 1 m*M* EDTA, 1% SDS, 200 µg/ml proteinase K). Pour solution back into wells of dish to lyse remaining attached cells. Incubate 37°C, 30 min.
3. Carefully pipette lysate into 1.5 ml microfuge tube, put on ice, add 0.25 ml 5 *M* NaCl, mix gently and store at 0 to 4°C for at least 4 h (or overnight).
4. Centrifuge at top speed in microfuge for 15 min. Transfer supernatant to 12 × 17 mm tube on ice. Extract twice with equal volume chloroform.
5. Take final aqueous supernatant and precipitate with ethanol. Resuspend in 40 µl 10 m*M* Tris, pH 8.0, 1 m*M* EDTA.
6. Analyze aliquots by gel electrophoresis in 1% agarose gel at room temperature.
7. Viral DNA can be detected by Southern blotting and hybridization with an appropriate ^{32}P-DNA probe.

References

1. Carter, B. J., *Curr. Opin. Biotechnol.,* 3, 533 (1992).
2. Muzyczka, N., *Curr. Top. Microbiol. Immunol.,* 158, 99 (1992).
3. Carter, B. J., in *Handbook of Parvoviruses,* Tjissen, P., Ed., CRC Press, Boca Raton, 1990, 155.
4. Srivastava, A., Lusby, E., and Berns, K., *J. Virol.,* 45, 555 (1983).
5. Samulski, R. J., Berns, K. I., Tan, M., and Muzyczka, N., *Proc. Natl. Acad. Sci. U.S.A.,* 79, 2077 (1982).
6. Laughlin, C. A., Tratschin, J. D., Coon, H., and Carter, B. J., *Gene,* 23, 65 (1983).
7. Tratschin, J. D., Miller, I. L., and Carter, B. J., *J. Virol.,* 51, 611 (1984).
8. Hermonat, P. L., Labow, M. W., Wright, R., Berns, K. I., and Muzyczka, N., *J. Virol.,* 51, 329 (1984).
9. Senapathy, P. and Carter, B. J., *J. Biol. Chem.,* 259, 4661 (1984).
10. Chejanovsky, N. and Carter, B. J., *J. Virol.,* 64, 1764 (1990).
11. Chejanovsky, N. and Carter, B. J., *Virology,* 173, 120 (1989).
12. Samulski, R., Srivastava, A., Berns, K., and Muzyczka, N., *Cell,* 33, 135 (1983).
13. Senapathy, P., Tratschin, J. D., and Carter, B. J., *J. Mol. Biol.,* 179, 1 (1984).
14. Im, D. S. and Muzyczka, N., *J. Virol.,* 63, 3095 (1989).

15. Im, D. S. and Muzyczka, N., *Cell,* 61, 447 (1990).
16. Owens, R. A., Trempe, J. P., Chejanovsky, N., and Carter, B. J., *Virology,* 184, 14 (1991).
17. Snyder, R. O., Im, D. S., and Muzyczka, N., *J. Virol.,* 64, 6204 (1990).
18. Snyder, R. O., Samulski, R. J., and Muzyczka, N., *Cell,* 60, 105 (1990).
19. Im, D. S. and Muzyczka, N., *J. Virol.,* 66, 1119 (1992).
20. McCarty, D. M., Ni, T. H., and Muzyczka, N., *J. Virol.,* 66, 4050 (1992).
21. Owens, R., Weitzman, M., Kyöstiö, S., and Carter, B. J., *J. Virol.,* 67, 997 (1993).
22. Yang, Q., Kadam, A., and Trempe, J. P., *J. Virol.,* 66, 6058 (1992).
23. Yang, Q. and Trempe, J. P., *J. Virol.,* 67, 4442 (1993).
24. Owens, R. A. and Carter, B. J., *J. Virol.,* 66, 1236 (1992).
25. Owens, R. A., Weitzman, M. D., Kyöstiö, S. R. M., and Carter, B. J., *J. Virol.,* 67, 997 (1993).
26. Weitzman, M. D., Kyöstiö, S. R. M., Kotin, R. M., and Owens, R. A., *Proc. Natl. Acad. Sci. U.S.A.,* 91, 5808 (1994).
27. Sedivy, J. M., Capone, J. P., RajBhandary, U. L., and Sharp, P. A., *Cell,* 50, 379 (1987).
28. Chejanovsky, N. and Carter, B. J., *Virology,* 171, 239 (1989).
29. Smuda, J. W. and Carter, B. J., *Virology,* 184, 310 (1991).
30. Mendelson, E., Miller, I. L., and Carter, B. J., *Virology,* 166, 154 (1988).
31. Yang, Q., Chen, F., and Trempe, J. P., *J. Virol.,* 68, 4847 (1994).
32. Holscher, C., Horer, M., Kleinschmidt, J., Zentgraf, H., Burkle, A., and Heilbronn, R., *J. Virol.,* 68, 7169 (1994).
33. Carter, B. J., Laughlin, C. A., delaMaza, L. M., and Myers, M. W., *Virology,* 92, 449 (1979).
34. Laughlin, C. A., Myers, M. W., Risin, D. L., and Carter, B. J., *Virology,* 94, 162 (1979).
35. Carter, B. J., Marcus-Sekura, C. J., Laughlin, C. A., and Ketner, G., *Virology,* 126, 505 (1983).

Chapter 14

Reconstitution of Adenovirus DNA Replication with Purified Replication Proteins

Job Dekker, Frank E. J. Coenjaerts, and Peter C. van der Vliet

Contents

0-8493-4412-3/96/$0.00+$.50

I. Introduction

In 1979 Challberg and Kelly[1] described the first *in vitro* DNA replication system employing nuclear extracts of HeLa cells infected with adenovirus type 5. Since then the viral and cellular components required for adenovirus DNA replication have been identified and, presently, replication can be reconstituted *in vitro* using highly purified proteins, often prepared by recombinant techniques.

Adenoviruses contain a linear double-stranded genome of 36 kb with inverted terminal repeats of approximately 100 bp containing the origins of DNA replication (Figure 14.1). A 55-kDa terminal protein is covalently attached to the 5′ ends of the DNA molecules. The origin contains a core region that is absolutely required for replication and an auxiliary region containing binding sites for the cellular transcription factors NFI and NFIII/Oct-1. Binding of these factors to the origin can stimulate replication 50 and 7 times, respectively.[2]

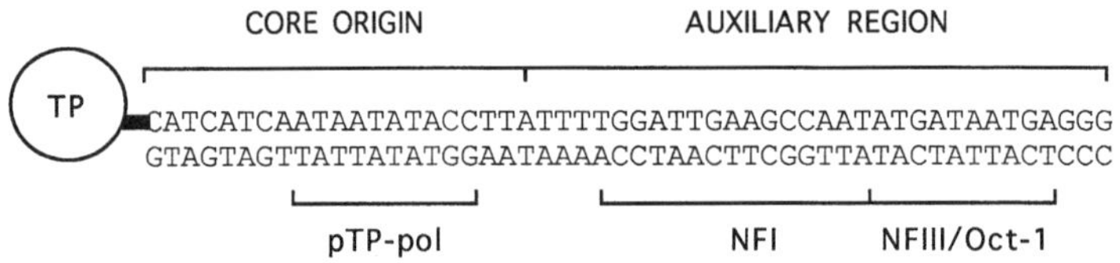

FIGURE 14.1
The adenovirus type 5 origin of replication. The terminal protein (TP) is linked to the first dCMP residue of the upper strand. Binding sites for the pTP-pol complex and for the transcription factors NFI and NFIII/Oct-1 are indicated.

Three viral proteins are required for replication: the adenovirus DNA polymerase (pol), the precursor of the terminal protein (pTP) that forms a tight complex with pol, and the DNA binding protein (DBP). DBP binds aspecifically to single-stranded DNA in a cooperative way, but also binds to double-stranded DNA, albeit not cooperatively. Besides these three proteins that are sufficient for initiation, several cellular proteins have been identified that stimulate the *in vitro* DNA replication reaction (Table 14.1). The DNA binding domains of the transcription factors NFI (NFI-BD, amino acids 1–236 of the rat protein[3]), and NFIII/Oct-1 (the POU domain, amino acids 284–436 of Oct-1) are sufficient for stimulation of replication.[4-6] NFI-BD stimulates the formation of a stable preinitiation complex via direct protein–protein interactions with the polymerase.[7-9] The way in which the POU domain of NFIII/Oct-1 stimulates replication is not known, but recently Coenjaerts et al.[10] showed that this protein also binds the pTP-pol complex. Thus, both transcription factors assist in binding and positioning of the pTP-pol complex to the origin. A similar role may apply to the parental TP.[11] NFII has topoisomerase I activity and is only required for full-length elongation.[12]

Replication starts via a protein priming mechanism. The pTP-pol complex binds the origin, presumably followed by opening of the helix. The first nucleotide, dCTP, is covalently coupled to serine 580 of pTP followed by two additional nucleotides leading to a pTP-CAT intermediate.[13] This will then serve as a primer for elongation.

TABLE 14.1
Proteins Involved in Adenovirus DNA Replication

Protein	Function	Expression System
pTP	Initiation	Vaccinia virus,[1] baculovirus[2]
pol	Initiation, elongation	Vaccinia virus,[1] baculovirus[2]
DBP	Initiation, elongation	Adenovirus-infected cells,[1] *E. coli*[1]
NFI	Initiation	Vaccinia virus,[3] *E. coli*[1]
NFII	Elongation	—
NFIII/Oct-1	Initiation	Vaccinia virus,[3] *E. coli*[1]

1. Described here.
2. Bosher, J., Clare-Robinson, E., and Hay, R. T., *New Biol.*, 2, 1083 (1990).
3. Mul, Y. M., Verrijzer, C. P., and van der Vliet, P. C., *J. Virol.*, 64, 5510 (1990).

During elongation the nontemplate strand forms a complex with DBP. The displaced strand becomes double stranded by renaturation with a displaced strand of opposite polarity or when replication has started on both ends of the same molecule before the strand was completely displaced. It is also possible that the ends of the displaced strand anneal to form a panhandle molecule with an intact double-stranded origin. Late in infection pTP is processed to TP by a viral-encoded protease. More details of the mechanism of adenovirus DNA replication can be found in several recent reviews.[14-16]

Here we give an overview of the purification procedures for the template DNA and the replication proteins. We describe the optimal conditions for reconstituted replication. Under these conditions replication is very efficient as replication rates are comparable to the rate *in vivo* and products of multiple rounds of replication are observed.[2] All purifications are performed at 4°C or on ice.

II. Purification of Replication Components

A. Purification of TP-DNA

Several templates can be used for *in vitro* DNA replication. The most efficient template is the TP-containing natural template (TP-DNA). Linearized plasmids containing the origin of replication at one of the ends can also be used,[17,18] as well as oligonucleotides, but the efficiency is low.[13,19,20] Here we describe the purification of TP-DNA by a procedure adapted from Reference 21.

1. Grow 2 l HeLa cells in suspension to a density of 5×10^5 cells/ml and infect them with Ad5 (10 pfu per cell).
2. Harvest the cells 44 to 48 h post-infection by centrifugation at $1000 \times g$ for 10 min.
3. Wash the cells in phosphate-buffered saline (PBS) and centrifuge at $1000 \times g$.
4. Resuspend the cells in 50 ml 10 m*M* Tris-Cl, pH 8.0, and sonicate three times for 30 sec on ice. Centrifuge for 20 min at $1000 \times g$ to remove cellular debris.

5. Apply the supernatant in 25 ml portions to a discontinuous CsCl gradient in 10 m*M* Tris-Cl, pH 8.0. The gradient composition is as follows: at the bottom 4 ml 1.4352 g/ml CsCl (η = 1.375), then 3 ml 1.3374 g/ml CsCl (η = 1.3657) and on top 3 ml 1.2387 g/ml CsCl (η = 1.3572). Centrifuge for 2 h in an SW25.1 rotor at 18,000 rpm. The virus particles accumulate in the fraction with a density of 1.3374 g/ml CsCl.
6. Collect the virus particles, apply to a continuous CsCl gradient in 10 m*M* Tris-Cl, pH 8.0, with a starting density of 1.3374 g/ml (η = 1.3657) and centrifuge for 18 h at 35,000 rpm in an SW41 rotor. The position of the virus particles is visible as a whitish band.
7. Collect the virus band, carefully add 0.5 volume 50 m*M* Tris-Cl, pH 7.2/1 m*M* ethylene diamine tetraacetic acid (EDTA)/100 m*M* NaCl and apply to a discontinuous CsCl gradient in the same buffer. The composition of the gradient is as follows: 4 ml 1.4352 g/ml CsCl (η = 1.375) at the bottom and 4 ml 1.2387 g/ml CsCl (η = 1.3572) at the top. Centrifuge in an SW25.1 rotor at 18,000 rpm for 2 h.
8. Collect the virus bands and dialyze against 10 m*M* Tris-Cl, pH 7.6/1 m*M* EDTA/0.1 m*M* phenyl methyl sulfonyl fluoride (PMSF).
9. Add an equal volume of 100 m*M* Tris-Cl, pH 7.6/1 m*M* EDTA/8 *M* guanidine hydrochloride and layer the suspension in 500 μl portions on a 12.5 ml sucrose gradient (5 to 20%) in 100 m*M* Tris-Cl, pH 7.6/1 m*M* EDTA/4 *M* guanidine hydrochloride. Centrifuge at 29,000 rpm in an SW41 rotor for 18 h.
10. Collect ten 1.3-ml fractions and screen for TP-DNA by agarose gel electrophoresis. TP-DNA will accumulate in the bottom half of the gradient.
11. Pool TP-DNA containing fractions and dialyze against (three times) 1 l of 10 m*M* Tris-Cl, pH 7.6/1 m*M* EDTA/0.1 m*M* PMSF. Add 20% glycerol during the last dialysis step. Add 500 μg/ml bovine serum albumin (BSA) to the TP-DNA solution before storage in aliquots at –80C.

B. Purification of DBP from Adenovirus-Infected HeLa Cells

DBP is expressed early in infection and accumulates to high levels (up to 2×10^7 molecules per cell). As an *in vitro* DNA replication reaction requires stoichiometric amounts of DBP, large amounts of DBP must be purified. To date, infected cells are the only source of DBP that provides sufficient protein. The purification takes advantage of the fact that the large DBP-containing nucleoprotein complexes that are formed can easily be isolated.[22] DBP expressed in *Escherichia coli* can be purified by denaturation-renaturation protocols yielding small amounts of protein with lower activity (see below).

1. Infect HeLa cells as described above and harvest the cells 44 to 48 h post-infection. Wash the cells in PBS and store them in 10 g aliquots at –80°C.
2. We start the purification routinely with 100 g cells. Lyse the cells by resuspending them in buffer A*/0.05% NP-40 (90 ml/10 g cells) and centrifuge at 1000 × *g*, followed

* Buffer compositions are listed in Section II.G.

by washing with buffer A without NP-40. Centrifuge the suspension at 1000 × *g* for 15 min. The nuclei are in the pellet; the supernatant contains cytoplasmic material and virus particles.

3. Wash the nuclei twice with buffer B and centrifuge at 2000 × *g*.
4. Wash the nuclei three times with buffer C and centrifuge at 3000 × *g*. During these washes the nuclear structure is destroyed and a gel-like pellet of approximately 400 ml is obtained.
5. Add buffer C to 500 ml and sonicate the suspension 10 times for 30 sec in 50 ml aliquots on ice. Make sure that the temperature stays below 6°C.
6. Add 5 *M* NaCl to 150 m*M* and remove insoluble material by centrifugation at 50,000 × *g* for 20 min.
7. Add 1 *M* Tris-Cl, pH 8.0, to the supernatant to a final concentration of 20 m*M*. Slowly add ammonium sulfate to 25% saturation (145 g/l) and remove precipitated material by centrifugation at 13,000 × *g* for 15 min.
8. Dissolve more ammonium sulfate to 60% saturation (290 g/l original volume) and centrifuge at 13,000 × *g*. Discard the supernatant and dissolve the pellet in 100 ml buffer D/50 m*M* NaCl.
9. Dialyze the solution against (two times) 2 l buffer C/50 m*M* NaCl. Remove any insoluble material by centrifugation at 50,000 × *g* for 30 min.
10. Apply the supernatant to a 100 ml phosphocellulose column equilibrated in buffer D/50 m*M* NaCl (100 ml/h). Wash the column with buffer D/150 m*M* NaCl and elute proteins with a linear gradient of D/150 m*M* NaCl to D/800 m*M* NaCl. DBP elutes between 320 and 460 m*M* NaCl.
11. Pool the DBP containing fractions, dilute with buffer E to 150 m*M* NaCl and apply to a 15 ml ssDNA cellulose column equilibrated in buffer D/200 m*M* NaCl (15 ml/h). The ssDNA cellulose column is prepared essentially as described in Reference 23 and contains at least 1 mg DNA per milliliter bed volume. Wash the column first with buffer D/200 m*M* NaCl (1 column volume) and then with buffer D/300 m*M* NaCl/200 m*M* potassium dextran sulfate (1 column volume). Remove potassium dextran sulfate by washing with buffer D/300 m*M* NaCl. Wash with 15 ml D/500 m*M* NaCl. Some DBP elutes, but this also contains DNA and is therefore not used for further purification. Elute with D/2 *M* NaCl. About 30 mg DBP elutes that is approximately 90% pure. This fraction is devoid of DNA.
12. Pool the DBP containing fractions and dilute with buffer F to 250 m*M* NaCl. Apply to an 8 ml heparin Sepharose column equilibrated in buffer F/250 m*M* NaCl. Wash the column with buffer F/370 m*M* NaCl. This removes any contaminating nucleases and phosphatases. Elute the column with F/440 m*M* NaCl. DBP elutes but contaminating ligase activity is retained on the column.
13. To concentrate the protein to about 2 mg/ml, a mono S column is used. Pool DBP containing fractions and dilute with buffer F to 100 m*M* NaCl. Apply to a mono S column equilibrated in buffer F/100 m*M* NaCl and elute DBP with buffer F/300 m*M* NaCl. The yield of 100 g cells is approximately 20 mg protein that is >95% pure as judged by Coomassie blue staining. On an SDS-gel faster migrating bands are present that represent breakdown products of DBP.

C. Purification of His-Tagged ΔN-DBP from *E. coli*

The DNA binding domain of DBP (amino acids 174–529, ΔN-DBP) is sufficient for DNA replication.[24] The crystal structure of this domain has recently been solved[25] and this makes it possible to design specific mutations in order to understand the way in which this domain binds DNA. A bacterial expression system greatly facilitates construction and purification of mutant proteins. Although high levels of ΔN-DBP can be expressed in *E. coli,* the protein is highly insoluble and more than 99% appears in inclusion bodies. Very small amounts of soluble GST-ΔN-DBP could be purified.[25] Renaturation of GST-ΔN-DBP or His-tagged ΔN-DBP by the well-known method of removing the denaturant via dialysis was not successful, as the protein rapidly precipitated. His-tagged ΔN-DBP can, however, be renatured while bound to Ni-NTA beads. This yields approximately 1 mg protein per liter cells. This is far more than the amount of soluble protein that can be purified.

The construct we use contains amino acids 174 to 529 of DBP cloned in the *Xho*I site of pET15b.[26] At the N-terminus, 6 histidines are present followed by a thrombin cleavage site. No additional amino acids are present at the C-terminus. We routinely use the *E. coli* strain BL21(DE3)/pLysS.[26] Expression is directed by T7 RNA polymerase that is encoded by the DE3 lysogen and is inducible with IPTG.

1. Grow 1 l cells in LB/50 μg/ml ampicillin/30 μg/ml chloramphenicol/0.4% glucose at 37°C until the A_{600} reaches 0.7 to 0.9. Add IPTG to a concentration of 0.5 m*M* and continue growth for 3.5 h.
2. Harvest the cells by centrifugation at 2000 × *g* for 25 min and dissolve the cells in 10 ml buffer G. Incubate the cell suspension for 30 min at room temperature.
3. Freeze the cells to –80°C in ethanol/dry ice and thaw quickly in a water bath at 16°C. The cells are now lysed and a viscous solution is obtained.
4. Sonicate the suspension three times for 20 sec in the presence of 0.2% NP-40 on ice.
5. Add DNAseI to a concentration of 20 μg/ml followed by 45 min incubation at room temperature. Collect inclusion bodies and cell debris by centrifugation at 10,000 × *g* for 20 min. The supernatant can be discarded.
6. Dissolve the pellet in buffer H/6 *M* guanidine hydrochloride. Remove insoluble material by centrifugation at 10,000 × *g* for 15 min.
7. Apply the supernatant to a 1 ml Ni-NTA column equilibrated in buffer H/6 *M* guanidine hydrochloride (0.1 ml/min).
8. Wash the column with a linear gradient of H/6 *M* guanidine hydrochloride to H/0 *M* guanidine hydrochloride for 2 h. During this wash renaturation occurs.
9. Wash the column with buffer H/20 m*M* imidazole. During this wash any aspecifically bound protein is eluted. Elute with a 20 ml gradient of H/20 m*M* imidazole to H/500 m*M* imidazole. His-ΔN-DBP elutes around 300 m*M* imidazole and is pure as judged by Coomassie blue staining. The yield is approximately 1 mg/l cells. The protein binds cooperatively to single-stranded DNA and is functional in adenovirus DNA replication. The affinity for single-stranded DNA is approximately six times lower compared to native DBP purified from adenovirus-infected HeLa cells. Note that the C-terminal fragment obtained by chymotrypsin digestion (containing amino acids 174–529[24]) has a fivefold lower affinity for single-stranded DNA compared to intact

DBP.[27] Renatured His-ΔN-DBP binds weakly to double-stranded DNA and can renature two single-stranded DNA molecules like DBP purified from infected HeLa cells.[28]

D. Purification of the pTP-pol Complex Using the Vaccinia Expression System

In infected cells the expression levels of pTP and pol are very low. The construction of recombinant vaccinia viruses containing the pTP and pol genes under the control of the T7 promoter[29] made it possible to purify the pTP-pol complex in relatively large amounts. HeLa cells are infected with three vaccinia viruses encoding pTP, pol, and T7 RNA polymerase.[30] The pTP-pol complex is readily formed in the cell and is purified as described below.

1. Grow 10 l HeLa cells in suspension to a density of 5×10^8 cells per liter.
2. Harvest the cells by centrifugation at 700 × *g* for 30 min and resuspend the cells in 1 l prewarmed SMEM. Incubate the cells at 37°C for 15 min.
3. Add 25×10^9 pfu of each virus (pTP, pol, and T7, m.o.i. = 5) and incubate for 1 h at 37°C.
4. Add 9 l prewarmed SMEM and proceed with incubation at 37°C for 22 h. Infection takes place and pTP and pol are expressed.
5. Harvest the cells at 1000 × *g* and resuspend the cells in 500 ml PBS.
6. Collect the cells by centrifugation at 1000 × *g* and wash with 200 ml buffer I. Centrifuge at 1500 × *g* for 10 min.
7. Resuspend the cells in 150 ml buffer I and leave on ice for 20 min.
8. Homogenize the cells using a Dounce homogenizer and add KCl to 300 m*M*. Leave the suspension on ice for 60 min.
9. Remove cellular debris by centrifugation at 2000 × *g* for 10 min. Centrifuge the supernatant at 100,000 × *g* for 40 min.
10. Add glycerol to a final concentration of 15%. Add dithiothreitol (DTT) to a final concentration of 1 m*M* and adjust to 500 ml with buffer J. The KCl concentration is now 100 m*M*.
11. Apply the lysate to a 50 ml DEAE column equilibrated in buffer J/100 m*M* KCl (100 ml/h), wash the column with buffer J/100 m*M* KCl (2 column volumes) and elute proteins with a 250 ml gradient of buffer J/100 m*M* KCl to J/500 m*M* KCl. Collect 5 ml fractions. To screen for pTP-pol activity, use 1 μl per fraction in a replication assay as described below. pTP-pol peaks around 200 m*M* KCl.
12. Pool pTP-pol containing fractions, dilute with buffer J to 100 m*M* KCl and apply to an 8 ml mono S FPLC column equilibrated in buffer J/100 m*M* KCl (0.5 ml/min). Wash the column with buffer J/100 m*M* KCl (4 column volumes) and elute proteins with an 80 ml gradient of J/100 m*M* KCl to J/600 m*M* KCl. Collect 3 ml fractions and screen for pTP-pol activity using a replication assay (0.1 μl per fraction). pTP-pol peaks around 360 m*M* KCl.
13. Pool pTP-pol containing fractions, dilute with buffer J to 150 m*M* KCl and apply to a 16 ml ssDNA cellulose column[23] equilibrated in buffer J/150 m*M* KCl (3 ml/h). Wash the column with 100 ml buffer J/150 m*M* KCl (0.5 ml/min) and elute with a 120 ml

gradient of J/150 m*M* KCl to J/600 m*M* KCl. Collect 3 ml fractions and screen for pTP-pol activity (0.1 μl per fraction) using a replication assay. Replication activity peaks around 280 m*M* KCl.

14. Pool pTP-pol containing fractions, dilute with buffer J to 100 m*M* KCl and apply to a 1 ml mono S column equilibrated in buffer J/100 m*M* KCl (0.5 ml/min). Wash the column with buffer J/100 m*M* KCl and elute with a 20 ml gradient of J/100 to J/600 m*M* KCl. This yields approximately 4 mg pTP-pol which is 95% pure as determined by silver staining.

E. Purification of His-Tagged NFI-BD from *E. coli*

The DNA binding domain of NFI expressed in *E. coli* is highly insoluble. More than 99% of His-tagged NFI-BD (His-NFI-BD) is present in inclusion bodies. GST-NFI-BD and His-NFI-BD can be purified by denaturation-renaturation protocols,[10] but the activity is low compared to NFI-BD purified from HeLa cells using the vaccinia virus expression system.[2] It is possible to purify approximately 0.5 mg soluble His-NFI-BD per liter cells. This soluble protein is as active in DNA binding and stimulation of adenovirus DNA replication as NFI-BD expressed with the vaccinia system. Here we describe the purification of soluble His-NFI-BD from *E. coli*.

The construct we use contains amino acids 1 to 236 of rat NFI[3] cloned in the *Bam*HI site of the pET15b vector.[26] At the N-terminus, six histidines are present and a thrombin cleavage site. The C-terminus contains five additional amino acids. His-NFI-BD is expressed in the *E. coli* strain BL21(DE3)/pLysS.[26]

1. Grow cells in 1 l LB/50 μg/ml ampicillin/30 μg/ml chloramphenicol/0.4% glucose at 37°C until the A_{600} reaches 0.3.
2. Add IPTG to a concentration of 0.5 m*M* and continue growth for 3.5 h.
3. Harvest cells by centrifugation at 2000 × *g* for 25 min and dissolve them in 10 ml buffer G. Allow lysis to occur for 30 min at room temperature.
4. Freeze the cell suspension at –80°C in dry ice/ethanol and thaw in cold water. A viscous solution is obtained.
5. Sonicate the suspension three times for 20 sec on ice in the presence of 0.2% NP-40.
6. Add DNAseI (20 μg/ml) and incubate at room temperature for 45 min. Remove insoluble material by centrifugation at 10,000 × *g* for 20 min.
7. Dialyze the supernatant two times against 2 l buffer K to remove DTT and EDTA. Remove any particulate material by centrifugation at 20,000 × *g* for 20 min.
8. Apply the supernatant to a 2 ml Ni-NTA column (0.3 ml/min), equilibrated in buffer K. Wash the column with buffer L until the A_{280} is less than 0.01.
9. Wash with buffer L/20 m*M* imidazole to remove aspecifically bound proteins (approximately 40 ml) and elute with a 20 ml gradient of L/20 m*M* imidazole to L/500 m*M* imidazole. His-NFI-BD elutes around 300 m*M* imidazole.
10. Pool His-NFI-BD containing fractions and dilute to 200 m*M* NaCl with buffer M. Apply to a fast flow S Sepharose column (bed volume 2 ml, 0.3 ml/min) equilibrated in buffer M/200 m*M* NaCl. Wash with 10 ml buffer M/200 m*M* NaCl and elute with

a 20 ml gradient of M/200 to M/600 m*M* NaCl. His-NFI-BD elutes around 500 m*M* NaCl.

11. Pool His-NFI-BD containing fractions and dilute with buffer N to 100 m*M* NaCl. Apply to a pKB67-88 DNA cellulose column (bed volume 2 ml, 0.1 ml/min) equilibrated in buffer N/100 m*M* NaCl. pKB67-88 is a plasmid containing 88 copies of the Ad2 origin (bp 1–67) including the NFI binding site.[31] Wash with 10 ml buffer N/100 m*M* NaCl and elute with a gradient of N/100 m*M* NaCl to N/1 *M* NaCl. His-NFI-BD elutes around 800 m*M* NaCl and is homogeneous as judged by silver staining. The yield is approximately 500 μg per liter cells.

F. Purification of His-Tagged Oct-1 POU Domain from *E. coli*

The Oct-1 POU domain (amino acids 284–436) is sufficient for DNA binding and stimulation of DNA replication.[6] Although efficient expression of the POU domain in HeLa cells using the vaccinia virus system has been described,[2] bacterial expression is now routinely used in our laboratory as high levels of soluble His-tagged POU domain can be obtained.

The construct we use contains amino acids 284–436 of Oct-1 cloned between the *Nde*I and *Bam*HI sites of the pET15b vector[26] and contains six histidine residues and a thrombin cleavage site at the N-terminus. No additional amino acids are present at the C-terminus. His-POU is expressed in the *E. coli* strain BL21(DE3)/pLysS.[26]

1. Grow 1 l cells in LB/50 μg/ml ampicillin/30 μg/ml chloramphenicol/0.4% glucose at 37°C until the A_{600} reaches 0.7.
2. Add IPTG to 1 m*M* and continue growth for 4 h.
3. Harvest cells by centrifugation at 2000 × *g* for 25 min and dissolve them in 10 ml buffer O.
4. Freeze the cell suspension at –80°C in dry ice/ethanol and thaw in cold water. A viscous solution is obtained.
5. Add lysozyme to a final concentration of 1 mg/ml and incubate for 30 min on ice.
6. Sonicate the suspension for 20 sec on ice.
7. Add DNAseI (20 μg/ml) and incubate on ice for 45 min. Remove insoluble material by centrifugation at 10,000 × *g* for 20 min.
8. Apply the supernatant to a 1.5 ml Ni-NTA column equilibrated in buffer O (0.3 ml/min). Wash the column with buffer P/l m*M* imidazole until A_{280} is below 0.01. Elute with a 25 ml gradient of P/20 m*M* imidazole to P/500 m*M* imidazole. His-POU elutes around 200 m*M* imidazole.
9. Pool His-POU containing fractions, dilute with buffer Q to 100 m*M* NaCl and apply to a 1.5 ml fast flow S Sepharose column (0.3 ml/min) equilibrated in buffer Q/100 m*M* NaCl. Wash the column with 10 ml buffer Q/100 m*M* NaCl and elute with a 30 ml gradient of Q/100 m*M* NaCl to Q/1 *M* NaCl. His-POU elutes around 400 m*M* NaCl. The yield is approximately 40 mg per liter cells and the protein is pure as determined by Coomassie blue staining. His-POU purified from *E. coli* is fully functional in DNA binding and stimulation of DNA replication.

G. Buffers

Buffer A:

10 m*M* $NaHCO_3$
150 m*M* NaCl
1 m*M* β-mercaptoethanol
1 m*M* PMSF
10 μg/ml TPCK

Buffer B:

25 m*M* NaCl
0.3% w/v EDTA
1 m*M* β-mercaptoethanol
0.5 m*M* PMSF
5 μg/ml TPCK

Buffer C:

10 m*M* Tris-Cl, pH 8.0
1 m*M* β-mercaptoethanol
0.2 m*M* PMSF
2 μg/ml TPCK
NaCl as indicated

Buffer D:

20 m*M* Tris-Cl, pH 8.0
1 m*M* β-mercaptoethanol
0.2 m*M* PMSF
10 μg/ml TPCK
NaCl as indicated

Buffer E:

10 m*M* Tris-Cl, pH 8.8
1 m*M* β-mercaptoethanol
0.2 m*M* PMSF
5 μg/ml TPCK

Buffer F:

25 m*M* HEPES-KOH, pH 8.0
1 m*M* DTT
0.1 m*M* PMSF
0.02% NP-40
20% glycerol
NaCl as indicated

Buffer G:

50 m*M* Tris-Cl, pH 8.0
250 m*M* NaCl
5 m*M* DTT
1 m*M* EDTA
5 m*M* $Na_2S_2O_5$
1 m*M* PMSF
0.7 mg/ml lysozyme

Buffer H:

50 m*M* Tris-Cl, pH 8.0
500 m*M* NaCl
3 m*M* β-mercaptoethanol
5 m*M* $Na_2S_2O_5$
15% glycerol
Guanidine hydrochloride as indicated
Imidazole as indicated

Buffer I:

20 m*M* HEPES-KOH, pH 7.5
5 m*M* KCl
0.5 m*M* $MgCl_2$
0.1% NP-40
10 μg/ml TPCK
5 m*M* $Na_2S_2O_5$

Buffer J:

25 m*M* MES, pH 6.2
1 m*M* DTT
1 m*M* EDTA
0.02% NP-40
0.1 m*M* PMSF
5% sucrose
20% glycerol
KCl as indicated

Buffer K:

50 m*M* Tris-Cl, pH 8.0
250 m*M* NaCl
3 m*M* β-mercaptoethanol
5 m*M* $Na_2S_2O_5$
0.5 m*M* PMSF

Buffer L:

50 Tris-Cl, pH 8.0
250 m*M* NaCl
3 m*M* β-mercaptoethanol
5 m*M* $Na_2S_2O_5$
0.5 m*M* PMSF
10% glycerol
Imidazole as indicated

Buffer M:

50 m*M* Tris-Cl, pH 8.5
1 m*M* DTT
5 m*M* $Na_2S_2O_5$
0.5 m*M* PMSF
10% glycerol
NaCl as indicated

Buffer N:

50 m*M* Tris-Cl, pH 8.0
1 m*M* DTT
0.01% NP-40
0.1 m*M* PMSF
15% glycerol
NaCl as indicated

Buffer O:

46.6 m*M* Na_2HPO_4
3.4 m*M* NaH_2PO_4
5 m*M* β-mercaptoethanol
300 m*M* NaCl
5 m*M* $Na_2S_2O_5$
0.1 m*M* PMSF
1 m*M* imidazole

Buffer P:

6 m*M* Na_2HPO_4
44 m*M* NaH_2PO_4
5 m*M* β-mercaptoethanol
300 m*M* NaCl
5 m*M* $Na_2S_2O_5$
0.1 m*M* PMSF
10% glycerol
Imidazole as indicated

Buffer Q:

50 m*M* Tris-Cl, pH 8.0
5 m*M* β-mercaptoethanol
0.5 m*M* DTT
0.1 m*M* PMSF
10% glycerol
NaCl as indicated

III. DNA Replication Assays

Adenovirus DNA replication can be reconstituted *in vitro* using the purified replication components described above. Assays have been developed to study initiation, partial elongation, and complete elongation. Recently an immobilized replication system was developed.[10]

A. The Initiation and Partial Elongation Reactions

Initiation involves the incorporation of the first dCMP residue and the formation of a pTP-dCMP intermediate. When ddATP is included in the assay no elongation occurs and pTP-dCMP is the only product. Partially elongated products are obtained when dCTP, dATP, dTTP, and ddGTP are added, because elongation will stop after the incorporation of ddGMP, which occurs at position 26. Higher levels of initiation and partial elongation are obtained when NFI-BD and/or the POU domain are added.

1. Add the following replication components together in a 25 μl reaction mixture: 50 ng pTP-pol, 1 μg DBP, 20 ng His-NFI-BD, 5 ng His-POU domain, and 50 ng TP-DNA in a buffer containing 25 m*M* HEPES-KOH, pH 7.5/1 m*M* DTT/1.5 m*M* $MgCl_2$. Add NaCl to a final concentration of 55 m*M*, taking into account the amount of NaCl already added through the replication proteins. For initiation reactions, add 40 μ*M* ddATP and 500 n*M* α^{32}P-dCTP (400 Ci/mmol). For partial elongation reactions add 40 μ*M* of dATP, dTTP, and ddGTP and 500 n*M* α^{32}P-dCTP.
2. Incubate mixtures for 1 h at 37°C. Stop the reaction by addition of 15 μl 200 m*M* sodium pyrophosphate/26 m*M* EDTA/260 ng/μl BSA.
3. Add 7.5 μl trichloroacetic acid (TCA) and incubate on ice for 30 min to precipitate proteins.
4. Centrifuge for 15 min in an Eppendorf centrifuge. Wash the pellet with 15% TCA and dissolve the pellet in Laemmli buffer.
5. Analyze the products on a 7.5% SDS-polyacrylamide gel followed by autoradiography. The pTP-dCMP initiation product migrates as an 80-kDa protein; the partial elongation product containing 26 nucleotides migrates as a 90-kDa protein. In the partial elongation assay another intermediate is also formed, pTP-CAT,[13] that is absent when higher dCTP concentrations are used,[32] e.g., 3 μ*M* (55 Ci/mmol).

B. The Elongation Reaction

Complete elongation of the entire chromosome of 36 kb requires, besides the components described above, an additional cellular protein: NFII. This protein contains topoisomerase I activity.[12] The need for this protein can be overcome by using smaller templates. For elongation reactions we use Ad5 TP-DNA digested with *Xho*I. This yields seven different fragments, named A to G. Fragments B (5.7 kb) and C (6.2 kb) contain the origins and are replicated. The K_m for dCTP is higher during elongation than during initiation;[32] therefore, more dCTP is added in the elongation assay.

1. Add together the following replication components in a 15 µl reaction mixture: 30 ng TP-DNA digested with *Xho*I, 1 µg DBP, 15 ng pTP-pol, 15 ng His-NFI-BD, and 3 ng His-POU domain in a buffer containing 25 m*M* HEPES-KOH, pH 7.5/0.4 m*M* DTT/1 m*M* $MgCl_2$. Add 40 µ*M* of dATP, dGTP, and dTTP and 3 µ*M* α^{32}P-dCTP (55 Ci/mmol). Add NaCl to a final concentration of 55 m*M*.
2. Incubate at 37°C for 1 h. Stop the reaction by adding 2 µl 30% sucrose/0.1% bromophenol blue/0.05% xylene cyanol/1% SDS.
3. Analyze the replication products on a 1% agarose gel containing 0.1% SDS followed by partial dehydration and autoradiography. Besides the B and C bands, extra bands will be seen that represent labeled single strands that are displaced during multiple rounds of replication.

C. Detection of Interactions Between the pTP-pol Complex and His-Tagged Proteins

By immobilizing a His-tagged protein on Ni-NTA beads it is possible to study interactions between this protein and other proteins by experiments analogous to those performed with GST fusion proteins (e.g., Reference 33). Here we describe a procedure to detect interactions between a His-tagged protein and the pTP-pol complex by immobilizing a His-tagged protein on Ni-NTA beads, followed by detection of the amount of bound pTP-pol by a replication assay. A major disadvantage of Ni-NTA beads is that they completely block replication when added to a replication reaction, probably by binding the template DNA. Although imidazole itself also inhibits replication, low levels of imidazole can partly counteract the inhibitory effect of Ni-NTA. Using this assay we found an interaction between the POU homeodomain and pTP-pol. This interaction was also found with a GST-homeodomain fusion protein.[10]

1. Mix 1 µg of His-tagged protein with 10 µl of a 50% suspension of Ni-NTA beads in 25 m*M* HEPES-KOH, pH 7.5/1 m*M* $MgCl_2$/55 m*M* NaCl/20 m*M* imidazole/3 m*M* β-mercaptoethanol. Incubate on ice for 15 min.
2. Wash the beads three times with the same buffer to remove unbound protein.
3. Add 1 µg pTP-pol, add buffer to 25 µl and incubate at room temperature for 30 min.
4. Wash the beads three times with the same buffer to remove unbound protein.
5. Detect pTP-pol with a replication assay using *Xho*I digested Ad5 TP-DNA as template, as described above. Perform the assay in the presence of 20 m*M* imidazole.
6. Stop the replication reaction by addition of 1/10 volume 30% sucrose/0.1% bromophenol blue/0.05% xylene cyanol/1% SDS and apply the suspension, including the Ni-NTA beads, to a 1% agarose gel containing 0.1% SDS.
7. Run the gel overnight at 30 mA and partly dehydrate it, followed by autoradiography.

IV. Concluding Remarks

Adenovirus DNA replication can efficiently be reconstituted *in vitro* using highly purified replication proteins. Three of the five replication proteins can now be expressed and purified in *E. coli,* which makes construction and purification of mutant proteins feasible. Initiation of replication requires the build up of a multicomponent nucleoprotein complex. Although some interactions have already been described, e.g., the interaction between pTP-pol and NFI-BD and the POU homeodomain, the interacting domains have not yet been characterized in detail. The three-dimensional structure of two of the replication proteins (DBP[25] and the POU domain[34]) have been solved. When more structural data are obtained from the other replication proteins it will be possible to understand the mechanism of adenovirus DNA replication in molecular detail.

Acknowledgments

We wish to thank Wim van Driel, Audrey King, Hans van Leeuwen, Wieke Teertstra, Roel Schiphof, Bram Essers, and Arjan van der Flier for useful comments. This work was supported in part by the Netherlands Foundation for Chemical Research (SON) and the Medical Research Council (G-MW) with financial support from the Netherlands Organization for Scientific Research (NWO).

References

1. Challberg, M. D. and Kelly, T. J., *Proc. Natl. Acad. Sci. U.S.A.*, 76, 655 (1979).
2. Mul, Y. M., Verrijzer, C. P., and van der Vliet, P. C., *J. Virol.*, 64, 5510 (1990).
3. Paonessa, G., Gounari, F., Frank, R., and Cortese, R., *EMBO J.*, 7, 3115 (1988).
4. Mermod, N., O'Neill, E. A., Kelly, T. J., and Tjian, R., *Cell,* 58, 741 (1989).
5. Gounari, F., De Francesco, R., Schmitt, J., van der Vliet, P. C., Cortese, R., and Stunnenberg, H., *EMBO J.*, 9, 559 (1990).
6. Verrijzer, C. P., Kal, A. J., and van der Vliet, P. C., *EMBO J.*, 9, 1883 (1990).
7. Bosher, J., Clare-Robinson, E., and Hay, R. T., *New Biol.*, 2, 1083 (1990).
8. Chen, M., Mermod, N., and Horwitz, M. S., *J. Biol. Chem.*, 265, 18634 (1990).
9. Mul, Y. M. and van der Vliet, P. C., *EMBO J.*, 11, 751 (1992).
10. Coenjaerts, F. E. J., van Oosterhout, J. A. W. M., and van der Vliet, P. C., *EMBO J.*, 13, 5401 (1994).
11. Pronk, R. and van der Vliet, P. C., *Nucleic Acids Res.*, 21, 2293 (1993).
12. Nagata, K., Guggenheimer, R. A., and Hurwitz, J., *Proc. Natl. Acad. Sci. U.S.A.*, 80, 4266 (1983).
13. King, A. J. and van der Vliet, P. C., *EMBO J.*, 13, 5786 (1994).
14. Salas, M., *Annu. Rev. Biochem.*, 60, 39 (1991).

15. van der Vliet, P. C., *Semin. Virol.,* 2, 271 (1991).
16. DePamphilis, M. L., *Annu. Rev. Biochem.,* 62, 29 (1993).
17. van Bergen, B. G. M., van der Ley, P. A., van Driel, W., van Mansfeld, A. D. M., and van der Vliet, P. C., *Nucleic Acids Res.,* 7, 1975 (1983).
18. Tamanoi, F. and Stillman, B. M., *Proc. Natl. Acad. Sci. U.S.A.,* 79, 2221 (1982).
19. Harris, M. P. G. and Hay, R. T., *J. Mol. Biol.,* 201, 57 (1988).
20. Dobbs, L., Zhao, L.-J., Sripad, G., and Padmanabhan, R., *Virology,* 178, 43 (1990).
21. Sharp, P. A., Moore, C., and Haverty, J. L., *Virology,* 75, 442 (1976).
22. Schechter, N. M., Davies, W., and Anderson, C. W., *Biochemistry,* 19, 2802 (1980).
23. Alberts, B. and Herrick, G., *Methods Enzymol.,* 21D, 198 (1971).
24. Tsernoglou, D., Tsugita, D. A., Tucker, A. D., and van der Vliet, P. C., *FEBS Lett.,* 188, 2813 (1985).
25. Tucker, P. A., Tsernoglou, D., Tucker, A. D., Coenjaerts, F. E. J., Leenders, H., and van der Vliet, P. C., *EMBO J.,* 13, 2994 (1994).
26. Studier, F. W., Rosenfeld, A. H., Dunn, J. J., and Dubendorff, J. W., *Methods Enzymol.,* 185, 60 (1990).
27. Kuil, M. E., van Amerongen, H., van der Vliet, P. C., and Grondelle, R., *Biochemistry,* 28, 9795 (1989).
28. Zijderveld, D. C., Stuiver, M. H., and van der Vliet, P. C., *Nucleic Acids Res.,* 21, 2591 (1993).
29. Nakano, R., Zhao, L.-J., and Padmanabhan, R., *Gene,* 105, 173 (1991).
30. Moss, B., Elroy-Stein, T., Mizukami, T., Alexander, W. A., and Fuerst, T. R., *Nature,* 348, 91 (1990).
31. Rosenfeld, P. J. and Kelly, T. J., *J. Biol. Chem.,* 261, 1398 (1986).
32. Mul, Y. and van der Vliet, P. C., *Nucleic Acids Res.,* 21, 641 (1993).
33. Verrijzer, C. P., van Oosterhout, J. A. W. M., and van der Vliet, P. C., *Mol. Cell. Biol.,* 12, 542 (1992).
34. Dekker, N., Cox, M., Boelens, R., Verrijzer, C. P., van der Vliet, P. C., and Kaptein, R., *Nature,* 362, 852 (1993).

Section III

Index

Index

A

B

C

D

E

F

G

H

I

J

K

L

O

P

Q

R

S

T

U

V

W

X

Z